# STATISTICAL DECISION THEORY IN PERCEPTION AND COGNITION

# STATISTICAL DECISION THEORY IN PERCEPTION AND COGNITION

## SIGNAL DETECTION AND GENERAL RECOGNITION THEORIES

F. GREGORY ASHBY

THE MIT PRESS CAMBRIDGE, MASSACHUSETTS LONDON, ENGLAND

The MIT Press
Massachusetts Institute of Technology
77 Massachusetts Avenue
Cambridge, MA 02139
mitpress.mit.edu



The MIT Press would like to thank the anonymous peer reviewers who provided comments on drafts of this book. The generous work of academic experts is essential for establishing the authority and quality of our publications. We acknowledge with gratitude the contributions of these otherwise uncredited readers.

This book was set in Stone Serif and Avenir by Westchester Publishing Services. Printed and bound in the United States of America.

Library of Congress Cataloging-in-Publication Data is available.

ISBN: 978-0-262-05251-1

EU Authorised Representative: Easy Access System Europe, Mustamäe tee 50, 10621 Tallinn, Estonia | Email: gpsr.requests@easproject.com

To Jim Townsend
who started me on my journey and inspired me along the way

# CONTENTS

# PREFACE

This book surveys signal detection theory (SDT) and its multidimensional generalization called general recognition theory (GRT). The two theories are based on identical assumptions and have different names only because they are usually applied to different tasks. The umbrella name "statistical decision theory," which dates back at least to Green and Swets (1966), is more descriptive than the names "signal detection theory" or "general recognition theory" because it emphasizes the two core fundamental assumptions of both theories—namely that all sensory information is noisy or statistical and all behaviors require a decision.

This is the first book to cover statistical decision theory comprehensively. Other excellent volumes have covered SDT in great depth (e.g., Green & Swets, 1966; Macmillan & Creelman, 2004; Wickens, 2002), but these books just introduce the basic conceptual framework of GRT, and make no attempt to supply readers with a deep enough understanding that they could use GRT in their own work. As a result, students and researchers who want to learn GRT have always had to collect and study the many original articles that developed the theory. This is a difficult task because these original articles were published in many different journals over the span of four or five decades, and were written by a variety of authors.

Readers of this volume will not only have all the knowledge they need to critically evaluate articles that use SDT or GRT but will also be able to use any current SDT- or GRT-based method of data analysis. The book was written in a style and at a level of mathematical rigor that is similar to my recent text on fMRI data analysis (Ashby,

**Table P.1** Road maps for readers

| Primary interest | Most relevant chapters |
|---|---|
| Introductory overview | 1, 2, 4, 5, 11 |
| Signal detection theory | 1, 2, 3, 10 – 10.2.1.1, 11.1 |
| General recognition theory | 1, 4 – 11 |
| Perceptual processes | 1, 2, 4, 6, 7 |
| Decision processes | 1, 2, 4, 5, 8 |

2019). Even so, this book requires a less sophisticated technical background to read and understand, compared to my fMRI volume. The only necessary background is a basic introduction to probability, and perhaps a one-semester course in calculus. But unlike my fMRI volume, almost no statistical background is needed, nor is any facility with matrix algebra (i.e., matrices only appear in one or two places in the book). Most results are given without proof but with a reference to an original article where the proof can be found. In the few cases where I included a proof, I did so because I thought it provided a deeper understanding of the result. Even so, none of the text depends on these proofs, so readers not interested in technical details can safely skip the proofs.

The book includes a number of worked-out examples that illustrate how the various methods are applied, and it also includes Matlab code (appendix B) that performs several key computations. All code included in appendix B is also posted online at: https://labs.psych.ucsb.edu/ashby/gregory/content/resources. Fortunately, there are a variety of freely available software packages that perform many of the more complex analyses. The text points readers at these packages in the appropriate places.

This book provides a comprehensive survey of all of statistical decision theory. The introduction to chapter 1 includes a brief overview of the topics that are covered and the approach that is taken. Of course, many readers will have a specialized interest in some subset of topics. Table P.1 provides road maps that point readers with a variety of different specialized interests to the chapters that are most relevant to those interests.

Finally, I want to thank Michael Miller, Philip Smith, Jim Townsend, Heidi Zetzer, and several anonymous reviewers for their helpful comments on earlier drafts of this book. Their comments and suggestions greatly improved the quality of this book.

# ACRONYMS

| | |
|---|---|
| AUC | area under the ROC |
| BIC | Bayesian information criterion |
| cdf | cumulative distribution function |
| FA | false alarm |
| GLC | general linear classifier |
| GQC | general quadratic classifier |
| GRT | general recognition theory |
| GRT-wIND | general recognition theory with individual differences |
| H | hit |
| iDBM | iterative decision-bound modeling |
| II | information integration |
| INDSCAL | individual differences multidimensional scaling |
| MDS | multidimensional scaling |
| MVPA | multivoxel pattern analysis |
| N | noise |
| PCA | principal component analysis |
| pdf | probability density function |
| RB | rule based |
| RJT | relative judgment theory |
| ROC | receiver operating characteristic |
| RT | response time |

**SCM** similarity-choice model
**SDT** signal detection theory
**SFT** systems factorial technology
**SIC** survivor interaction contrast
**SN** signal + noise
**SPC** striatal pattern classifier
**SPRT** sequential probability ratio test
**SVM** support vector machine

# 1

# INTRODUCTION

In 2002, Estes referred to signal detection theory (SDT) as "the most towering achievement of basic psychological research in the last half century" (p. 15). Similarly, Wixted suggested that "it should not be possible to earn a Ph.D. in experimental psychology without having some degree of proficiency in signal detection theory" (Wixted, 2020, p. 225). SDT is, by far, the most dominant model in psychophysics, and it is unrivaled as a method for deciding whether an effect is primarily perceptual or decisional. Its multidimensional generalization, called general recognition theory (GRT), is the default approach for defining and studying perceptual interactions, and provides among the most widely used foundations from which to study decision processes with multidimensional stimuli.

The name "signal detection theory" came from initial applications of the theory to simple experiments in which an observer was attempting to detect a signal embedded in noise. The only relevant dimension in these detection tasks was signal intensity. Tasks that require attention to more than one stimulus dimension typically require a decision more complex than detection—for example, the participant may be required to identify the presented stimulus or assign it to a predetermined category. In such cases, the same statistical model is more appropriately called general recognition theory. Both theories assume that observers make decisions in the face of noisy sensory and perceptual representations, and therefore both can be considered special cases of a more general approach that has been called *statistical decision theory* (Ashby & Wenger, 2023; Green & Swets, 1966).

Every year, thousands of published articles use SDT or GRT as a key method of data analysis. These papers span an enormous range of fields. A partial list includes vision, audition, and all other areas of perception, all types of memory, both individual and group decision-making, eyewitness testimony, response time modeling, face perception, visual search, categorization, perceived similarity, preference, stereotyping, implicit learning, fMRI data analysis, and food science.

This book surveys both SDT and GRT, with an emphasis on the underlying logic, the assumptions, and the strengths and weaknesses of these important theories. The focus is on how SDT and GRT account for performance in a wide variety of different tasks. In the case of SDT, the focus is on identification tasks with two stimuli and two responses, like the simple detection task. However, other related tasks will also be considered, including tasks where observers are asked to rate their confidence that a signal was presented, tasks where observers are given multiple looks at the stimulus, old-new recognition memory tasks, visual-search tasks, eyewitness-identification tasks, and group-decision-making tasks. In the case of GRT, the focus is on identification tasks with four or more stimuli and categorization tasks with many stimuli. Chapter 4 explains clearly why identification is the task of choice if the primary interest is in sensory and perceptual processes, whereas categorization is the task of choice if the primary interest is in decision processes. In addition, however, chapter 9 also covers applications of GRT to same-different tasks, tasks where the observer judges the similarity of pairs of stimuli, and tasks where the observer is asked to make preference judgments about the stimuli. Chapter 10 reviews extensions of SDT and GRT that can account for response time as well as response accuracy, and chapter 11 considers extensions of both theories into neuroscience.

## 1.1 THE AXIOMS OF STATISTICAL DECISION THEORY

Statistical decision theory rests on two fundamental axioms: one directed at sensory and perceptual processes and one directed at decisions and response selection.

### 1.1.1 THE AXIOM OF PERCEPTUAL VARIABILITY

The first axiom of statistical decision theory concerns the relationship between the stimulus and the percept it elicits.

**Axiom of perceptual variability** There is trial-by-trial variability in the perceptual information obtained from every object or event.

In other words, the percept changes across trials even if the stimulus does not. The evidence for this is overwhelming (for a review, see, e.g., Faisal et al., 2008). The inherent variability emanates from many sources. First, physical stimuli are intrinsically variable, and this variability begins at the quantum level. For example, the number of photons emitted by a constant light source during a fixed time interval is approximately Poisson distributed (e.g., Geisler, 1989; Wyszecki & Stiles, 1982). In a Poisson distribution, the mean equals the variance, so the standard deviation of the number of photons reaching the cornea increases as the square root of the stimulus luminance. Therefore, intense stimuli vary more across repeated presentations than threshold-level stimuli.

Second, variability occurs after the stimulus enters the sensory system and before transduction (called perireceptor noise). For example, in vision, variable amounts of light are scattered or absorbed while it passes through the lens and the aqueous and vitreous humors. In fact, H. B. Barlow (1977) estimated that somewhere between 67 and 89 percent of the photons that strike the cornea are never absorbed by a photoreceptor.

Third, virtually all neurons in the brain exhibit spontaneous activity. For example, spontaneous isomerization of photopigment occurs frequently enough to be called "dark light" (H. B. Barlow, 1957), and retinal ganglion neurons can have spontaneous firing rates as high as 100 Hz (Robson, 1975). Statistical decision theory predicts that dark light should affect the resulting percept and, therefore, also detection decisions. As a strong test of this prediction, Aho et al. (1988) reported that spontaneous isomerization of photopigment increases with body temperature. Furthermore, they noted that if the assumptions of statistical decision theory are correct, then animals with higher resting body temperatures should therefore experience more visual noise and have higher detection thresholds for light flashes than animals with lower body temperatures. As strong support for this prediction, Aho et al. (1988) reported that frogs, with a body temperature of 59° F, have much lower detection thresholds than humans.

Fourth, there is also variability in the chemical events that occur at every neural synapse and across the neuronal membrane. There are many causes, including variability in the number of neurotransmitter molecules released presynaptically, variability in the time it takes neurotransmitter to diffuse across the synapse, and variability in the number of neurotransmitter molecules that bind to a postsynaptic receptor. Collectively, these stochastic processes produce significant variability in the postsynaptic response to identical presynaptic stimulation (e.g., Kleppe & Robinson, 2006).

Furthermore, due to thermodynamic effects, voltage-gated ion channels open and close randomly, producing random variability in electrical currents (J. A. White et al., 2000). This "channel noise" might be one of the most important sources of biochemical noise at the neuronal membrane (Faisal, 2012).

Finally, statistical decision theory assumes that all decisions depend on memory. Even the simplest detection task requires observers to remember the criterion they used on previous trials to decide whether or not a signal is present. Under most conditions, variability in the remembered criterion, called *criterial noise,* can not be estimated separately from perceptual noise. Instead, all that can be estimated is the sum of stimulus, perceptual, and criterial noise (see section 2.2.3 for details). Therefore, any factor that increases criterial noise (e.g., such as lengthening the delay between trials) will appear to increase perceptual noise.

Thurstone was the first to propose that the effects of all this noise could be modeled by assuming that the sensory magnitude produced by a stimulus on any given psychological dimension could be represented on each trial as a point on a continuum (Link, 1992; Thurstone, 1927c). The variability implies that for any given stimulus, sensory magnitude is a random variable, and therefore, the actual magnitude experienced by an observer on any single trial could be modeled as a random sample from the population of all possible sensory magnitudes elicited by that stimulus. Thurstone also had the important insight that this same representation could be used for any psychological construct, rather than just for sensory magnitudes that could be associated with some scale where physical measurement is possible, such as weight, length, or brightness. For example, Thurstone (1927b) used this idea to model the seriousness of various crimes and the beauty of handwriting samples.

### 1.1.2 THE AXIOM OF UNIVERSAL DECISIONS

The second axiom of statistical decision theory, which is considerably more recent, concerns the relationship between perception and behavior.

**Axiom of universal decisions** Every behavior is preceded by a decision that is under the observer's control.

Before statistical decision theory, psychophysics was dominated by the classical concept of an absolute threshold (e.g., Link, 1992; Swets, 1961). According to this hypothesis, there is a fixed level of stimulation that determines whether or not a stimulus is detected. Any stimulus that induces stimulation above this threshold causes a conscious sensation, whereas there is no sensation of a stimulus when stimulation falls below the threshold. As a result, no decision is needed to detect a signal or stimulus. Whether or not an object or event is detected is purely a matter of how much stimulation it induces.

Note that the assumption of a fixed threshold is logically unrelated to the axiom of perceptual variability. In other words, a fixed threshold could exist regardless of whether or not there is universal perceptual variability. In fact, adding the axiom of perceptual variability to the notion of a fixed threshold is necessary to account for the well-known phenomenon that threshold-level stimuli are not always detected, and largely for this reason, this noisy, fixed threshold model was adopted by Fechner (1860; see also Link, 1992).

The theory of absolute fixed thresholds encountered difficulties almost immediately. For example, Fechner (1860) proposed several methods for estimating the threshold. In one of these, called the method of limits, the stimulus intensity is gradually increased from an initial unnoticeable value or gradually decreased from an initial easily noticed value until the detection threshold is reached. Classical threshold theory predicts that it should not matter whether stimulus intensity is increased or decreased—both methods should produce the same threshold estimate. But they don't. Instead, perceptual hysteresis is common. The threshold measured when stimulus intensity is decreased is typically lower than the threshold measured when stimulus intensity is increased. Another well-known result that is problematic for fixed-threshold theories is that detection behavior can be modified significantly by varying the costs and rewards that are associated with the possible outcomes of a simple YES-NO detection task (e.g., Green & Swets, 1966; see section 2.2.1 for details).

The axiom of universal decisions rejects the idea of a fixed threshold. Instead it assumes that every behavior is preceded by a decision. In general terms, a perceptual decision is a form of statistical inference (e.g., Gold & Shadlen, 2007). The observer has some number of alternative hypotheses about the state of the world, and must use the available noisy perceptual information to select among these alternatives. For example, in a detection task, the two hypotheses are $H_1$: *there is a signal present* and $H_2$: *there is noise, but no signal*. So every detection response is based on a decision about whether the sensory evidence justifies a YES or NO response. This assumption separates sensory and decision processes. It essentially assumes that even the simplest behaviors include two serial stages. In the first stage, the stimulus is coalesced by sensory and perceptual systems into a percept. In the second stage, the percept is used to make a decision about how to behave. And this decision is based, only in part, on the percept. It also could depend on the observer's statistical knowledge of the environment (e.g., stimulus base rates), the value of any rewards and punishments that might be expected from each of the available behavioral options, and even on more arbitrary factors, such as the observer's personality or recent reward history. Chapter 11 reviews considerable neuroscience evidence that supports this two-stage model (see section 11.1.1).

The axiom of universal decisions caused an enormous and profound shift in how even the simplest behaviors were viewed. A change in any behavior—even one as simple as stimulus detection—could now be caused by a change in sensory processing, by a change in how the decision was made to respond, or by both. This more flexible approach immediately opened many new lines of research and allowed a deeper understanding of a wide variety of behavioral phenomena. For example, if hypnotized participants have altered detection thresholds, is this change because hypnosis improves sensory processing? Or does hypnosis just change the criteria that participants use to report the presence of a stimulus? Deciding between these two alternatives fundamentally changes how hypnosis is viewed.

Even seemingly obvious lines of research could benefit from this approach. For example, consider a simple experiment in which detection behavior is studied as stimulus intensity is increased. Obviously, an increase in stimulus intensity will tend to increase the magnitude of the resulting sensation. But what about the decision process? Ideally, if nothing else in the experiment changes, then increasing stimulus intensity should only change sensory processing (i.e., see later discussions of the principle of correspondent change). But logically, it could also change the decision process. For example, when signal intensity is low, some participants might be reluctant to report that a signal is present because they might worry that a false alarm could be interpreted by the experimenter as a hallucination, causing these participants to judge the cost of a false alarm to be higher than the cost of a miss and thereby causing them to adopt a strict criterion for responding that YES a signal is present. As signal intensity increases, they might view this as less of a concern, causing them to relax this criterion. In fact, there is considerable individual difference in the tendency of people to shift between liberal and conservative response criteria (e.g., M. B. Miller & Kantner, 2020). Without the axiom of universal decisions, criterion shifts of this type would be interpreted as changes in sensory processing.

## 1.2 THE HISTORICAL ROOTS OF STATISTICAL DECISION THEORY IN PERCEPTION AND COGNITION

The historical antecedents of statistical decision theory in psychophysics, perception, and cognition date back more than a century. For excellent and more detailed treatments—especially on the history of SDT—see Link (1994) or Wixted (2020).

The first application of SDT was to the YES-NO detection task, in which a sequence of signal-present and signal-absent trials are randomly intermixed, and the observer's task is to respond YES or NO on each trial, depending on whether or not they believe a signal was present. The axiom of perceptual variability predicts that signal-absent trials can sometimes elicit sensations that are identical to those produced on signal-present trials. In other words, it predicts that a sensation of light can occur even when no light is presented. In fact, evidence supporting this assumption has been known for centuries. Perhaps the most famous example is phosphenes, which are created when mechanical pressure on the eye causes the subjective experience of light. Helmholtz (1867) experimented with phosphenes, but curiosity about this phenomenon predates Helmholtz by many years. For example, sketches of phosphenes can be found in Newton's notes.[1] The axiom of perceptual variability also predicts that a presented light can sometimes fail to elicit a sensation of light. This prediction was less controversial since, for example, it is also made by absolute-threshold theory.

These two insights influenced the work of both Fechner and Thurstone, and were a primary motivation behind the development of two-alternative tasks, in which there are two different trial types and each type is associated with a unique response (Fechner, 1860; Link, 1994; Wixted, 2020). The YES-NO detection task is perhaps the simplest example. If sensation was a perfect predictor of stimulus presence, there would be no need for signal-absent trials. Instead, one could just vary stimulus intensity and study how this affects sensory magnitude. In fact, the importance of including signal-absent trials was only universally appreciated fairly recently. For example, Wixted (2020) notes that in the classic study of Hecht et al. (1942), signal-absent trials were omitted, which led directly to their astonishing conclusion that the sensation of light required the absorption of only six or more photons. Shortly thereafter, however, the inclusion of signal-absent trials became standard practice, and it was quickly noted that, especially with weak signals, signal-absent and signal-present trials could elicit similar sensations (H. B. Barlow, 1956). This insight considerably complicates the question of how many photons are required for signal detection.

Another critical insight, which led directly to the axiom of perceptual variability, was that the sensation produced by a stimulus may be corrupted by noise. The earliest scientific worry about this problem occurred centuries ago in the field of astronomy, where it arose from the ubiquitous observation that repeated measurements of celestial objects at the same distance never produced exactly the same result. Gauss proposed the normal distribution as a model of measurement error (Dunnington et al., 2004), while Fechner (1860), Cattell (Fullerton & Cattell, 1892), and Thurstone (1927a, 1927b) all proposed that such noise is an important determinant of variation in subjective human experience.

The formulation and eventual acceptance of the axiom of universal decisions did not occur until the twentieth century. The first critical step was to formalize a theory of how decisions should be made in the face of inherent randomness. The theory of decision-making adopted by statistical decision theory was developed in the 1920s and 1930s—first by R. A. Fisher (1922), who introduced the notions of a *p* value and a hypothetical population of infinite size, and then by Neyman and Pearson (1933), who formalized the distinction between type I and type II errors. The second step was to apply this statistical model of decision-making to real-world problems. This apparently happened first in decisions that were based on radar or sonar—for example, this approach was used in World War II to decide whether to drop a bomb or a depth charge on a possible enemy submarine (Marcum, 1947). The third step was to realize that this same decision-making model might also account for human perceptual decisions, even the simplest possible perceptual decisions—namely, decisions about whether a signal is or is not present. As noted by Wixted (2020), this insight dates back at least to the 1940s and Rose's proposal that the probabilistic behavior of photographic film and television tubes might provide a model for the human visual system (Rose, 1942, 1948). The final step was to put this all together to create SDT in what is essentially its modern form. This appears to have occurred independently and at about the same time at MIT and the University of Michigan (Creelman, 2015; Peterson & Birdsall, 1953; Van Meter & Middleton, 1954). The first publications that introduced SDT to psychology appeared in 1954 (Peterson et al., 1954; Tanner Jr. & Swets, 1954).

The development of a multidimensional version of statistical decision theory lagged far behind the one-dimensional theory. In fact, the modern form of GRT was not developed for another 30 years (Ashby & Townsend, 1986). It is true that the first attempt to generalize SDT to multiple stimulus dimensions did actually appear just a few years after SDT was first proposed (Tanner, 1956). Even so, this early generalization made many simplifying assumptions that caused it to be more similar to traditional multidimensional scaling than to modern GRT. Because of these simplifying assumptions, it failed to exploit the power of the full multidimensional model. As a result, this early generalization never had much influence. For example, in the first 30 years after its publication, the Tanner (1956) paper averaged fewer than three citations per year.

There are several reasons why the development of GRT lagged so far behind the development of SDT. The most important though is probably that at the time when SDT was proposed, the fields of psychophysics and perception had not yet developed to the point where there was much attention paid to interactions between perceptual dimensions. The focus instead was on simpler questions that mostly concerned single dimensions. For example, Weber's law was still an area of active research during the 1950s.

Interest in interactions among multiple sensory dimensions emerged for several reasons, but arguably, one of the most important was the seminal work of Hubel and Wiesel (1962). The discovery of simple and complex neurons in the visual cortex of cats earned the pair a Nobel prize and inspired an avalanche of new research on human pattern recognition, which was focused initially on how people identify letters in their native alphabet. Much of this work explored the hypothesis that pattern recognition is a feature-analytic process, and most proposals were that these hypothetical features were similar to the lines and bars that seemed to maximally excite the simple and complex cells identified by Hubel and Wiesel.

Although there was disagreement about what set of features mediated human pattern recognition, all of the proposals during the first 15 or 20 years of this research assumed that all of these putative features were processed independently. In other words, the default assumption was that there were no perceptual interactions. In this case, there is no need for GRT. Simpler models will suffice. For example, if some stimulus includes two features $f_1$ and $f_2$, and on trials when this stimulus is presented, the probability that feature $f_1$ is perceived equals $\alpha_1$ and the probability that feature $f_2$ is perceived equals $\alpha_2$, then all models at that time assumed that the probability that features $f_1$ and $f_2$ are both perceived must equal $\alpha_1 \times \alpha_2$. Townsend and Ashby (1982) called this assumption *feature-sampling independence*.

The first projects I worked on in graduate school, which were all designed by Jim Townsend, tested feature-sampling independence. In these experiments, on each trial, a single letter or letter-like character was briefly displayed to an observer using a tachistoscope. This was a device, rarely seen today, that could limit display durations to just a couple of milliseconds. The observer's task in these experiments was to identify the presented stimulus. The short exposure durations guaranteed errors, and we used the frequencies of the different types of errors made by observers to test feature-sampling independence.

The stimuli used in my first project with Jim are described in panel (a) of figure 1.1 (Townsend & Ashby, 1982). In this experiment, one of the four capital letters A, E, F, or H was displayed on each trial. Note that each letter was constructed from some subset of the five line segments shown in the bottom of panel (a). Before the experiment began, all observers were shown the four letters and the five line segments, and they were taught the numerical labels shown in the figure that identified each of the lines. On each trial, the observer gave two responses: the letter they saw and the list of line segments they saw. Both reports were forced choice. For example, given the five line segments, there were 32 possible feature reports that were allowed. Specifically, each line segment could be reported as present or absent, so the number of possible reports was $2^5 = 32$. The data from this experiment were summarized in three different matrices. Each matrix described different types of errors that observers could make, so, in the field, these are known as confusion matrices. First, there was a $4 \times 4$ overall confusion matrix in which row $i$ and column $j$ contained the frequency with which the observer responded that they saw letter $j$ on trials when the stimulus was letter $i$. Second, there was a $4 \times 32$ feature-sampling matrix in which row $i$ and column $j$ contained the frequency with which the observer responded that they saw feature list $j$ on trials when the stimulus was letter $i$. Finally, there was a $32 \times 4$ decision matrix in which row $i$ and column $j$ contained the frequency with which the observer responded that they saw letter $j$ on trials when the reported feature list was $i$.

Much of my first year in graduate school was spent fitting a wide variety of computational models to these three confusion matrices. In general, we found models that gave good accounts of the overall confusion matrices and of the decision matrices. But all models that we fit to the feature-sampling matrices performed poorly. To make sure that these poor fits weren't because we were using the wrong feature list, we fit models that assumed the primary features were some other, more complex combination of the five line segments. For example, perhaps lines 1 and 3 together formed a single feature (call this feature 13), so that the correct feature list is 13, 2, 4, 5. Altogether, there are 75 possible alternative feature lists of this type, and we tested all of these. They all fit poorly. Although we fit many, many different models to these feature-sampling confusion matrices, they all assumed feature-sampling independence. So we suspected that the problem with the models was that independence

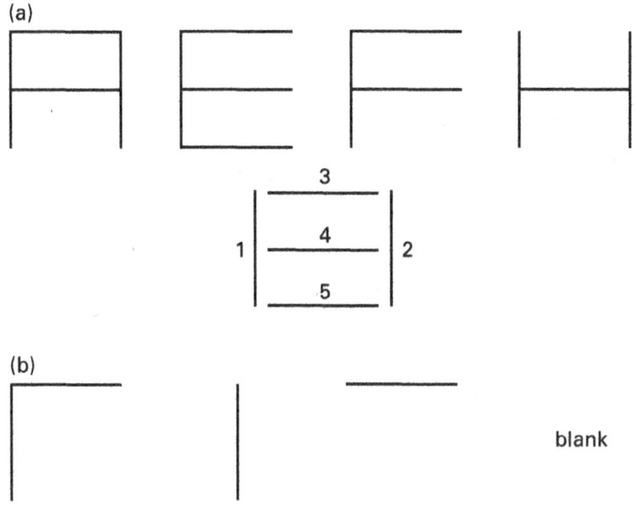


1.1 (a) The four letters used as stimuli in the experiment of Townsend and Ashby (1982) and the five line segments used to construct each stimulus. The observer's task on each trial was to report the letter and features that they saw. (b) The four stimuli used in the experiment of Townsend, Hu, and Ashby (1980).

failed—that is, that there was some type of perceptual interaction going on that none of our models was capturing.

Together with Gary Hu, we then ran a follow-up, even simpler experiment that provided the most rigorous test of feature-sampling independence that we could design—more accurately, that Jim could design (Townsend et al., 1980).[2] The four stimuli used in this experiment are shown in panel (b) of figure 1.1. These stimuli were so simple that a five-millisecond exposure duration was required to observe errors. Note that this experiment includes all possible combinations of features 1 and 3 from the earlier experiment. This simpler experiment also rejected feature-sampling independence (see section 6.1.2 for a description of this test). It seems that perceptual interactions are present even with such simple and obvious features.

Unfortunately, during the late 1970s when these experiments were run, there were no available computational methods for modeling perceptual interactions. So we had run into a dead end. Fortunately, though, at about this same time I took a course from the Electrical Engineering Department on campus called Statistical Pattern Recognition that was taught by Keinosuke Fukunaga and used a text that he wrote with the same title. This course introduced me to much of the mathematics that came to form the foundations of GRT. It described the use of the multivariate normal distribution as a model of noisy observations of some important class of objects, and it described optimal and near-optimal methods for classifying objects that are sampled from one of several different classes.

I was stunned by how well this approach seemed to map onto our pattern recognition research. It was immediately obvious to me that the covariances in a multivariate normal distribution might provide a powerful, yet straightforward mechanism via which to model perceptual interactions. Even so, many problems remained. First, in the typical engineering application, every sample from each distribution is observable (i.e., as a vector containing measured values on each relevant dimension), whereas, in our research, the samples and distributions were unobservable. All we could observe was the true class (i.e., the stimulus) and the judged class (i.e., the observer's response) of each observation. How could such impoverished data ever provide information about the many parameters of a multivariate normal distribution? Second, it wasn't clear that covariances were enough to model all the different types of perceptual interactions and noninteractions that were mentioned in the perception literature. The list of these seemed almost endless. The most prominent included perceptual independence, separability, integrality, holism, redundancy gain, performance parity, dimensional orthogonality, and report independence. Finally, engineering applications focused on optimal, or near-optimal classification schemes, and there was abundant evidence in psychology at that time that human decision-making was often fundamentally suboptimal (e.g., Tversky & Kahneman, 1974). Modeling any such suboptimality would require a different approach from any of those that existed in the engineering literature. Answering these questions turned out to be a major focus of my research for the next twenty years.

# 2

# SIGNAL DETECTION THEORY

## 2.1 THE YES-NO DETECTION TASK

The original applications of statistical decision theory were to tasks in which there was only one relevant sensory dimension. Furthermore, the original development was restricted almost exclusively to YES-NO detection tasks, in which noise-only trials, denoted as N trials, were randomly intermixed with trials in which some signal was embedded in the noise, denoted as SN trials. The observer's task on each trial was to respond YES or NO, depending on whether they thought a signal was present or absent. In other words, the observer's task was to detect the presence of a signal, and, for this reason, the theory was called signal detection theory.

For example, in the classical auditory detection task, white noise, which sounds like the static one hears when a television or radio is not receiving any signal, was used for the N stimulus. The signal used on SN trials was typically a pure tone—similar to the sound produced when a tuning fork is struck. So SN stimuli would be constructed by embedding a pure tone in white noise, and the observer's task on each trial was to respond YES or NO, depending on whether or not the presented stimulus on that trial included a pure tone. The intensity of the noise and tone were typically adjusted beforehand to guarantee that the observer made errors.

The four possible outcomes to this experiment are illustrated in table 2.1. The two possible correct responses are to respond NO on noise-only N trials, which is called a correct rejection, and to respond YES on SN trials, which is called a hit. The two

**Table 2.1** Possible outcomes of a YES-NO detection task

| | NO | YES |
|---|---|---|
| N | Correct Rejection | False Alarm |
| SN | Miss | Hit |

possible errors are to respond YES on N trials, called a false alarm, and NO on SN trials, which is called a miss.

The accuracy data from this experiment can be summarized by recording the frequencies of the four different possible outcomes described in table 2.1. The resulting $2 \times 2$ matrix is an example of a *confusion matrix* because it describes the frequency with which the observer confuses SN and N. Note that the two row sums of this matrix must equal the total number of N and SN trials, respectively. Because the experimenter sets these values before the experiment begins, each row of the confusion matrix includes only one unknown value. For example, once we know the number of correct rejections, the number of false alarms must equal the total number of N trials minus the number of correct rejections. Therefore, the data collected in this experiment include only two degrees of freedom (one for each row). This means that the entire confusion matrix, and therefore the complete results for an observer, can be described by reporting only two numbers—one frequency from each row. It makes no difference which number in each row is reported, but the convention is to report the numbers in the second column. So results of a YES-NO detection task are usually reported as the frequency or proportion of false alarms and hits.

## 2.2 THE SDT MODEL OF THE YES-NO DETECTION TASK

The standard SDT model for this YES-NO detection task is illustrated in figure 2.1. The model assumes that performance in this task is based on a single continuous sensory value, denoted by $X$. In the YES-NO detection task where the stimuli are N and SN, this value is sensory magnitude—for example, loudness with auditory stimuli, or brightness with visual stimuli. So the model assumes that when a stimulus is presented—either N or SN—the observer experiences a sensory magnitude that can be represented by a single value $X$. As described in chapter 1, a fundamental

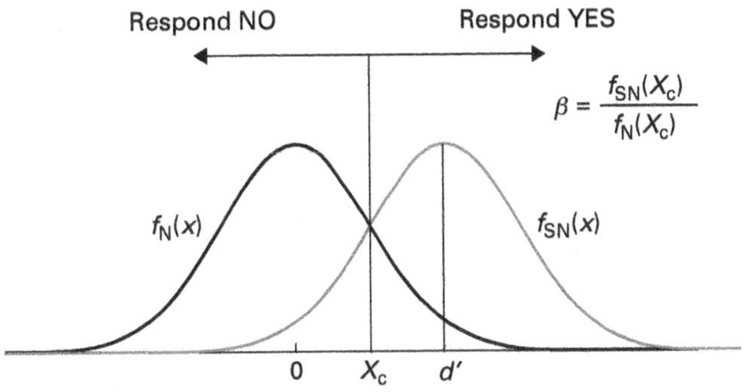


**2.1** The normal, equal-variance, SDT model.

assumption of statistical decision theory, called the axiom of perceptual variability, is that all sensations are inherently noisy. Therefore, SDT assumes that $X$ varies trial by trial, with N trials eliciting one set of possible values of $X$ and SN trials eliciting a different set. On average, we expect the observer to experience larger sensory magnitudes on SN trials than on N trials, but SDT accounts for imperfect accuracy by assuming that the two sets may overlap. Even so, as signal intensity increases, sensory magnitude will also increase, making the magnitudes on N and SN trials more widely separated.

Each set of sensory magnitudes is described by a probability distribution, or more specifically by a probability density function (pdf). The pdf describing the distribution of sensory magnitudes on N trials is denoted by $f_N(x)$, and the pdf on SN trials is denoted by $f_{SN}(x)$. Technically, $f_N(x)$ and $f_{SN}(x)$ describe the likelihoods that the sensory magnitude $X$ on a random N or SN trial, respectively, equals the specific numerical value $x$, for all possible values of $x$. In other words, $f_N(x)$ equals the likelihood that $X=x$ on N trials.[1] In figure 2.1, both of these distributions are normal with the same variance. This normal, equal-variance model is the most commonly used model in signal detection analysis, but any distributions are possible.

The normal, equal-variance model assumes that adding a signal to noise increases mean sensory magnitude, but has no effect on sensory variability, or on any other distributional property. As a result, only one parameter is needed to describe the sensory effects of the experiment—namely, the mean sensory magnitude on SN trials, which is denoted by $d'$. SDT assumes an interval scale of measurement, so the 0-point and unit of measurement are arbitrary. The convention is to set 0 as the mean of the N distribution, and the assumption that both standard deviations equal 1 fixes the unit of measurement. Note that under these conditions, $d'$ also equals the standardized distance between the N and SN distributions (i.e., the number of standard deviations between the two means).

Another fundamental assumption of SDT, called the axiom of universal decisions, is that there is no fixed threshold on sensation that determines whether or not an observer will detect a signal (see section 1.1.2). Instead the observer is assumed to set a criterion value, denoted by $X_C$, and then to respond YES on a trial if the sensory magnitude is greater than this criterion value. This is equivalent to using the following decision rule:

$$\text{Respond YES if } X > X_C\text{; Otherwise respond NO.} \tag{2.1}$$

So as one would expect, the greater the sensory magnitude, the more likely it is that the observer responds YES.

A false alarm occurs in this model if a random sample $X$ from the N distribution is greater than $X_C$ and the probability that this occurs equals the area under $f_N(x)$ to the right of $X_C$. More formally, the normal, equal-variance SDT model predicts that the probability of a false alarm, denoted $P(\text{FA})$, equals

$$\begin{aligned} P(\text{FA}) &= P(X > X_C \mid \text{N}) \\ &= 1 - P(X \leq X_C \mid \text{N}) \\ &= 1 - P\left(Z \leq \frac{X_C - 0}{1} \,\middle|\, \text{N}\right) \\ &= 1 - \Phi(X_C) \\ &= \Phi(-X_C), \end{aligned} \tag{2.2}$$

where $Z$ is a random variable that has a standard normal distribution (i.e., a Z distribution)—that is, a normal distribution with mean 0 and variance 1—and $\Phi$ is the cumulative distribution function of a Z distribution. In other words, $\Phi(z)$ is the area under the Z distribution to the left of the numerical value $z$. These are just the values listed in a Z table. More specifically, if the Z table lists areas under the Z distribution to the left of each Z value, then if we look up $-X_C$ in this table, the associated area will equal $P(\text{FA})$. Similarly a hit occurs if a random sample from the SN distribution is greater than $X_C$. By a derivation that is identical to equation 2.2, the probability of a hit, denoted $P(\text{H})$, equals

$$\begin{aligned} P(\text{H}) &= P(X > X_C \mid \text{SN}) \\ &= 1 - \Phi(X_C - d') \\ &= \Phi(d' - X_C). \end{aligned} \tag{2.3}$$

Note that increasing the criterion for responding YES—that is, the numerical value of $X_C$—makes YES responses less likely. As a result, any experimental manipulation that causes $X_C$ to increase will decrease the probability of a false alarm and the probability of a hit. Responding is said to be more conservative because the observer is more cautious about claiming a signal is present. Note that increasing $X_C$ in equations 2.2 and 2.3 clearly reduces both probabilities. On the other hand, note that decreasing $X_C$ has the opposite effect—that is, it increases the probability of both a hit and a false alarm. So increasing or decreasing $X_C$ always involves a trade-off. Either type of change causes one type of error to increase and the other to decrease. Importantly though, SDT assumes that the value of the criterion $X_C$ is under the observer's control. In other words, SDT assumes that observers are free to set $X_C$ to any value they choose.

### 2.2.1 ALTERNATIVE DECISION RULES

Note that $X_C$ is a criterion on action, not on perception. Classical psychophysical theory includes the notion of a fixed absolute threshold. Stimuli with intensities below this value produce no sensation, whereas stimuli with intensities above the threshold produce a conscious sensation. SDT breaks from this classical account by rejecting thresholds on sensation. Instead, SDT assumes that sensory magnitude is a continuum. SDT includes a threshold, but the threshold is on action, not sensation, and furthermore, this threshold is under the observer's control. For example, suppose you are in the shower and you think you hear something that may or may not be your phone ringing. The two actions you can take are to stay in the shower or get out and check your phone (analogous to NO and YES responses, respectively, in a YES-NO detection task). If you are expecting an important call, you are much more likely to exit the shower based on a weak sensation, and SDT accounts for this change in behavior by assuming that the expectation of an important call reduces the response criterion $X_C$.

Changing $X_C$ changes the hit and false alarm probabilities, and therefore changes the observer's accuracy. One specific value of $X_C$ results in the highest possible accuracy. This optimal decision strategy depends on the likelihood ratio

$$L(x) = \frac{f_{SN}(x)}{f_N(x)}, \tag{2.4}$$

which provides a measure of the likelihood that a given sensory value $x$ was generated by stimulus SN rather than by stimulus N. A value of $L(x) = 1$ means that the sensory sample was equally likely to have been generated by either stimulus. This occurs at the point where $f_N(x)$ and $f_{SN}(x)$ cross, which in figure 2.1 is exactly where the criterion $X_C$ is set. Values of $L(x)$ greater than 1 indicate that the stimulus was more likely to be SN than N, whereas values less than 1 signal the opposite—that is, that N was more likely.

If N and SN trials are equally likely and the goal is to maximize accuracy, then the optimal decision rule is to

$$\text{Respond YES if } L(X) > 1; \text{ Otherwise respond NO.} \tag{2.5}$$

In other words, the optimal strategy is to respond YES if the current sensory magnitude was more likely generated by SN than by N, and to respond NO if the opposite is true. An observer who uses the optimal strategy is called an *ideal observer*. Humans are rarely exactly optimal, however, and so typically the criterion is not set exactly at $L(x) = 1$. Because the ideal observer acts on the likelihood ratio rather than directly on sensory magnitude, it is common to express the criterion that best describes a human observer's performance in terms of $L(x)$, rather than $x$. As a result, instead of expressing the SDT decision rule as in equation 2.1, it is common to write the rule as:

$$\text{Respond YES if } L(X) > \beta; \text{ Otherwise respond NO,} \tag{2.6}$$

where $\beta$ is the response criterion. So $\beta$ and $X_C$ play similar roles; $X_C$ is the criterion on sensory magnitude to respond YES, and $\beta$ is the criterion on likelihood ratio to respond YES. With the normal, equal-variance model, $L(x)$ increases monotonically with $x$. In other words, this model agrees with our intuition that the greater the sensory magnitude, the more likely a signal was presented. As a result, in this model, the equations 2.1 and 2.6 decision rules are equivalent when

$$\beta = L(X_C) = \frac{f_{SN}(X_C)}{f_N(X_C)}. \tag{2.7}$$

Next, consider a YES-NO detection task in which N trials are more frequent than SN trials. Now suppose that the sensory magnitude on some trial is equivocal, so that $L(x) = 1$. Objectively, this sensation does not favor either response alternative. But since N trials are more likely, if we have to guess, it makes more sense to guess N than SN. As a result, the optimal strategy is to respond NO on this trial. So the ideal observer shifts its strategy depending on the N and SN base rates. More specifically, in experiments where base rates might be unequal, the strategy that maximizes accuracy is to

$$\text{Respond YES if } L(X) > \frac{P(\text{N})}{P(\text{SN})}; \text{ Otherwise respond NO,} \tag{2.8}$$

where $P(\text{N})$ and $P(\text{SN})$ are the probabilities that N and SN, respectively, are presented on each trial (i.e., the stimulus base rates). Note that if there are more N trials than SN trials, then $P(\text{N})/P(\text{SN}) > 1$, and therefore the equation 2.8 decision rule indicates that equivocal evidence, in which $L(x) = 1$, calls for a NO response. For example, suppose N trials are twice as likely as SN trials—in other words, that signals are rare events. If N trials are twice as likely, then $P(\text{N}) = \frac{2}{3}$ and $P(\text{SN}) = \frac{1}{3}$, and as a result, the equation 2.8 decision rule reduces to

$$\text{Respond YES if } L(X) > 2;\ \text{Otherwise respond NO.} \tag{2.9}$$

So the optimal strategy is to respond NO unless the sensory magnitude is so large that SN is at least twice as likely as N.

In some applications, the different types of errors may incur different penalties and the different types of correct decisions may bring different benefits. Let $V_{\text{CR}}$ and $V_{\text{H}}$ denote the values of a correct rejection and hit, respectively, and let $V_{\text{FA}}$ and $V_{\text{M}}$ denote the values of a false alarm and miss, respectively. In most cases, we would expect $V_{\text{CR}}$ and $V_{\text{H}}$ to be positive values and $V_{\text{FA}}$ and $V_{\text{M}}$ to be negative—that is, we would typically expect correct decisions to be rewarded, and errors to incur a cost. Regardless of the values though, the decision strategy that maximizes value is (e.g., Green & Swets, 1966):

$$\text{Respond YES if } L(X) > \frac{(V_{\text{CR}} - V_{\text{FA}})P(\text{N})}{(V_{\text{H}} - V_{\text{M}})P(\text{SN})};\ \text{Otherwise respond NO.} \tag{2.10}$$

If the goal is to maximize value or payoffs, then equation 2.10 shows that the criterion for responding YES depends on the costs of the two types of errors (i.e., misses and false alarms), the benefits of the two types of correct responses (i.e., hits and correct rejections), and on the N and SN base rates (i.e., the relative frequency of N and SN trials). As an example of this decision rule, suppose, as in our shower scenario, that a miss is more costly than a false alarm. To make matters concrete, suppose the base rates are equal, both correct decisions earn a reward of +1, a miss causes a loss of 5 and therefore has a value of $V_{\text{M}} = -5$, and a false alarm has a value of $V_{\text{FA}} = -1$. Then the equation 2.10 decision rule reduces to:

$$\text{Respond YES if } L(X) > \frac{1-(-1)}{1-(-5)} = \frac{1}{3};\ \text{Otherwise respond NO.}$$

This is a liberal decision rule because it responds YES to equivocal evidence (i.e., when the likelihood ratio equals 1). In the shower example, missing an important call is probably more costly than exiting the shower and discovering that the phone did not actually ring. So the equation 2.10 decision rule prescribes adopting a more liberal decision strategy to avoid missing a call, despite the fact that this will increase the false alarm rate.

Despite this example, it is important to note that SDT does not assume that observers will always set the criterion to a value that maximizes accuracy or payoffs. Some people are inherently more conservative than others, and SDT is flexible enough to account for individual differences of this kind. In general, in a typical SDT analysis, $X_C$ is estimated to take whatever value is most consistent with the false alarm and hit rates produced by the observer.

### 2.2.2 ESTIMATING $d'$ AND $X_C$

The normal, equal-variance SDT model shown in figure 2.1 has two free parameters that must be estimated from the data—namely, $d'$ and $X_C$. The data are the observed false alarm and hit rates—that is, the observed proportion of N and SN trials, respectively, on which the observer responds YES. Denote these proportions by $\hat{P}(\text{FA})$ and $\hat{P}(\text{H})$, respectively. Then the standard estimate of $X_C$ is obtained by solving equation 2.2 for $X_C$ to produce:

$$\hat{X}_C = -\Phi^{-1}\left[\hat{P}(\text{FA})\right], \tag{2.11}$$

where $\Phi^{-1}(z)$ is the Z value that has area to the left equal to $z$. So $\hat{X}_C$ equals the negative of the Z value with area to the left equal to $\hat{P}(\text{FA})$. In other words, to estimate $X_C$, we look up $\hat{P}(\text{FA})$ in a Z table that lists areas to the left of each tabled Z value, and our $X_C$ estimate is just the negative of the Z value associated with this proportion. Our next goal is to estimate $d'$. Note from equation 2.3 that

$$\Phi^{-1}\left[\hat{P}(\text{H})\right] = \hat{d}' - \hat{X}_C, \tag{2.12}$$

and therefore,

$$\hat{d}' = \Phi^{-1}\left[\hat{P}(\text{H})\right] + \hat{X}_C. \tag{2.13}$$

So to estimate $d'$, we find the Z value that has area to the left equal to $\hat{P}(\text{H})$ and then add this Z value to our estimated value of $X_C$.

In summary, the data from the YES-NO detection task has two degrees of freedom [i.e., $\hat{P}(\text{FA})$ and $\hat{P}(\text{H})$ are both free to vary] and the normal, equal-variance SDT model has two free parameters ($X_C$ and $d'$). After we estimate these two parameters from the data, there are no degrees of freedom left to test the validity of the model. In fact, just as a line can perfectly fit any two data points, the normal, equal-variance SDT model can perfectly fit any observed values of $P(\text{H})$ and $P(\text{FA})$. Given this fact, then an obvious question is why fit this model to two-stimulus identification data? Thousands of applications have confirmed that the SDT model is highly successful at separating perceptual and decisional effects. In particular, manipulations that should only affect sensory magnitude—such as increasing or decreasing signal intensity—mostly cause $d'$ to change but not $X_C$, whereas manipulations that should only affect the observer's decision about how to act on their sensory experience—such as changing the costs and benefits associated with the various possible outcomes—mostly cause $X_C$ to change but not $d'$. In contrast, any of these changes is likely to cause accuracy to change, so without SDT, it is generally impossible to know whether a change in accuracy is due to a change in perception or a change in decision strategy. SDT offers a highly effective method for solving this problem.

As a concrete example, suppose we want to investigate the mechanisms that cause hypnosis to affect performance. Toward this end, we run identical YES-NO auditory detection tasks on a volunteer under two different conditions—control and hypnosis. Now suppose that the results show that the participant's accuracy was 74% correct under control conditions but 84% correct when hypnotized. Based on these results, it is tempting to conclude that hypnosis improved the participant's hearing. However, suppose that a closer analysis showed that, under control conditions, $\hat{P}(\text{H}) = 0.50$ and $\hat{P}(\text{FA}) = 0.02$, whereas under hypnosis, $\hat{P}(\text{H}) = 0.82$ and $\hat{P}(\text{FA}) = 0.14$. Inserting these values into equations 2.11 and 2.13 produces an identical estimate of $\hat{d}' = 2$ in both conditions but an estimate of the response criterion that equals $\hat{X}_C = 2$ under control conditions and $\hat{X}_C = 1.1$ under hypnosis. So the SDT analysis suggests that the improved accuracy under hypnosis has nothing to do with a change in hearing. Instead, it appears that hypnosis makes the participant less conservative about responding YES, and that this more liberal response strategy improved overall accuracy.

### 2.2.3 CRITERIAL NOISE

The equation 2.1 decision rule assumes that the observer uses the exact same value of the response criterion $X_C$ on every trial. Of course, in the case of any biological organism, this value will vary from trial to trial because of noise in the memory systems that store the value. This criterial noise can be modeled by revising equation 2.1 as follows:

$$\text{Respond YES if } X > X_C + \epsilon_C\text{; Otherwise respond NO,} \tag{2.14}$$

where $\epsilon_{\rm C}$ is a random variable with mean 0 and variance $\sigma_{\rm C}^2$. The most common assumption is that $\epsilon_{\rm C}$ is normally distributed.

Consider a version of the normal, equal-variance SDT model that includes criterial noise. Denote the common perceptual variance that characterizes the N and SN distributions as $\sigma_{\rm p}^2$. Suppose that on N trials, sensory magnitude $X$ is normally distributed with mean 0 and variance $\sigma_{\rm p}^2$, whereas on SN trials $X$ is normally distributed with mean $d'$ and variance $\sigma_{\rm p}^2$. Therefore, the predicted probability of a false alarm equals

$$\begin{aligned} P(\text{FA}) &= P(X > X_{\rm C} + \epsilon_{\rm C} \mid \text{N}) \\ &= P(X - \epsilon_{\rm C} > X_{\rm C} \mid \text{N}) \\ &= P\left[\frac{(X-\epsilon_{\rm C}) - 0}{\sqrt{\sigma_{\rm p}^2 + \sigma_{\rm C}^2}} > \frac{X_{\rm C} - 0}{\sqrt{\sigma_{\rm p}^2 + \sigma_{\rm C}^2}} \,\middle|\, \text{N}\right] \\ &= P\left(Z > \frac{X_{\rm C}}{\sqrt{\sigma_{\rm p}^2 + \sigma_{\rm C}^2}}\right). \end{aligned} \tag{2.15}$$

The last equality holds because under the assumption that $X$ and $\epsilon_{\rm C}$ are independent, then $X - \epsilon_{\rm C}$ is normally distributed with mean 0 and variance $\sigma_{\rm p}^2 + \sigma_{\rm C}^2$. In contrast, on SN trials, note that $X - \epsilon_{\rm C}$ is normally distributed with the same variance, but now the mean is $d'$ (i.e., on SN trials, $X$ has mean $d'$ and $\epsilon_{\rm C}$ has mean 0). As a result, the probability of a hit equals

$$\begin{aligned} P(\text{H}) &= P\left[\frac{(X-\epsilon_{\rm C}) - d'}{\sqrt{\sigma_{\rm p}^2 + \sigma_{\rm C}^2}} > \frac{X_{\rm C} - d'}{\sqrt{\sigma_{\rm p}^2 + \sigma_{\rm C}^2}} \,\middle|\, \text{SN}\right] \\ &= P\left(Z > \frac{X_{\rm C} - d'}{\sqrt{\sigma_{\rm p}^2 + \sigma_{\rm C}^2}}\right). \end{aligned} \tag{2.16}$$

Therefore, note that both $P(\text{H})$ and $P(\text{FA})$ depend in exactly the same way on the sum $\sigma_{\rm p}^2 + \sigma_{\rm C}^2$. Any combination of $\sigma_{\rm p}^2$ and $\sigma_{\rm C}^2$ values that produce the same sum will make exactly the same predictions for $P(\text{H})$ and $P(\text{FA})$. Therefore, the best we can do with this model is estimate the sum $\sigma_{\rm p}^2 + \sigma_{\rm C}^2$. There is no way to estimate unique values of $\sigma_{\rm p}^2$ or $\sigma_{\rm C}^2$. In other words, in SDT, perceptual and criterial noise are nonidentifiable. So the standard SDT model that does not include an explicit criterial noise term makes exactly the same predictions as the criterial noise version if we set the common variance in the standard model, denoted as $\sigma^2$, to $\sigma^2 = \sigma_p^2 + \sigma_c^2$ (i.e., the two models are mathematically equivalent), and furthermore, without loss of generality we can set $\sigma^2 = 1$ (which fixes the unit of measurement).

One important implication of this equivalence is that the noise in the standard SDT model can not be assumed to be purely perceptual. Rather, it should be assumed to reflect all sources of trial-by-trial variability—including noise in the stimulus, in the perception of the stimulus, and in the memory of the criterion used to select a response.

## 2.3 THE RECEIVER OPERATING CHARACTERISTIC

A standard way to summarize the results of a collection of YES-NO detection experiments is via the receiver operating characteristic (ROC), which plots $P(\mathrm{H})$ on the ordinate against the probability of a false alarm $P(\mathrm{FA})$ on the abscissa. The different points in this plot could be empirical or theoretical. Empirical estimates of $P(\mathrm{FA})$ and $P(\mathrm{H})$ could be computed from separate experimental conditions or from separate participants in some YES-NO detection task, whereas theoretical values could be generated by deriving predictions from some version of SDT. Making inferences from an ROC is an example of *state-trace analysis*, which is a method for determining the number of underlying parameters or latent variables that are varying across two or more dependent variables (Ashby & Bamber, 2022; Bamber, 1979). The key result from state-trace analysis is that under very weak conditions, any model in which $r$ free parameters are varying across $N$ dependent variables, where $N \geq r$, predicts that the resulting state-trace plot will be $r$ dimensional (Ashby & Bamber, 2022). Since an ROC curve is a two-dimensional state-trace plot, we therefore know that varying any single parameter in any SDT model will always result in a one-dimensional ROC curve.

For example, with the normal, equal-variance SDT model, three different types of one-dimensional ROC curves are possible. One type of curve is generated when we vary $X_C$ (or $\beta$) and hold $d'$ fixed at some constant value. Examples are shown in the left panel of figure 2.2. Because the value of $d'$ is the same for every point on any one curve, these are isosensitivity contours. A qualitatively different type of curve is generated when we vary $d'$ and hold $X_C$ fixed at some constant value. Examples are shown in the middle panel of figure 2.2. These are isocriterion curves. They are all vertical line segments because $P(\mathrm{FA})$ depends on $X_C$ but not on the value of $d'$ (i.e., see equation 2.2). A third, but related type of ROC is generated when we again vary

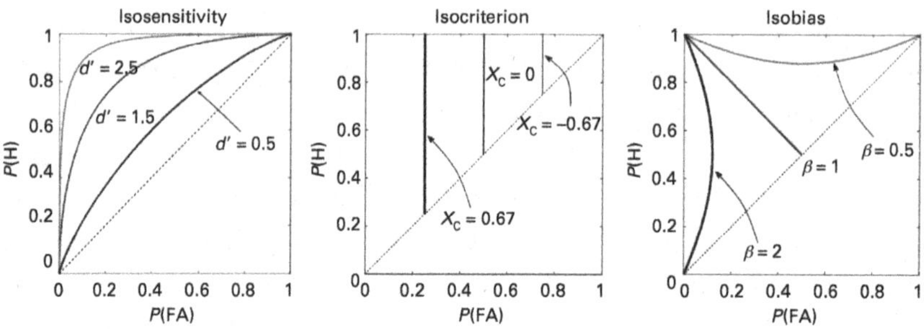


**2.2** The left panel is an ROC showing isosensitivity contours predicted by the normal, equal-variance SDT model for three different values of $d'$. The middle panel is an ROC showing isocriterion contours predicted by the same model for three different values of $X_C$. The right panel is an ROC showing isobias contours for three different values of $\beta$.

$d'$ but now we hold $\beta$ fixed at some constant value. The rightmost panel of figure 2.2 shows examples of these isobias curves. Isosensitivity curves are so much more common than isocriterion or isobias curves that when an article or book refers simply to an ROC curve, they almost always mean an isosensitivity curve. Even so, all three types of curves illustrated in figure 2.2 are ROC curves, since they are all plots of $P(\mathrm{H})$ against $P(\mathrm{FA})$.

In any YES-NO detection task, our expectation is that adding a signal to noise will increase mean sensory magnitude. In an auditory detection task, on average, SN should be louder than N. If so, then $d'$ should always be a positive value. For any positive value of $d'$, it must be true that $P(\mathrm{H}) > P(\mathrm{FA})$ and therefore, all isosensitivity, isocriterion, and isobias curves should fall completely in the upper left half of the plot. The main diagonal, on which $P(\mathrm{H}) = P(\mathrm{FA})$ (denoted by the dotted lines in figure 2.2) corresponds to $d' = 0$. An observer who guesses randomly on every trial will produce a $P(\mathrm{H})$ and $P(\mathrm{FA})$ that fall on this diagonal. Any point below this diagonal indicates a negative $d'$. Since pure guessing should produce $d' = 0$, a (significantly) negative $d'$ should only occur either because of participant deception or because the participant used a highly suboptimal decision rule.

As we will see, different SDT models predict different shapes for their isosensitivity curves. Figure 2.2 shows that the normal, equal-variance SDT model predicts isosensitivity curves that are symmetric with respect to the negative diagonal [i.e., the line that runs through the points (0,1) and (1,0)]. Other models predict asymmetric isosensitivity curves. Therefore, to choose the best SDT model for a task, it is helpful to know what the empirical isosensitivity curves look like in that task.

### 2.3.1 ESTIMATING EMPIRICAL ROCS

Estimating empirical isosensitivity ROC curves is a two-step process. The first step is to run an experiment that collects enough data to estimate multiple points in ROC space that should all fall on the same isosensitivity curve, and the second step is to fit some curve to these points. There are several popular experimental designs that are used to collect the relevant data. One approach is to include a variety of conditions in which the stimulus characteristics remain fixed, but different payoffs are used to encourage observers to change their criterion for responding YES.

Another approach, which uses the same N and SN trials but is experimentally more efficient, is to ask observers to rate their confidence that a signal is present on each trial. For example, suppose a seven-point rating scale is used. So on each trial, rather than respond YES or NO, the observer responds with a number in the range from 1 to 7, where 1 indicates that the observer is extremely confident that no signal was presented, 4 indicates complete uncertainty, and 7 indicates extreme confidence that the signal was present. The confusion matrix that results from this experiment has order $2 \times 7$, with one row for N trials and one row for SN trials, and one column for each of the seven possible rating responses. So the entries in column 5, for example, give the frequency with which the observer rated their confidence that a signal was present as "5" on N trials (row 1) and on SN trials (row 2).

The standard normal, equal-variance SDT model of this experiment is shown in figure 2.3. Given an $r$-point rating scale, the model assumes that observers construct $r-1$ criteria, denoted by $X_1, X_2, \ldots, X_{r-1}$, and that they respond with rating $i$ if and only if $X_{i-1} < X \leq X_i$, where $X_0 = -\infty$ and $X_r = \infty$. Figure 2.3 shows hypothetical values of the six criteria an observer might use when asked to respond on a seven-point rating scale.

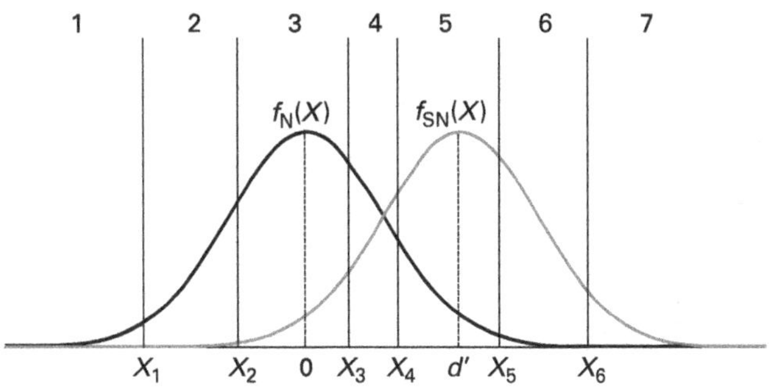


**2.3** The normal, equal-variance SDT model of a rating experiment with a seven-point rating scale.

The first step in our two-step process is to estimate $r-1$ points on an ROC from these ratings data, and the second step is to fit a model to these $r-1$ points. To estimate the $r-1$ points, the criteria that separate different rating responses are interpreted as different values of the SDT response criterion $X_C$. Therefore, a seven-point rating-scale experiment produces six points on a hypothetical ROC. Each point is an ordered pair that includes a different estimate of $P(\mathrm{H})$ and $P(\mathrm{FA})$. Denote the $i$th of these pairs by $[\hat{P}_i(\mathrm{H}), \hat{P}_i(\mathrm{FA})]$ and let $R$ denote the observer's rating on a randomly selected trial.

For example, consider a model of the YES-NO detection task with N and SN distributions like those shown in figure 2.3, in which the criterion $X_C$ is set to $X_C = X_1$, where $X_1$ is the value shown in the figure. Note that in this model, any rating response of 2 or larger indicates a sensory value greater than $X_1$, whereas a rating of 1 indicates that the sensory value was less than $X_1$. So a rating response of 2 or greater corresponds to a hit in the analogous YES-NO model, whereas a rating response of 1 corresponds to a miss. As a result, our first empirical ROC point is estimated as:

$$\begin{aligned}\hat{P}_1(\mathrm{H}) &= \hat{P}(2 \leq R \leq 7|\mathrm{SN})\\ \hat{P}_1(\mathrm{FA}) &= \hat{P}(2 \leq R \leq 7|\mathrm{N}),\end{aligned} \tag{2.17}$$

where $\hat{P}(2 \leq R \leq 7|\mathrm{SN})$, for example, is the proportion of SN trials on which the observer provided a rating of 2 or higher. The other five points are generated in a similar fashion. For example, to generate the second point, we repeat this analysis exactly except this time we assume that the YES-NO criterion was set to $X_C = X_2$. In this case, a hit corresponds to a rating of 3 or higher, and, as a result, we end up with the points:

$$\begin{aligned}\hat{P}_2(\mathrm{H}) &= \hat{P}(3 \leq R \leq 7|\mathrm{SN})\\ \hat{P}_2(\mathrm{FA}) &= \hat{P}(3 \leq R \leq 7|\mathrm{N}).\end{aligned} \tag{2.18}$$

Continuing in this fashion produces the other points:

$$\begin{aligned}\hat{P}_3(\mathrm{H}) &= \hat{P}(4 \leq R \leq 7|\mathrm{SN})\\ \hat{P}_3(\mathrm{FA}) &= \hat{P}(4 \leq R \leq 7|\mathrm{N})\\ \hat{P}_4(\mathrm{H}) &= \hat{P}(5 \leq R \leq 7|\mathrm{SN})\\ \hat{P}_4(\mathrm{FA}) &= \hat{P}(5 \leq R \leq 7|\mathrm{N})\\ \hat{P}_5(\mathrm{H}) &= \hat{P}(6 \leq R \leq 7|\mathrm{SN})\\ \hat{P}_5(\mathrm{FA}) &= \hat{P}(6 \leq R \leq 7|\mathrm{N})\end{aligned}$$

$$\hat{P}_6(\mathrm{H}) = \hat{P}(R = 7|\mathrm{SN})$$
$$\hat{P}_6(\mathrm{FA}) = \hat{P}(R = 7|\mathrm{N}). \quad (2.19)$$

Plotting these six points produces our empirical isosensitivity data.

The next step is to fit the model—for example, the normal, equal-variance SDT model—to these six points. With a seven-point rating scale, the model includes seven free parameters: $d'$, and the six response criteria shown in figure 2.3. Note that the sum of entries in row 1 of the confusion matrix from this experiment equals the number of N trials and the sum in row 2 equals the number of SN trials, so this $2 \times 7$ confusion matrix includes 12 degrees of freedom (six in each row). Because the model includes seven parameters, the data therefore include enough degrees of freedom to estimate the seven parameters and test the validity of the model with the five remaining degrees of freedom. For example, as figure 2.2 illustrates, the normal, equal-variance model always predicts a symmetric isosensitivity ROC, whereas there is no reason that the six points estimated via equations 2.17–2.19 must produce a symmetric curve.

The parameters of the normal, equal-variance model could be estimated by successively substituting the $[\hat{P}_i(\mathrm{FA}), \hat{P}_i(\mathrm{H})]$ pairs that we estimated from equations 2.17–2.19 into equations 2.11 and 2.13. But note that this would produce six different estimates of $d'$. For this reason, all estimated ROC points must be fit simultaneously with a single SDT model that includes one $d'$ parameter and $r - 1$ criteria (when an $r$-point rating scale is used). A number of software packages are available that automate this process. For example, The ROC Toolbox is a free toolbox written in Matlab that fits a variety of different models—including the normal, equal-variance SDT model—to ROC points estimated from ratings data (Koen et al., 2017). It returns maximum likelihood estimates of each model parameter, the standard error of each estimate, and a variety of different goodness-of-fit measures that can be used for model comparison and selection (e.g., AIC, BIC; see chapter 7 for details).

One well-used and relatively simple method for fitting any normal SDT model to points in an ROC space is to first transform each $\hat{P}(\mathrm{H})$ and $\hat{P}(\mathrm{FA})$ to a Z score (e.g., Wickens, 2002). Denote these by $Z_{\hat{P}(\mathrm{H})}$ and $Z_{\hat{P}(\mathrm{FA})}$, respectively. So for example, $Z_{\hat{P}(\mathrm{H})}$ is the Z score that has area to the left under the standard, normal Z distribution equal to $\hat{P}(\mathrm{H})$. The advantage of this extra step is described in the next result.

**Theorem 2.1** Consider an SDT model in which the N distribution is normal with mean 0 and variance 1 and the SN distribution is normal with mean $d'$ and variance $\sigma^2$. Then when plotted in Z space, the model predicts that the isosensitivity ROC curve is a line with slope equal to $1/\sigma$ and intercept equal to $d'/\sigma$. More specifically, the model predicts that

$$Z_{P(\text{H})} = \frac{1}{\sigma} Z_{P(\text{FA})} + \frac{d'}{\sigma}. \tag{2.20}$$

*Proof* See section 2.8. ■

Therefore, all normal SDT models predict that when plotted in Z space, isosensitivity curves must always be linear with slope equal to the ratio of the N to SN standard deviations and intercept proportional to $d'$. Furthermore, note that the normal, equal-variance model predicts that the slope must always equal +1 and the intercept is always equal to $d'$.

An example is shown in figure 2.4. Panel (a) shows standard isosensitivity curves predicted by the normal, equal-variance model for three different values of $d'$. The intermediate black curve is the isosensitivity curve predicted by the model when $d' = 2$, and the two gray curves correspond to $d' = 1$ and $d' = 3$. The three points on the black curve represent hypothetical outcomes of three different conditions—all in which $d' = 2$ but where participants were induced to use different response criteria. The highest point was generated with $X_C = 0$—so exactly at the mean of the N distribution. The intermediate point was generated with $X_C = 1$, which is the optimal criterion if the base rates and payoffs are equal. Finally, the lowest (and leftmost) point was generated with $X_C = 2$—so exactly at the mean of the SN distribution.

Panel (b) of figure 2.4 shows exactly the same three points and exactly the same three predicted isosensitivity curves, except plotted in Z space—that is, on a graph where the abscissa is $Z_{P(\text{FA})}$ and the ordinate is $Z_{P(\text{H})}$. Note that the three curves in panel (a) are now all lines with a slope of +1. So to fit the normal, equal-variance SDT model to a set of $[\hat{P}(\text{FA}), \hat{P}(\text{H})]$ points one could proceed as follows.

*Step 1*: Transform each $[\hat{P}_i(\text{FA}), \hat{P}_i(\text{H})]$ point to a point in Z space. In other words, transform each point $[\hat{P}(\text{FA}), \hat{P}(\text{H})]$ to the point $[Z_{\hat{P}(\text{FA})}, Z_{\hat{P}(\text{H})}]$.

*Step 2*: Plot these points in Z space.

*Step 3*: Find the line with slope +1 that best fits the points from step 2 (e.g., using methods from standard linear regression).

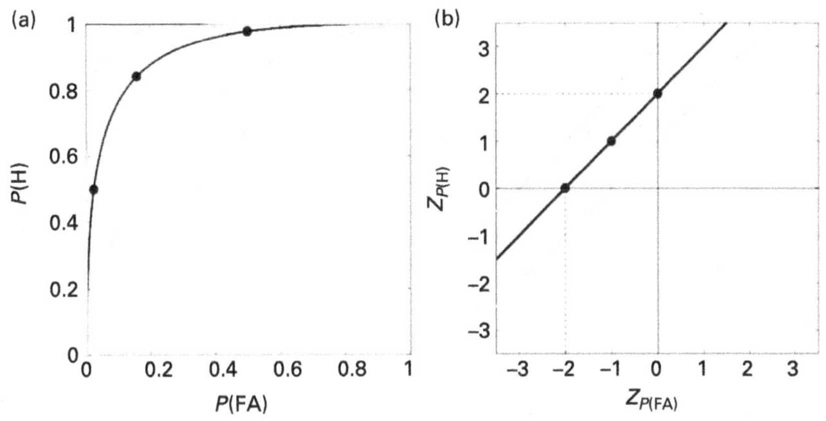


**2.4** (a) Standard isosensitivity curves predicted by the normal, equal-variance SDT model for three values of $d'$ (i.e., $d' = 1, 2$, and 3). (b) The same three isosensitivity curves as in panel (a) except plotted in Z space.

A scatterplot of points in Z space that is best fit by a line with some slope other than +1 is evidence that the N and SN perceptual distributions are normal, but with unequal variances (more on this below). In contrast, a best fit to the points by some nonlinear curve is evidence that the distributions are not normal.

Figure 2.4b also illustrates that the intercept of the linear ROC curves predicted by the normal, equal-variance model equals $d'$. More specifically, the linear ROC will always cross the two Z axes at $\pm d'$. The black line in figure 2.4b was generated from the normal, equal-variance SDT model with $d' = 2$. Note that it crosses the vertical axis at +2 and it crosses the horizontal axis at −2. As a result, after step #3 in this fitting algorithm, one can estimate the single value of $d'$ that is common to all points simply by noting the values of the two intercepts.

### 2.3.2 THE SHAPE OF THE ROC

Because of the important role that the likelihood ratio plays in SDT, it is useful to examine how it changes as sensory magnitude grows. Our intuition is that the likelihood ratio should increase monotonically as sensory magnitude increases. This would mean that the more intense the sensation, the greater the likelihood that a signal was presented. Despite this intuition however, it turns out that not all SDT models make this prediction, and, as a result, it is important to test whether empirical likelihood ratios have this feature. But how can we learn about empirical likelihood ratios without fitting some model that makes assumptions about the likelihood ratio? As it turns out, we can learn how the empirical likelihood ratio changes with sensory magnitude simply by studying empirical ROCs. The following result describes the relationship between the likelihood ratio and the ROC curve.

**Theorem 2.2** Consider any isosensitivity ROC curve that is generated from the equation 2.1 decision rule and is differentiable (i.e., smooth with no vertical segments). Then the likelihood ratio

$$L(x) = \frac{f_{\mathrm{SN}}(x)}{f_{\mathrm{N}}(x)} \tag{2.21}$$

is a monotonically increasing function of $x$ (i.e., sensory magnitude) if and only if the ROC curve is concave down (i.e., its derivative is monotonically decreasing, or alternatively, all points on a line segment that connects any two points on the curve lie below the curve).

*Proof* See section 2.8. ■

Because our intuition suggests that the likelihood ratio should increase monotonically with sensory magnitude, we expect that empirical ROCs should be concave down. In fact, the evidence strongly supports this prediction (Green & Swets, 1966). As a result, we should favor SDT models that have this property. As it turns out though, many alternative SDT models predict ROC curves that are not concave. Perhaps the best-known model in this class is the normal, unequal-variance model, which is considered below in section 2.5.2.

## 2.4 MODEL FLEXIBILITY CONTRIBUTED BY $d'$ VERSUS $X_C$

In some cases, it might be unclear whether a scatterplot of [$\hat{P}$(FA), $\hat{P}$(H)] pairs estimated in some experiment is an isosensitivity curve or an isocriterion curve. For example, consider a large database of jury trials in which each case includes the jury's verdict and some subsequent evidence that becomes available after the trial concludes, which either confirms or disconfirms the verdict (e.g., DNA evidence). Now suppose we separate the cases into different categories based on the type of crime that was suspected and we estimate the $P$(H) and $P$(FA) associated with each category (where, e.g., $\hat{P}$(H) is the proportion of cases in the category for which the jury verdict was guilty and subsequent evidence confirmed this verdict). The different categories might include murder, other violent crimes, drug offenses, and embezzlement and other white-collar crimes. When we examine all the $[\hat{P}(\text{FA}), \hat{P}(\text{H})]$ pairs that result from this analysis, we notice that some are strikingly different from others. The question is, why? One possibility is that juries use very different criteria for deciding guilt across the different crime categories that we studied. In this case, the scatterplot is an isosensitivity curve. Another possibility however, is that the police and district attorney's offices investigate crimes more thoroughly that they consider to be more serious. In this case, the scatterplot is an isocriterion curve because the primary difference across categories is in the value of $d'$.

How could we test whether a scatterplot of $[\hat{P}(\text{FA}), \hat{P}(\text{H})]$ pairs is an isocriterion or isosensitivity curve? The standard approach would be to fit two different versions of the normal, equal-variance, signal detection model to these data. In one version, $X_C$ varies across points but $d'$ remains constant, whereas in the second version $d'$ varies across points but $X_C$ remains constant. So, in both cases, we are fitting a model with one free parameter to the data. If the former model fits better than the latter model then we conclude that the curve is an isosensitivity curve, whereas in the opposite case, we conclude that the curve is an isobias curve.

Unfortunately, this practice is severely biased in favor of the conclusion that the points are on an isosensitivity curve, rather than an isocriterion curve (Ashby, 2023). To see this, note from figure 2.2 that isosensitivity curves are longer than isocriterion curves. In fact, on average, isosensitivity curves tend to be at least four times longer than isocriterion curves (Ashby, 2023). All isosensitivity curves vary between the points (0,0) (when $X_C = \infty$) and (1,1) (when $X_C = -\infty$). The shortest path between these two points, which occurs when $d' = 0$, has a length of $\sqrt{2}$, and the longest path, which occurs when $d' = \infty$ has a length of 2. In contrast, all isocriterion curves are vertical line segments that begin on the major diagonal (when $d' = 0$) and end at $P(\text{H}) = 1$ (when $d' = \infty$). So note that the shortest isocriterion curve has length 0 and the longest has length 1. As a result, isosensitivity curves are substantially longer than isocriterion curves and, therefore, fill up more of ROC space. Because of this extra space-filling property, a set of randomly generated points in ROC space are collectively all more likely to fall near some single isosensitivity curve than near some single isocriterion curve, which suggests that some single isosensitivity curve should fit random or highly noisy data better than some single isocriterion curve.

To test this prediction, Ashby (2023) simulated 100,000 replications of two different experiments—one with five separate conditions (or participants) and one with 10. In both experiments, every data point from each condition was an ordered pair [$P$(FA),$P$(H)] that was generated by randomly sampling a point (from a uniform distribution) in the upper left-half region of ROC space—that is, in the region corresponding to $d' \geq 0$. The random sampling produced a scatterplot of either five or 10 points (in simulated experiments 1 and 2, respectively), all that fell on or above and to the left of the main diagonal in ROC space. Next, two different normal, equal-variance, SDT models were fit to each resulting scatterplot—one model in which all conditions shared the same $d'$ but differed in their value of $X_C$ and one in which all conditions shared the same value of $X_C$ but differed in the value of $d'$. As predicted, the model that assumed all conditions shared the same $d'$ but had different values of $X_C$ was the clear winner, and its dominance increased with the number of data points. With ten randomly generated data points, this model fit better than the model that assumed all conditions shared the same $X_C$ but had different values of $d'$ more than 97% of the time.

In summary, applying the SDT model to random data almost always leads to the conclusion that all the points share the same value of $d'$ but were generated under different values of $X_C$. Therefore, in our hypothetical example, a decision about whether juries use different criteria for deciding guilt across different crime categories or whether the relevant investigative agencies investigate more serious crimes more thoroughly should not be based simply on goodness-of-fit alone. If this is the only criterion used, then it is highly likely that the conclusion will be that juries use different criteria for deciding guilt, regardless of whether or not this is true.

## 2.5 OTHER MODELS

The normal, equal-variance SDT model is the most widely used SDT model. It is conceptually and computationally simple, it seems highly effective at separating sensory from decision processes, and it correctly predicts that all isosensitivity ROC curves are concave down. Even so, it makes one problematic prediction—namely, that all isosensitivity curves are symmetric around the negative diagonal. Some empirical ROCs have this property, but many are skewed (e.g., Green & Swets, 1966), and for this reason, a variety of other SDT models have been proposed. This section describes three of these.

### 2.5.1 THE SIMILARITY-CHOICE MODEL

Luce (1963) proposed a simple model of signal detection that he derived by applying the similarity-choice model (SCM) to the YES-NO detection task. We will consider the SCM in some detail in chapters 7 and 9. This section only considers the YES-NO detection version of the model. In the YES-NO detection task, the SCM predicts that

$$P(\mathrm{FA}) = \frac{\eta\gamma}{(1-\gamma)+\eta\gamma}, \tag{2.22}$$

and

$$P(\mathrm{H}) = \frac{\gamma}{\eta(1-\gamma)+\gamma}. \tag{2.23}$$

Like the normal, equal-variance SDT model, the SCM has two parameters: $\eta$ and $\gamma$. The parameter $\eta$ is a measure of the similarity between the N and SN stimuli. The weaker the signal, the greater the similarity, so $\eta$ is inversely related to the $d'$ parameter of SDT. The parameter $\eta$ is constrained to the interval [0, 1], with $\eta = 1$ signaling maximal similarity between N and SN and therefore corresponding to $d' = 0$ and chance performance. In contrast, $\eta = 0$ signals minimal similarity between N and SN, or equivalently, maximal dissimilarity, and therefore $\eta = 0$ corresponds to $d' = \infty$ and perfect performance. Similarity is assumed to be symmetric, which means that the similarity of N to SN and the similarity of SN to N both equal $\eta$. Furthermore, the similarity of N to N and SN to SN both equal 1.

The parameter $\gamma$ is a measure of the observer's bias to respond YES. The bias to respond NO equals $1-\gamma$, and $\gamma$ is also constrained to the interval [0, 1]. $\gamma$ is purely a decision parameter, in the sense that it should be unaffected by changes in SN or N. As a result, it plays the same role as the SDT response criterion $X_C$ or criterial likelihood ratio $\beta$.

The SCM has its origins in Luce's Choice Axiom (Luce, 1959), rather than in the axioms of statistical decision theory that we considered in chapter 1. Despite this very different theoretical rationale, and the very different equations for $P(\mathrm{H})$ and $P(\mathrm{FA})$ that result, the SCM and the normal equal-variance SDT model predict almost identical isosensitivity curves. This similarity is documented in the first row of figure 2.5. Note that both models predict symmetric curves that visually appear nearly identical. As a result, if the only concern is the shape of the predicted isosensitivity curves, then there is little reason to prefer one model over the other.

On the other hand, the models make different isobias predictions. These are shown in the second row of figure 2.5. First, note that the isobias curves predicted by the SCM are much more similar to the normal, equal-variance model's isobias curves than to its isocriterion curves (i.e., compare figures 2.5 and 2.2). As a result, figure 2.5 compares the SCM's isobias curves to the isobias curves predicted by the normal, equal-variance model. Even so, note that the isobias curves predicted by these two

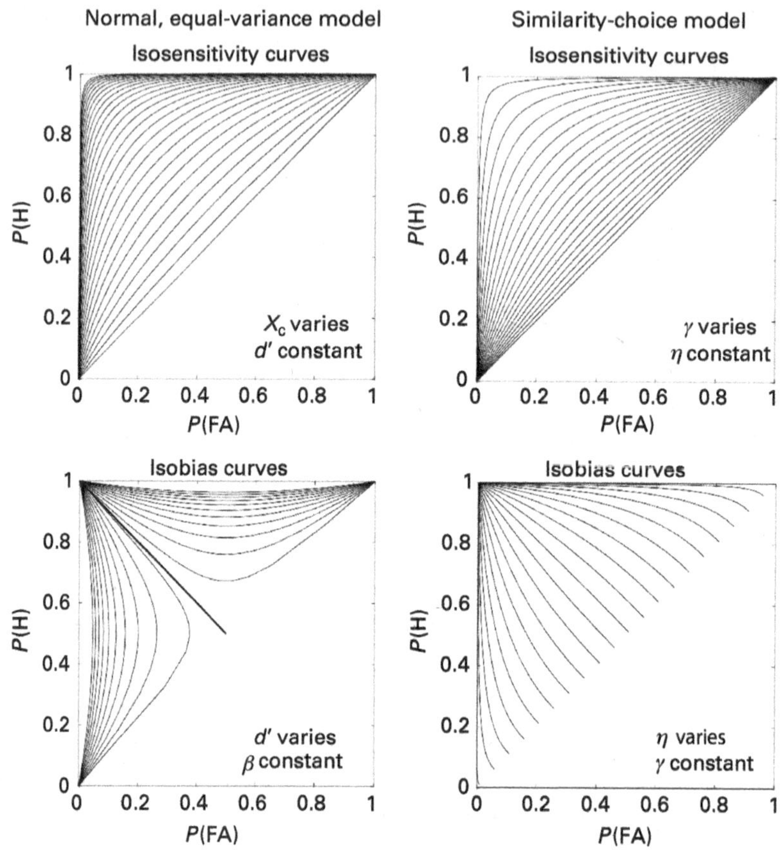


**2.5** Predictions of the normal, equal-variance SDT model (left column) and the SCM (right column) in the YES-NO detection task. The first row shows predicted isosensitivity curves, and the second row shows predicted isobias curves.

models have qualitatively different shapes. The curves predicted by the SCM are bowed out, whereas the normal, equal-variance isobias curves are bowed in. These different isobias predictions have been known for more than half a century. Even so, to my knowledge, they have never been contrasted empirically. The needed experiment is conceptually simple. Each observer would complete say, five or more blocks of YES-NO detection trials. The N stimuli would be identical in each block, but the signal strength on SN trials would vary across blocks. Observers would be encouraged to maintain the same criterion for responding YES in all blocks—either through instruction or via experimental design (e.g., via payoffs). The critical empirical question is whether the resulting scatterplot of $[\hat{P}(\text{FA}), \hat{P}(\text{H})]$ values is bowed in or bowed out.

### 2.5.2 THE NORMAL, UNEQUAL-VARIANCE MODEL

Perhaps the most popular alternative model retains all assumptions of the normal, equal-variance model except that the N and SN distributions have equal variance. This normal, unequal-variance model is illustrated in figure 2.6. Panel (a) shows an N distribution with small variance and two alternative SN distributions, both with larger variances. Panel (b) shows the ROC curves predicted by the model for both SN distributions shown in panel (a) and under the assumption that the observer uses the equation 2.1 decision rule—that is, under the assumption that the observer responds YES to any sensation $x > X_C$. Panel (c) shows the same two isosensitivity curves as in panel (b), except this time plotted in Z space. Note that the curves are linear, but both slopes are less than 1.0 because the SN distribution has larger variance than the N distribution (i.e., see theorem 2.1). The broken line in panel (c) shows predictions of the normal, equal-variance model with $d' = 1.5$.

This normal, unequal-variance model has several attractive properties. First, note that it assumes the variance of the SN distribution is greater than the variance of the N distribution. This makes sense, especially if SN can be modeled as signal plus noise. Adding two nonnegative random variables always increases both the mean and the variance. Furthermore, it is almost a universal law of nature that the variance of any empirically produced random variable increases with its mean. For example, the number of photons emitted from a constant light source that reach a photosensitive

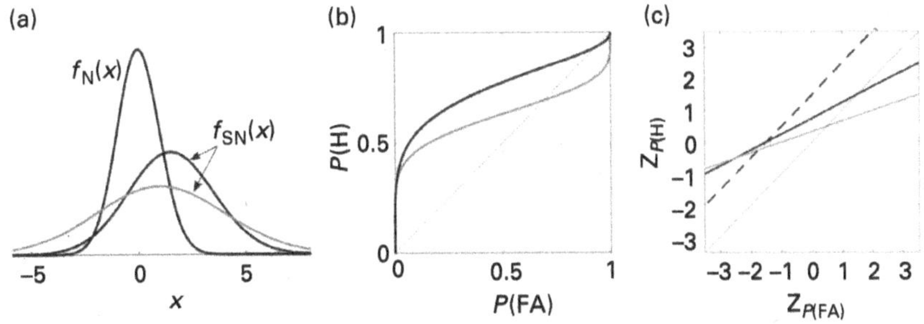


**2.6** (a) The normal, unequal-variance model of SDT. The N distribution is normal with mean 0 and variance 1. Two alternative SN distributions are shown. The pdf in black is normal with mean 1.5 and standard deviation 2, whereas the pdf in gray is normal with mean 1 and standard deviation 3. (b) The ROC showing the isosensitivity contours predicted by the two models shown in panel (a). Both curves assume the N distribution is normal with mean 0 and variance 1. The black curve assumes the SN distribution has mean 1.5 and standard deviation 2, whereas the gray curve assumes the SN distribution has mean 1 and standard deviation 3. (c) The same curves as in panel (b), except plotted in Z space. The broken line shows predictions of the normal, equal-variance model with $d' = 1.5$.

plate is approximately Poisson distributed. The variance of a Poisson distribution equals its mean, so if the task requires observers to detect a light signal embedded in a noisy light field, then at the stimulus level, the SN stimulus will have greater variance than the N stimulus. And not surprisingly, there is substantial evidence that in single neurons, the variance of the spike count tends to increase in proportion to the mean (e.g., Tolhurst et al., 1983). The normal, unequal-variance model nicely accounts for these effects, whereas the normal, equal-variance model does not.

Second, note that the normal, unequal-variance model predicts isosensitivity curves that are skewed, rather than symmetric around the negative diagonal. In fact, many empirical ROCs are skewed in this same manner (Green & Swets, 1966), and the fact that it predicts skewed ROC curves is the main reason that the normal, unequal-variance model is so popular.

Despite these attractive attributes, the normal, unequal-variance model also makes several problematic predictions. First, almost all empirical isosensitivity curves are concave down, whereas the isosensitivity curves predicted by the model violate concavity. As is evident in figure 2.6b, the upper right portion of the curves displays a pronounced violation of concavity. Second, note that both ROCs shown in figure 2.6 dip below the main diagonal, which, as mentioned earlier, reflects suboptimal decision-making. In this case, the suboptimality occurs on trials when the sensory magnitude is well below the mean of the N distribution. In the case of loudness, this would indicate a trial when the observer experiences a loudness that is less than the loudness typically experienced on noise-only trials. The equation 2.1 decision rule prescribes a NO response on these trials because the sensory magnitude is so small. But note that figure 2.6a shows that because the SN variance is larger than the N variance, such small magnitudes are more likely to be generated on SN trials, and so the optimal response is actually to respond "YES a signal is present," even though the brightness or loudness is much less than on typical noise-only trials. In other words, panel (a) shows that the model predicts that although large sensory magnitudes strongly favor the presence of a signal, so do extremely small sensory magnitudes (i.e., magnitudes well below the mean of the N distribution). Thus, given these N and SN distributions, the optimal strategy is to respond YES to small and large sensory magnitudes and NO only for magnitudes of intermediate value. However, the predicted ROC curves shown in figure 2.6 were generated under the assumption that the observer is using the equation 2.1 decision rule, which is highly suboptimal for small sensory magnitudes.

There are two different approaches for resolving these problematic predictions. One approach is to assume that the observer uses the equation 2.6 optimal decision rule based on the likelihood ratio, rather than the decision rule of equation 2.1 that bases the decision solely on sensory magnitude. This approach is illustrated in figure 2.7, which shows the ROC curve predicted by the gray-curve, figure 2.6a, normal, unequal-variance model for the two different types of decision rules. Both curves assume that the sensory effects of N are normally distributed with mean 0 and standard deviation 1, whereas the sensory effects of SN are normally distributed with mean 1 and standard deviation 3. The gray curve is the same as the gray curve in figure 2.6b, so it assumes the observer uses the equation 2.1 decision rule in which decisions are based solely on sensory magnitude—that is, the observer responds YES if the sensory magnitude is large and NO if it is small. In contrast, the black curve assumes optimal responding (i.e., the equation 2.6 decision rule), in which the observer responds YES if the likelihood ratio is large and NO if it is small.

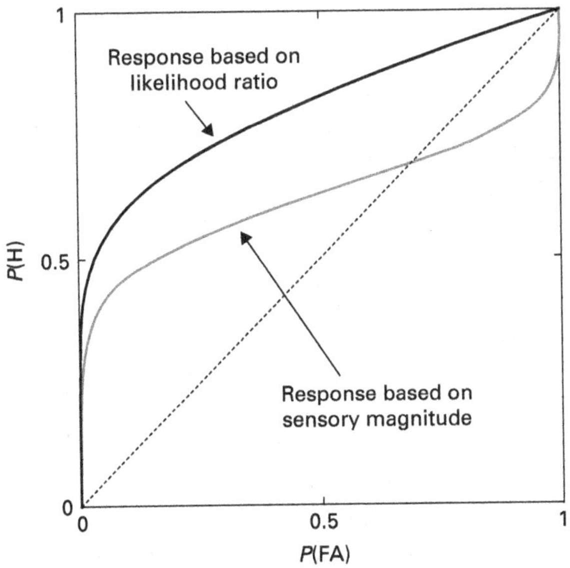


**2.7** Two ROC curves predicted by the normal, unequal-variance model shown in figure 2.6. The gray curve is the same as the gray curve in figure 2.6. The black curve assumes the same N and SN distributions but assumes the observer uses the equation 2.6 likelihood-ratio decision rule. Both curves assume the N distribution is normal with mean 0 and variance 1, and that the SN distribution has mean 1 and standard deviation 3.

The black curve ROC in figure 2.7 is consistent with many empirical ROC curves, since it is concave down and skewed in the same direction as many empirical curves. Even so, as a model of human signal detection it has two weaknesses. First, because it assumes the same underlying N and SN distributions as the gray curve, note that it therefore also still assumes that the smallest sensory magnitudes that an observer experiences are more likely to be generated by SN than by N. This seems highly counterintuitive. It is difficult to imagine how adding a signal to noise could increase the likelihood of extremely small perceived magnitudes.

The second potential problem with the normal, unequal-variance model that predicts the black curve ROC in figure 2.7 is that it not only assumes that the smallest sensory magnitudes occur on SN trials but also assumes that observers are aware of this and therefore respond YES on these trials. This prediction also seems implausible. It is difficult to imagine any human observer responding "YES a signal is present" when the perceived loudness or brightness is considerably less than one would expect on a noise-only trial.

A second approach to avoiding the problematic predictions of the normal, unequal-variance model is more practical than theoretical. Specifically, the practical approach is to note that all problems with the model diminish as the ratio of the SN-to-N standard deviations decreases. For example, consider the case in which the SN standard deviation is 1.25 (i.e., when the N standard deviation equals 1). If the SN distribution has a mean of 1.5 and the N and SN base rates are equal, then the optimal decision rule, which is to respond YES whenever the likelihood ratio exceeds 1, is as follows:

$$\text{Respond YES if } X \leq -6.2$$
$$\text{Respond NO if } -6.2 < X \leq 0.7$$
$$\text{Respond YES if } X > 0.7.$$

So a decision rule that responds NO whenever sensory magnitude falls below a criterion value of 0.7 and YES whenever sensory magnitude is greater than 0.7 is only suboptimal on trials when the sensory magnitude is less than –6.2. On both N and SN trials, the probability of a sensory magnitude less than –6.2 is less than 0.0000001. In other words, such small sensory magnitudes are essentially impossible in this model, so the primary weakness of the normal, unequal variance SDT model is eliminated under these conditions. The one possible drawback of this model is that because the ratio of the standard deviations is so close to 1, this version of the normal, unequal-variance model predicts isosensitivity curves that, albeit skewed, are not skewed very much. This property of the model is illustrated in figure 2.8, which shows isosensitivity curves predicted by two SDT models that are identical except for the standard deviation of the SN distribution. The gray curve shows the predictions of the normal, equal-variance model, and the black curve was generated from a model in which the SN standard deviation is 1.25. Note that the curves differ by only a small amount for larger values of *P*(FA). Of course, empirical isosensitivity curves could show such modest violations of symmetry, in which case the normal, unequal-variance SDT model is an excellent choice for a candidate model. In fact, we will see one such example in chapter 3 (see section 3.1). However, if the empirical curves are more skewed than in figure 2.8, then other models might be preferred.

Finally, note that the standard measure of sensitivity, namely *d′*, is not defined in the normal, unequal-variance SDT model. Traditionally, *d′* is defined as the distance between the N and SN means divided by the common standard deviation. In the normal, unequal-variance model, however, there is no common standard deviation, so the traditional *d′* is undefined. This is also a common problem with multivariate extensions of SDT. Chapter 6 considers this issue in some detail.

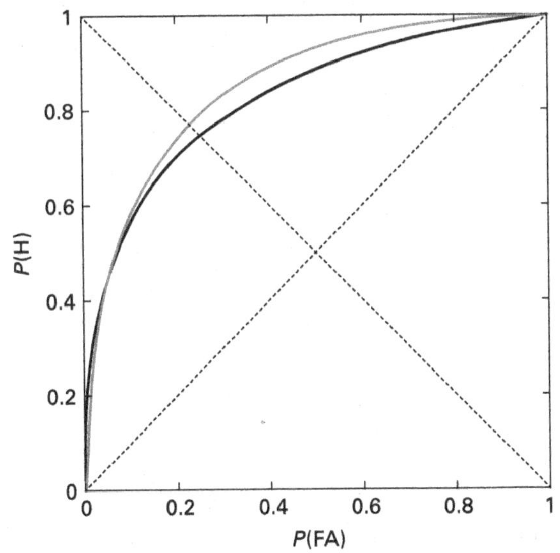


**2.8** (a) Isosensitivity curves predicted by the normal, unequal-variance model of SDT when the standard deviation of the SN distribution equals 1.25 (in black) and when it equals 1 (in gray).

### 2.5.3 THE GAMMA MODEL

Empirical isosensitivity curves are either symmetric about the negative diagonal or skewed in the direction shown in figure 2.6. Furthermore, all of these empirical curves are concave down, and evidence suggests that observers make detection decisions on the basis of sensory magnitude, rather than likelihood ratio (Balakrishnan & Ratcliff, 1996). These properties suggest that the likelihood ratio of the empirical N and SN distributions increases monotonically with sensory magnitude. The likelihood ratio of the normal, equal-variance model is monotonically increasing, so this model provides a good description of symmetric isosensitivity curves. As it turns out, though, so do an infinite number of other SDT models (Killeen & Taylor, 2004).[2] All of these, however, fail to account for skewed ROC curves.

The normal, unequal-variance model accounts for skewed ROCs, but its likelihood ratio is not monotonic with sensory magnitude. So are there any SDT models with monotonic likelihood ratio that predict skewed ROC curves? It turns out that there are many such models. For example, models in which the N and SN distributions are both exponential, gamma, or Rayleigh distributions all have these properties. Figure 2.9 describes a model in which the N and SN perceptual representations are both gamma distributions with shape parameter equal to 2. The N distribution in this figure has mean 1 and the SN distribution has mean 3. Note that the likelihood ratio increases monotonically with sensory magnitude and the isosensitivity curve is skewed in the direction most commonly seen in empirical data. Therefore, this is a good model of tasks in which the empirical ROC is skewed.

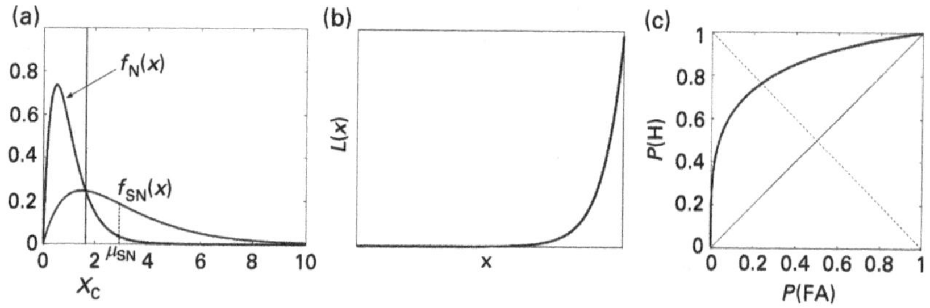


**2.9** The gamma distribution model of SDT. The N distribution has a gamma distribution with a mean of 1, whereas the SN distribution is gamma with a mean of 3. Both distributions have a shape parameter equal to 2. (a) N and SN distributions. (b) The likelihood ratio. (c) The ROC.

## 2.6 APPLICATIONS TO TASKS WITH MULTIPLE STIMULI

Although the original applications of SDT in psychology were to YES-NO detection tasks, the theory has also been applied to many other tasks. Several of these are reviewed in detail in the next chapter. This section considers applications to tasks that are structurally similar to the YES-NO detection task, but that include more stimuli. First though, consider a two-stimulus, identification task with any two arbitrary stimuli that are not N and SN (as, e.g., in our application of SDT to jury trials). Applications of SDT to any such task are identical to the YES-NO detection task except for relabeling of the stimuli and responses. Denote the two stimuli in this task by "A" and "B" and their identifying responses as "a" and "b." If A and B are different stimuli, then they must differ in some way. Stevens (1957) popularized the distinction between prothetic and metathetic stimulus dimensions. With prothetic dimensions, such as loudness and brightness, the stimuli can be ordered by sensory magnitude. As stimulus intensity increases, so do brightness and loudness. In contrast, with metathetic dimensions, such as pitch and hue, increases in stimulus value primarily change the quality of the sensation, rather than the magnitude. Increasing wavelength causes the quality of light to change, but not necessarily the magnitude of the visual sensation. So there are two possibilities. First, the two stimuli A and B could differ on some single prothetic dimension. In this case, we can still interpret the underlying sensory dimension as sensory magnitude and then use the standard SDT model in which the stimulus with the smaller sensory magnitude is modeled as the N stimulus in a YES-NO detection task and its associated response (i.e., a or b) is modeled as a NO response. The second possibility is that the stimuli differ on some single metathetic dimension or they differ on multiple stimulus dimensions. In either of these cases, the association of A and B to N and SN is arbitrary. Computationally, this model is isomorphic to the YES-NO detection model. Conceptually, though, the interpretation of the SDT sensory dimension changes from sensory magnitude to something like "degree of B'ness"—that is, to an abstract quantitative measure in which increasing values signal increasing evidence that the stimulus was B (i.e., assuming A is assigned the role of N and B is assigned the role of SN). For example, in our application to jury trials, the SDT sensory dimension would represent something like "evidence of guilt." In either case though, the model is computationally identical to the YES-NO detection model that we have already considered.

In addition, SDT has been applied to a variety of different types of detection and identification experiments that include multiple stimuli. The most widely used is probably the two-sample, two-alternative forced-choice task, On each trial, two stimuli are presented—one N and one SN (or one A and one B), and the observer's task is to identify which one is SN (or, e.g., B). SDT assumes that exposure to the two stimuli produces two sensory magnitudes—one that is a random sample from the N distribution and one randomly sampled from the SN distribution—and that the observer identifies the larger of these as the SN.

The fact that the observer gets two sensory magnitudes in two-sample, two-alternative forced-choice experiments makes this an easier task than the standard YES-NO detection task, in which the observer must make a decision on the basis of a single sample. The standard model assumes that the observer compares the two samples by subtracting the second magnitude from the first. If the stimuli are auditory, then note that this difference will be positive if the first sample is louder than the second, and negative if the second sample is louder. So the obvious decision rule is to respond 1 if the difference is positive to signal that SN was the first of the two stimuli presented and to respond 2 if the difference is negative.

The normal, equal-variance SDT model of this task is described in figure 2.10a. Let $X_1$ and $X_2$ denote the sensory magnitudes of the first and second stimuli, respectively. Then the statistic that is used for decision-making is $Y = X_1 - X_2$. On trials when N is presented first and SN second, note that $Y$ is normally distributed with mean $-d'$ and variance 2 (because the variances add), whereas on trials when SN is presented first, $Y$ is normally distributed with mean $d'$ and variance 2. The typical approach is to standardize these distributions so that both variances are set to 1 and the smaller mean is set to 0. The first step is to transform the distribution with negative mean to a Z distribution via

$$Z = \frac{Y - (-d')}{\sqrt{2}} = \frac{Y + d'}{\sqrt{2}}, \tag{2.24}$$

which, like all Z transformations, produces a distribution with mean 0 and variance 1. The second step is to apply this same transformation to the distribution with positive mean. This produces a new normally distributed random variable with mean

$$\mu = \frac{\mu_Y + d'}{\sqrt{2}} = \frac{d' + d'}{\sqrt{2}} = \sqrt{2}d' \tag{2.25}$$

and variance

$$\sigma^2 = \text{Var}\left(\frac{Y + d'}{\sqrt{2}}\right) = \left(\frac{1}{\sqrt{2}}\right)^2 \text{Var}(Y) = \left(\frac{1}{2}\right) 2 = 1. \tag{2.26}$$

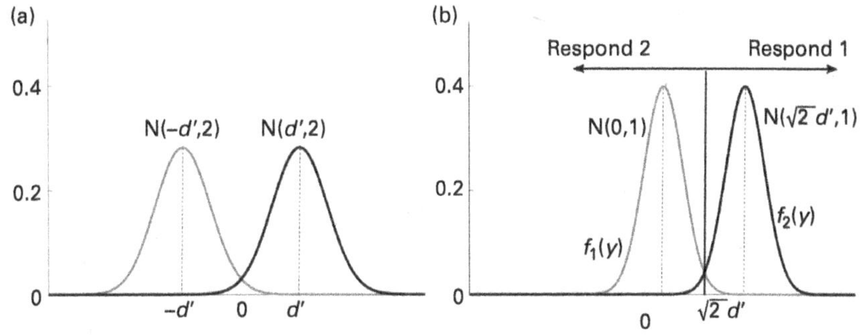


**2.10** Normal, equal-variance SDT model of the two-sample, two-alternative forced-choice task. (a) The sensory distributions of the difference statistic $X_1 - X_2$ on the two trial types (i.e., either N followed by SN, or vice versa). (b) The model after standardizing.

This standardized model is shown in figure 2.10b. This version is convenient for comparing predictions of the model in the two-sample, two-alternative forced-choice task to predictions in the YES-NO detection task when the same N and SN stimuli are used in both tasks. Figure 2.10b shows that the model predicts that the $d'$ in the two-sample task should be larger than the $d'$ in the one-sample YES-NO task by a factor of $\sqrt{2}$.

Another well-known result, described in the following theorem, relates the isosensitivity ROC from the YES-NO detection task, and specifically the area under this curve, to accuracy in the two-sample, two-alternative forced-choice task (Green & Swets, 1966; Green, 1964).

**Theorem 2.3** Regardless of the sensory distributions of the N and SN stimuli, the area under the isosensitivity ROC curve from the YES-NO detection task, denoted AUC, equals the probability correct that would be expected from the same observer in the two-sample, two-alternative forced-choice task, assuming the same stimuli are used in both tasks, and that the observer responds optimally in the two-alternative forced-choice task.

*Proof* See Green and Swets (1966). ∎

This theorem is widely used, not so much to test predictions of SDT, but rather as justification for using the area under the YES-NO ROC curve, frequently denoted as AUC, as an alternative performance measure to $d'$. Compared to $d'$, it has a number of distinct advantages. Perhaps the most important is that AUC is a nonparametric measure that makes no assumptions about the underlying N and SN sensory distributions. In contrast, $d'$ is unambiguously defined only when the N and SN distributions have variances that are equal. Largely for this reason, AUC is used more widely than $d'$ to measure performance in the fields of data science and machine learning.

The two-sample, two-alternative forced-choice task is closely related to multiple-look experiments, in which the observer is presented with $r$ independent samples of either N or SN on each trial. As in the YES-NO detection task, the observer's task is to respond YES or NO, depending on whether the $r$ samples were all SNs or Ns. Another well-known result relates the performance of an ideal observer in the multiple-look experiment to performance in the YES-NO detection task.

**Theorem 2.4** Suppose an ideal observer with perfect memory participates in a multiple-look experiment in which $r$ independent samples of N or SN are presented on each trial. Denote the $d'$ of this observer in the YES-NO detection task as $d'_{\text{YN}}$ and the $d'$ in the multiple-look experiment as $d'_r$. Then the normal, equal-variance model predicts that

$$d'_r = \sqrt{r}\, d'_{\text{YN}}. \tag{2.27}$$

*Proof* See section 2.8. ■

Note that when $r = 2$, theorem 2.4 states that $d'_2 = \sqrt{2} d'_{\text{YN}}$, which means that the relationship between the two-sample, two-alternative forced-choice experiment and the YES-NO detection task is the same as the relationship between the two-look experiment and the YES-NO detection task (i.e., see figure 2.10). This means that SDT predicts that it doesn't matter whether the two stimuli are different (i.e., one N and one SN) or the same (i.e., either both Ns or both SNs). The critical point is that, in both cases, the observer gets two independent samples of data, rather than the one sample provided in the YES-NO detection task. Of course, this better performance is only possible if the observer is explicitly instructed that the two stimuli will always either be different or the same. Without this knowledge, the task is more difficult because there would then be four logical possibilities, rather than only two [i.e., (N,N), (N,SN), (SN,N), and (SN,SN)].

Experiments with human observers that have tested the theorem 2.4 prediction show that $d'_r$ increases with $r$, but more slowly than predicted by equation 2.27 (Green & Swets, 1966). The most likely reason is that human observers do not have perfect memory, and thus are unable to take full advantage of all $r$ stimulus samples.

Theorem 2.4 describes predictions of an ideal observer. Human performance is rarely as good as the performance of an ideal observer, however, so why derive such predictions? There are many advantages, but one is that the ideal observer establishes a benchmark against which to compare human performance (for other advantages, see Geisler, 2011). When modeling human data in some new task, a common approach is to begin with the ideal-observer model and then build in limitations that are known to characterize human perception and cognition (e.g., limited working memory capacity). This process continues until a model is created that accounts for the human performance. Identifying the limitations that seem to characterize the human data often provides deep insights into perceptual processing (e.g., Geisler, 1989).

## 2.7 ACCOUNTING FOR RESPONSE TIMES

Marr (1982) famously proposed the hierarchical classification of mathematical models as computational, algorithmic, and implementational. Briefly, computational models account for behavior in a task while making only minimal assumptions about psychological process. Algorithmic models describe the psychological processes that produce the behavior, and implementational models describe the hardware that implements those processes. In mathematical psychology, Marr's algorithmic-level models are often referred to as process models. SDT provides a computational-level description of decision-making, since it makes no attempt to describe the underlying algorithms or perceptual or cognitive processes that mediate decision-making. During the 1970s, great efforts were devoted to developing process models of decision-making, and currently there are several different process interpretations of SDT. This topic is considered in detail in chapter 10. Briefly though, arguably the most popular process interpretation is provided by the drift-diffusion model (Link & Heath, 1975; Ratcliff, 1978), which is illustrated in figure 2.11. The idea is that instead of representing the sensory effects of the stimulus on each trial with a single random sample from the N or SN distributions, as in classical SDT, the observer is assumed to repeatedly sample the presented stimulus as long as it is available. Each sample $X$ is compared to the criterion $X_C$ by computing the difference $X - X_C$, and these differences are accumulated. The sampling and accumulating processes continue until the resulting sum (or integral) first exceeds an upper criterion A or falls below a lower criterion –B. Sampling terminates with a YES response in the former case, and with a NO response in the latter case.

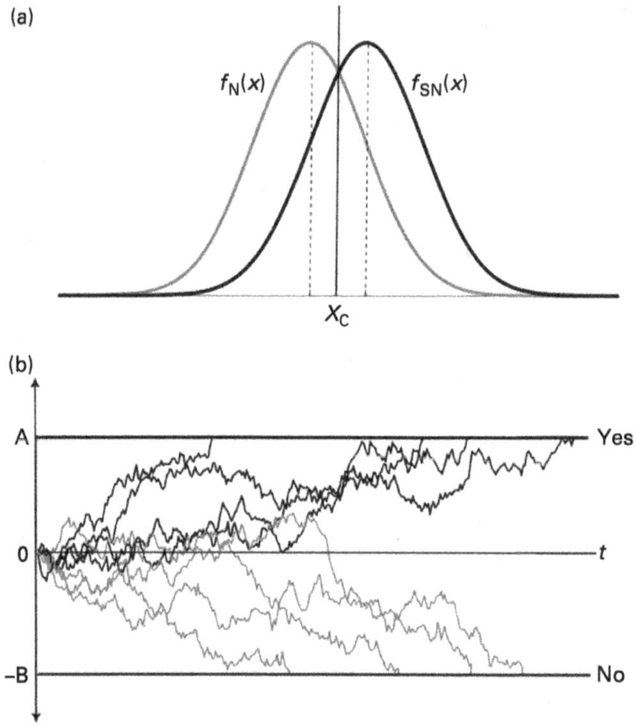


**2.11** (a) The normal, equal-variance model in which $d' = 2$. (b) A drift-diffusion model in which the drift is determined by random sampling from the N or SN distribution. Samples larger than $X_C$ push the drift up, whereas samples smaller than $X_C$ push it down. Sample paths are shown for eight hypothetical trials—four SN trials (in black) and four N trials (in gray).

This version of the drift-diffusion model includes the $d'$ and $X_C$ parameters of SDT plus the response criteria A and B. However, in addition to predicting accuracy data, the diffusion model also predicts response times (RTs) because closed-form expressions exist for first passage times (i.e., time when the process first crosses a response threshold). Chapter 10 considers the RT predictions of this model in some detail (i.e., see section 10.2.1). So the purpose of this section is simply to introduce the model and point out some of its advantages over classical SDT. First, because the drift-diffusion model makes predictions about both accuracy and response time, there are more data to fit, and therefore more degrees of freedom available for parameter estimation and model testing. Several computer packages are available that automate this parameter estimation process (Vandekerckhove & Tuerlinckx, 2007; Wiecki et al., 2013).

Second, note that the drift-diffusion model can represent a response bias in two different ways. One is to place $X_C$ at some point where the likelihood ratio is different from 1 (assuming equal base rates and payoffs), and another is to set A $\neq$ B. Of course, the classical SDT model can account for bias only by adjusting $X_C$. Consider a condition in which the observer adopts a conservative criterion and therefore is biased toward responding NO. Therefore, according to SDT, $X_C$ is set at some point where the likelihood ratio is greater than 1 (i.e, $\beta > 1$). Now consider trials in that condition where the sensory value falls at some point where the likelihood ratio is greater than 1 but less than $\beta$. According to SDT, the observer will respond NO on this trial, even though the evidence objectively favors a YES response (because the likelihood ratio is greater than 1). Balakrishnan (1999) presented evidence against this prediction. In particular, he described results of several experiments that suggested that observers always respond with the alternative that is most likely to be correct, even if they are biased toward one response and against the other. Unfortunately, there is no way to represent this state of affairs in classical SDT. In contrast, the drift-diffusion model offers an elegant resolution to this apparent paradox. Balakrishnan's results suggest that $X_C$ is set at the point for which $\beta$ is approximately 1.0 in all applications (e.g., as in figure 2.11). A bias toward a NO response can then be implemented by setting A > B. Thus, according to this account, the evidence is always judged objectively. Evidence that objectively favors SN always makes a YES response more likely, and evidence that favors N always makes a NO response more likely. Therefore, a bias toward responding NO does not color the observer's view of the world. Instead, the observer is simply willing to stop and respond NO on the basis of less overall evidence then they are willing to stop and respond YES. This more reasonable view of response bias is among the greatest advantages that the drift-diffusion model provides over and above classical SDT.

## 2.8 PROOFS

### 2.8.1 PROOF OF THEOREM 2.1

Recall from equations 2.2 and 2.3 that

$$P(\text{FA}) = \Phi(-X_C) \tag{2.28}$$

and

$$P(\text{H}) = \Phi\left(\frac{-X_C + d'}{\sigma}\right), \tag{2.29}$$

where $\Phi(z)$ is the cumulative distribution function of the Z distribution—that is, $\Phi(z)$ equals the area under the Z distribution to the left of the value $z$. Note that these two equalities imply that

$$Z_{P(\mathrm{FA})} = -X_\mathrm{C} \tag{2.30}$$

and

$$Z_{P(\mathrm{H})} = \frac{-X_\mathrm{C} + d'}{\sigma}. \tag{2.31}$$

As a result,

$$Z_{P(\mathrm{H})} = \frac{Z_{P(\mathrm{FA})} + d'}{\sigma}, \tag{2.32}$$

from which the result directly follows. ■

### 2.8.2 PROOF OF THEOREM 2.2

By definition, a differentiable function is concave down if and only if its slope is monotonically decreasing. The slope of the ROC curve is

$$\frac{dP(\mathrm{H})}{dP(\mathrm{FA})} = \frac{d[1 - F_{\mathrm{SN}}(X_\mathrm{C})]}{d[1 - F_\mathrm{N}(X_\mathrm{C})]} = \frac{f_{\mathrm{SN}}(X_\mathrm{C})}{f_\mathrm{N}(X_\mathrm{C})} = L(X_\mathrm{C}). \tag{2.33}$$

An isosensitivity curve sweeps across all possible values of $X_\mathrm{C}$. For example, in all SDT models considered in this chapter, continuously decreasing $X_\mathrm{C}$ in the equation 2.1 decision rule from $+\infty$ to $-\infty$ generates an isosensitivity curve that increases continuously from the point (0,0) to the point (1,1). Therefore, because the slope of the isosensitivity curve equals the likelihood ratio, if that slope monotonically decreases between the points (0,0) and (1,1), then

1. the isosensitivity curve is concave down (by definition of concave down);
2. the likelihood ratio monotonically decreases as $X_\mathrm{C}$ decreases; and therefore
3. the likelihood ratio monotonically increases as $X_\mathrm{C}$ increases. ■

### 2.8.3 PROOF OF THEOREM 2.4

In the multiple-look experiment, each of the $r$ N or SN samples generates its own sensory value. Denote the $i$th of these by $X_i$, and the collection of all $r$ by the vector $\underline{X}' = [X_1, X_2, \ldots, X_r]$. Under the assumptions of the theorem, note that on N trials, $\underline{X}$ has an $r$-dimensional multivariate $Z$ distribution, and on SN trials it has an $r$-dimensional multivariate normal distribution with mean vector $\underline{\mu}' = [d'_{\text{YN}}, d'_{\text{YN}}, \ldots, d'_{\text{YN}}]$ and variance-covariance matrix equal to the identity. Since the variance equals 1 in all directions, the standardized distance between the N and SN means equals

$$\begin{aligned} d'_r &= \sqrt{(d'_{\text{YN}} - 0)^2 + (d'_{\text{YN}} - 0)^2 + \ldots + (d'_{\text{YN}} - 0)^2} \\ &= \sqrt{r\, d'^{2}_{\text{YN}}} \\ &= \sqrt{r}\, d'_{\text{YN}}. \end{aligned} \tag{2.34}$$

■

# 3

# SDT MODELS OF RECOGNITION MEMORY, VISUAL SEARCH, EYEWITNESS IDENTIFICATION, AND GROUP DECISION-MAKING

SDT has been used to model a great many phenomena. A review of all of these is well outside the scope of any single chapter. Instead, this chapter selectively focuses on three popular applications of SDT—as a model of: recognition memory; visual search and eyewitness identification; and group decision-making. I chose these applications partly because they are each supported by large and widely read literatures, and partly because they each include some novel mathematical generalizations of the standard SDT model. My focus will be on this latter feature of each application. I have no particular expertise in any of these empirical domains, and in each case there are already a number of excellent reviews of each literature. I'll try to point the interested reader to some of these in each of the following sections. The goal of this chapter, then, is not to review each of these large and vibrant research areas, but rather to describe the use of SDT in each area, with a special emphasis on describing any mathematics that might be needed to transition from a chapter 2–level understanding of SDT to these specialized literatures.

## 3.1 RECOGNITION MEMORY

The first applications of SDT, of course, were to YES-NO detection tasks in which the observer was trying to detect the presence of a signal embedded in noise. But the basic SDT model was applied to recognition memory almost immediately (Bernbach, 1967; Egan, 1958). In a typical experiment, called old-new recognition, a participant is presented with a list of words and then sometime later presented with a series of single words and asked to respond whether each word is old or new—that is, participants are asked to indicate whether each word was or was not on the studied list.

A straightforward SDT interpretation of this task is that each word elicits a memory strength, the participant sets a criterion on this strength, responds OLD if the memory strength of the current word exceeds this criterion, and otherwise responds NEW. New items are often called "lures," and old items are called "targets." Lures and targets will both generate a range of memory strengths, but on average we expect targets to generate larger memory strengths than lures. As a result, a natural approach would be to model responding in this task exactly as in the YES-NO detection task, except for relabeling. In the SDT model of the old-new recognition-memory task, the abscissa becomes memory strength (instead of sensory magnitude), the N distribution becomes the lure distribution, and the SN distribution is relabeled as the target distribution.

This simple SDT model was challenged almost immediately by researchers who argued that recognition memory could succeed for multiple reasons—for example, because of either high familiarity or recollection. Specifically, I can decide that I've previously seen a person that I encounter in a coffee shop either because I recollect that this is a person I met on my last group bicycle ride or simply because they look familiar. Several different models have been proposed on how best to account for these processes. The next three subsections review three of the most prominent approaches.

### 3.1.1 THE NORMAL UNEQUAL-VARIANCE SDT MODEL

Empirical ROCs (i.e., isosensitivity curves) estimated from confidence judgments collected in the old-new recognition-memory task are curved and when plotted in Z space, the ROCs are approximately linear, most typically with a slope less than 1.0 (e.g., Glanzer et al., 1999; Ratcliff et al., 1992). These results are consistent with the predictions of a normal SDT model in which the target distribution has a larger variance than the lure distribution. A typical value for the slope in Z space is about 0.8 (Ratcliff et al., 1992; Wixted, 2007), which is consistent with a model in which the target standard deviation is 1.25, given a lure standard deviation of 1.0 (i.e., 0.8 = 1/1.25; see theorem 2.1).

For these reasons, the normal, unequal-variance SDT model is a popular model of performance in the old-new recognition-memory task (Wixted, 2007). Mathematically, the model suffers from the same weaknesses that we reviewed in chapter 2—in particular, it makes the nonsensical prediction that items eliciting the weakest possible memory strengths are more likely to be targets than lures. However, as discussed in section 2.5.2, this weakness has no practical consequence when the ratio of the target-to-lure standard deviations is only 1.25. With a ratio this modest, it is essentially impossible for the target distribution to produce a weak memory strength that is so weak that an ideal observer would respond OLD. As a result, the primary liability of the normal, unequal variance SDT model is eliminated under these conditions.

Note that the normal, unequal-variance model assumes that all OLD responses depend only on a single memory-strength value. Thus, this is a single-process model because it does not assume separate recollection and familiarity processes, or at least that these separate processes do not both contribute to old-new recognition responses. In contrast, dual-process models assume that judgments that an item is old depend both on recollection and familiarity (e.g., Yonelinas, 1994; Yonelinas et al., 1998). Several proposals have tried to reconcile this conflict by arguing that recognition depends only on the value of a single continuous memory strength, but that this memory strength increases with both familiarity and recollection (Rotello et al., 2004; Wixted & Stretch, 2004). The basic idea is as follows. Suppose the before-study memory strengths elicited by the population of items that might be used in the old-new recognition-memory task are normally distributed with mean $\mu_b$ and variance $\sigma_b^2$. Next suppose that studying the list of target items causes the memory strength of each item to be increased by two processes—familiarity and recollection—and further that these two sources are additive. Suppose the increase in memory strength due to familiarity is normally distributed with mean $\mu_f$ and variance $\sigma_f^2$ and that the increase in memory strength due to recollection is normally distributed with mean $\mu_r$ and variance $\sigma_r^2$. According to this model, memory strength on lure trials is normally distributed with mean $\mu_b$ and variance $\sigma_b^2$, whereas memory strength on target trials is normally distributed with mean $\mu_b + \mu_f + \mu_r$ and variance $\sigma_b^2 + \sigma_f^2 + \sigma_r^2$ (assuming the three sources are statistically independent). So this model is formally equivalent to the normal, unequal-variance SDT model. The lure and target distributions are both normal, the difference in means equals $\mu_f + \mu_r$ and the target distribution has a greater variance than the lure distribution.

### 3.1.2 THE DUAL-PROCESS SDT MODEL

Dual-process models assume that recognition memory depends on separate recollection and familiarity processes (Yonelinas, 1994; Yonelinas et al., 1998). The most common assumption is that OLD responses in the old-new recognition-memory task are a mixture. Some OLD responses occur because the participant recollects that the item is old and other OLD responses occur because recollection fails but the item appears highly familiar. Yonelinas (1994) assumed that recollection is an all-or-none process that operates according to classical threshold theory. On target trials, there is some probability $p$ that recollection is successful and on these trials the participant always responds OLD. If recollection fails, which occurs with probability $1-p$, it completely fails in the sense that there is no partial recollection value that can be used to select a response. Instead, the choice between responding OLD versus NEW depends completely on familiarity. In the model proposed by Yonelinas (1994), judgments based on familiarity are modeled by the normal, equal-variance SDT model.

If we let R denote successful recollection and F indicate that the response is selected based on the familiarity of the presented item, then the dual-process model proposed by Yonelinas (1994) assumes that the probability of an OLD response on target trials—that is, the probability of a hit—equals

$$\begin{aligned} P(\text{OLD}|\text{target}) &= P(\text{R}|\text{target})P(\text{OLD}|\text{R}) + [1 - P(\text{R}|\text{target})]P(\text{OLD}|\text{F \& target}) \\ &= (p \times 1) + (1-p)[1 - \Phi(X_\text{C} - \mu_\text{T})] \\ &= p + (1-p)\Phi(\mu_\text{T} - X_\text{C}), \end{aligned} \tag{3.1}$$

where $\mu_\text{T}$ is the mean familiarity strength of target items, $X_\text{C}$ is the criterion on familiarity for responding OLD, and as before, $\Phi$ is the standard-normal cumulative distribution function. Note that $P(\text{OLD}|\text{R}) = 1$, whereas if recollection fails then the probability of responding OLD depends on familiarity in exactly the way predicted by the normal, equal-variance model [i.e., $P(\text{OLD}|\text{F \& target}) = \Phi(\mu_\text{T} - X_\text{C})$]. The model also assumes that recollection is impossible on lure trials, so $P(\text{R}|\text{lure}) = 0$. As a result, the probability of responding OLD on lure trials—that is, the probability of a false alarm—is exactly the same as in the normal, equal variance SDT model. In other words,

$$P(\text{OLD}|\text{lure}) = 1 - \Phi(X_\text{C}) = \Phi(-X_\text{C}). \tag{3.2}$$

Note that this model has three free parameters: $p$, $\mu_\text{T}$, and $X_\text{C}$—the same number as the normal, unequal variance model.

Figure 3.1 shows ROC curves predicted by this model. Panel (a) shows isorecollection curves because every point on the same curve has the same probability of successful recollection, and panel (b) shows isofamiliarity curves because every point of each of these curves is associated with the same mean familiarity strength. Panels

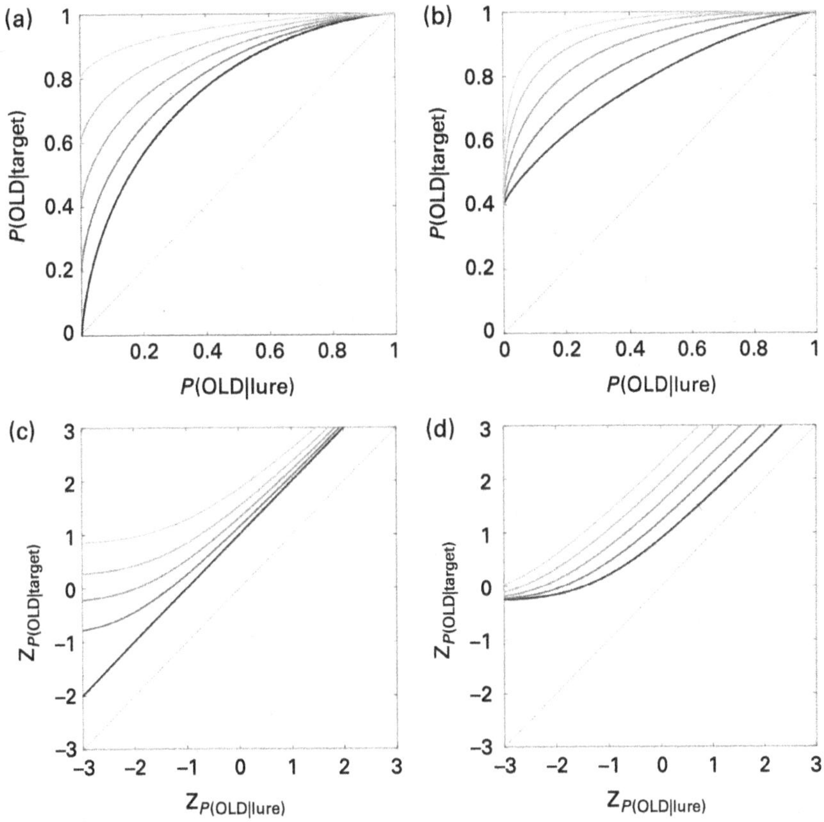


**3.1** ROC curves predicted by the dual-process SDT model of recognition memory. (a) ROC curves predicted by the model for five values of $p$ (i.e., $p=0, .2, .4, .6$, and $.8$). $\mu_T=1$ for all curves. (b) ROC curves predicted by the model for five values of $\mu_T$ (i.e., $\mu_T=.5, .9, 1.3, 1.7$, and $2.1$). $p=.4$ for all curves. (c) Same curves as in panel (a), except plotted in Z space. (d) Same curves as in panel (b), except plotted in Z space.

(c) and (d) are the same as (a) and (b), except plotted in Z space. First, note that in panel (a), the curves intersect the ordinate at $P(\text{OLD}|\text{target})=p$. This intersection occurs when $P(\text{OLD}|\text{lure})=0$; that is, when the probability of a false alarm equals zero, which requires that $X_C=\infty$. Under these conditions $\Phi(-X_C)=0$, but when $X_C=\infty$, it is also true that $\Phi(\mu_T-X_C)=0$, and therefore from equation 3.1, note that $P(\text{OLD}|\text{target})$ must always equal $p$. Note that this prediction is fundamentally different from predictions of the normal, unequal-variance model. In that model, if the probability of a false alarm equals zero, then the probability of a hit must also equal zero, because both require that the criterion is set to $X_C = \infty$. In contrast, the dual-process model predicts that the probability of a hit can take on any value, even when the probability of a false alarm equals zero. The model predicts that the conditions under which this will occur are that there is some nonzero probability that recollection succeeds and on trials when it fails, participants are so conservative that they never respond YES. Although this is an interesting prediction that represents a fundamental psychological difference between the dual-process and the normal, unequal-variance SDT models, it is not clear that it is empirically useful, because it would be extremely difficult to set up experimental conditions that would cause participants to be so conservative on trials when their decisions are based on familiarity.[1]

Second, note that because all curves in figure 3.1 terminate at the point (1,1), this property guarantees that the ROC curves predicted by the model are all skewed in the same direction as empirical ROC curves from the old-new recognition-memory task (that is, when $p > 0$).

Later versions of the model added a fourth parameter: a threshold that came into play only on lure trials (e.g., Koen et al., 2017). The idea was that if the memory strength of an item on a lure trial was below this threshold then the participant would respond NEW without assessing familiarity. This threshold did not affect target trials because it was assumed that no target item would ever generate such a low memory strength. As a result, equation 3.1 still described predictions for $P(\text{OLD}|\text{taget})$ in this generalized model. However, predictions on lure trials did change. In particular,

$$P(\text{NEW}|\text{lure}) = q + (1-q)\Phi(X_C), \tag{3.3}$$

where $q$ is the probability that the memory strength of a lure falls below the newness threshold. As a result,[2]

$$P(\text{OLD}|\text{lure}) = 1 - P(\text{NEW}|\text{lure}) = (1-q)\Phi(-X_C). \tag{3.4}$$

### 3.1.3 THE MIXTURE SDT MODEL

The mixture SDT model of recognition memory is similar to the dual-process model, in the sense that both models assume target trials include a mixture of responses that have different statistical properties (DeCarlo, 2002, 2010; Koen et al., 2017). The dual-process model assumes that some OLD responses on target trials are the result of recollection and the others are because recollection failed but the presented item elicited a strong sense of familiarity. The mixture SDT model is similar but assumes the two types of OLD responses on target trials are due to different levels of attention during the encoding of the target item at the time of initial study (DeCarlo, 2002). The idea is that as the study items are initially presented, the participant's attention waxes and wanes. The target items presented when attention is high will elicit a large memory strength when later presented during the testing phase, whereas target items initially presented when attention is low will elicit a weak memory strength during test.

The mixture SDT model (DeCarlo, 2002) accounts for memory strengths elicited by lures exactly as the N distribution is modeled in the normal, equal-variance SDT model. As a result,

$$P(\text{OLD}|\text{lure}) = \Phi(-X_C).$$

In contrast, the model assumes that the memory strengths elicited by targets are a binary mixture of large and small memory strengths, each normally distributed with variance 1, but with different means. More specifically, the model assumes that

$$P(\text{OLD}|\text{target}) = p\Phi(d_1 - X_C) + (1-p)\Phi(d_2 - X_C), \tag{3.5}$$

where $p$ is the probability that a target item received a high level of attention during study, $d_1$ is the mean memory strength when attention was high, and $d_2$ is the mean memory strength when attention was low. In many applications, it is common to assume $d_2 = 0$ – that is, that the mean memory strength during low attention is the same as for the lures, or in other words, that with probability $1-p$, the participant completely ignores the target item during study. One advantage of this simplifying assumption is that it reduces the number of free parameters in the model from 4 to 3 (i.e., $p$, $d_1$, and $X_C$), and as a result, this version of the mixture model has the same number of free parameters as the normal, unequal-variance and dual-process SDT models.

Figure 3.2 shows some isoattention ROC curves predicted by the model. Each point on every curve is associated with the same probability that a target item received a high level of attention during study. Note that these curves are similar to the ROC curves predicted by the dual-process SDT model (i.e., compare figures 3.1 and 3.2). This is not a surprise because both models make identical predictions for $P(\text{OLD}|\text{lure})$ and similar predictions for $P(\text{OLD}|\text{target})$ (i.e., compare equations 3.1 and 3.5).

### 3.1.4 THE FIXED-POINT PROPERTY OF BINARY MIXTURES

The normal, unequal-variance model, the dual-process model, and the mixture model all make similar predictions about ROC curves (but not identical predictions; see, e.g., Wixted, 2007). For example, a comparison of figures 2.6, 3.1, and 3.2 show that

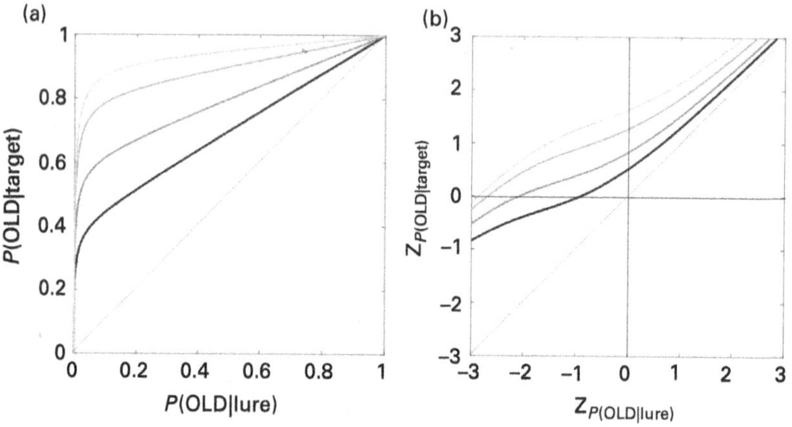


**3.2** ROC curves predicted by the mixture SDT model of recognition memory. (a) ROC curves predicted by the model when $p = .4, .6, .8$, and $.9$. (b) The same curves as in panel (a), except plotted in Z space. In all curves, $d_1 = 3$ and $d_2 = 0$.

all of these models predict, or can predict, skewed isosensitivity curves that generally have a slope in Z-ROC space that is less than 1.0—properties that are present in virtually all empirical ROCs estimated from the old-new recognition-memory task. Despite these similar predictions, the models make fundamentally different assumptions about the nature of processing on target trials. The normal, unequal-variance model predicts that this is a unitary process in the sense that responding on every trial is mediated by the same cognitive processes. In contrast, both the dual-process and mixture models assume that responding on target trials is a mixture of two qualitatively different types of trials. It turns out that this prediction—that responding to targets is a mixture of two types of trials—has a distinct empirical signature that, to my knowledge, has never been tested empirically. That signature is described in this subsection.

Mixture models have been proposed in a variety of different domains. For example, one of the earliest and most widely known mixture models is the fast-guess model (Yellott Jr., 1971), which accounts for fast error responses in a wide variety of cognitive tasks by proposing that observable RTs are a probability mixture of two types of responding. With some probability $p$, the participant ignores the stimulus and just randomly guesses a response. The lack of perceptual or cognitive processing on these trials causes accuracy to be low and RT to be fast. With probability $1 - p$, the participant fully processes the stimulus, causing responding to be accurate and slow.

All binary-mixture models have a unique empirical signature that was discovered by Falmagne (1968), which he called the *fixed-point property*. Specifically, consider a set of experimental conditions in which the mixture probability varies but the component distributions do not. For example, in the case of the fast-guess model, the conditions might vary speed stress. The fast-guess model predicts that the more that speed is emphasized, the more likely it is that the participant will ignore the stimulus and make a quick guess. So the model predicts that the primary effect of increasing speed stress will be to increase the guessing probability $p$. Falmagne (1968) showed that in any experiment with different conditions that affect the mixture probability $p$, but not the component distributions, all binary-mixture models predict that the pdfs predicted for each experimental condition must all intersect at exactly the same point. This property is described more formally in the following result.

**Theorem 3.1** *The Fixed-Point Property of Binary Mixtures* Consider a model that predicts that some relevant pdf in experimental condition $i$, denoted $f_i(x)$, is a binary mixture of two component pdfs $g_1(x)$ and $g_2(x)$, with mixture probability $p_i$. More specifically, suppose the model predicts that

$$f_i(x) = p_i g_1(x) + (1 - p_i) g_2(x). \tag{3.6}$$

Now consider a set of conditions in which the mixture probability $p_i$ varies, but $g_1(x)$ and $g_2(x)$ do not. Then if the pdfs predicted for each condition intersect, they must all intersect at the same fixed point.

*Proof* See Falmagne (1968) or Ashby (2024). ■

The fixed-point property holds for any random variable that is a binary mixture. As we saw, the mixture SDT model predicts that on target trials, the decision variable that drives old-new recognition judgments is a binary mixture of two normal distributions (i.e., see equation 3.5). As a result, it predicts that across a set of conditions in which the mixture probabilities vary but the component pdfs do not, the set of all target distributions predicted by the mixture model must all intersect at the same point. The dual-process SDT model is also a binary mixture model, but its predicted familiarity distributions are not constrained by the fixed-point property because there is no $g_1(x)$ pdf to intersect with $g_2(x)$. The big-picture question here though is why any of this should matter since memory strength distributions are not observable. The fixed-point property is useful only if the mixture variable is some observable dependent measure. If it was observable, then a strong test of the model would be to estimate the mixture distributions and check whether they all intersect at the same point. Although memory strength is not an observable variable, the dual-process and mixture models predict that several dependent measures depend directly on memory strength and therefore are candidates for tests of the fixed-point property. Two widely used dependent variables come immediately to mind—RTs and ROC curves. Ashby (2024) showed that both the dual-process and mixture models make strong fixed-point predictions for both of these dependent variables.

**3.1.4.1 Response times** By themselves, none of the recognition-memory models we have considered so far make any assumptions or predictions about the RTs we might expect in the old-new recognition-memory task. Even so, as we saw in the previous chapter, there are generalizations of SDT that do make RT predictions. Section 2.7 briefly considered the drift-diffusion model as one such model. Chapter 10 considers the drift-diffusion model in much more detail, as well as a variety of other generalizations of SDT (and GRT) that make RT predictions. Among the oldest and simplest methods for deriving RT predictions from SDT is to add a straightforward assumption to SDT, called the RT-distance hypothesis, that simply assumes that RT decreases with the distance between the percept and the response criterion. Chapter 10 considers the RT-distance hypothesis in detail, but briefly, the idea is that if decisions are made by comparing a percept or memory strength to a criterion, then the greater the distance between the two, the easier, and hence the faster the decision. Theoretical predictions of the generalized SDT model that include this assumption were worked out by Murdock (1985) and empirical support for the RT-distance hypothesis was first reported by Emmerich et al. (1972) and Gescheider et al. (1969) and later, within the multidimensional context of GRT, by Ashby et al. (1994). Furthermore, Murdock (1985) showed that this generalized SDT model gives good accounts of the RTs that are observed in recognition-memory experiments. In addition, as we will see in chapter 10, the RT predictions of the drift-diffusion model, as well as all other current statistical decision theory models that make RT predictions, are all in general agreement with the RT-distance hypothesis. So adding the RT-distance hypothesis to any of the recognition-memory models considered in this chapter represents a theoretically minimal approach to extending these models to the RT domain.

The following result, due to Ashby (2024), shows that adding the RT-distance hypothesis to any model that predicts memory strengths on target trials are a binary mixture causes that model to also predict that the RT pdfs on target trials must satisfy the fixed-point property.

**Theorem 3.2** Consider any recognition-memory model that assumes OLD responses in the old-new recognition-memory task are a binary mixture (e.g., including the dual-process and mixture SDT models). Suppose we extend this class of models to the RT domain by assuming that

$$RT = T + T_0, \tag{3.7}$$

where $T$ is the decision time that is computed by adding the RT-distance hypothesis to the memory model and $T_0$ is any random motor time that is statistically independent of $T$. Now consider an old-new recognition experiment that includes a set of conditions in which the mixture probability $p$ varies across conditions, but the component distributions do not. Then all RT pdfs predicted by these models on target trials when the participant responds correctly must satisfy the fixed-point property—that is, if the RT pdfs predicted for each condition intersect, they must all intersect at the same fixed point.

*Proof* See Ashby (2024). ■

A test of this strong prediction of all binary mixture models requires an old-new recognition experiment that includes at least three conditions that the model predicts should all be identical, except for the numerical value of the mixture probability $p$. The dual-process model assumes that $p$ is the probability that recollection is successful, so a candidate experiment would manipulate some independent variable that selectively influences the probability of recollection. In contrast, the mixture model assumes $p$ is the probability that the item received a high level of attention during study, so in this case, a candidate experiment would manipulate an independent variable that selectively influences the amount of attention available for encoding the target list during initial study. One way to do this might be to require the participant to perform a simultaneous dual task during encoding that varies across conditions in memory load. For example, the experiment might include four conditions in which the participant is required to study the target items while holding in working memory a list of 0, 2, 4, or 6 digits, respectively.

Theorem 3.2 is illustrated in figure 3.3, which shows RT pdfs predicted by the dual-process SDT model (panel a), the mixture SDT model (panel b), and the normal, unequal-variance SDT model (panels c and d) in a hypothetical experiment of this type. In all cases, predicted RT pdfs are shown for correct OLD responses in four different hypothetical experimental conditions. In all cases where the decision was based on an SDT model in which the memory strength was compared to a response

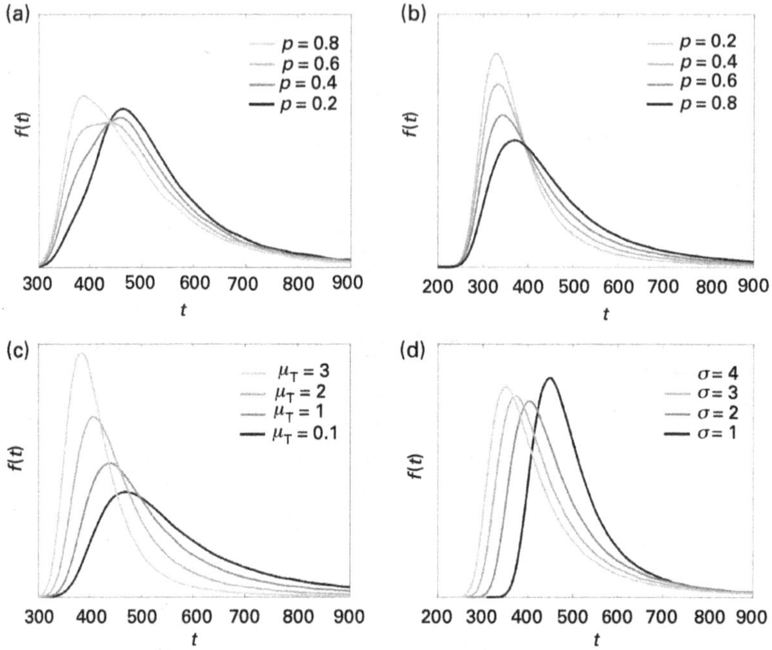


**3.3** RT pdfs predicted by recognition memory models on target trials of an old-new recognition-memory task on which the participant responds correctly. (a) Predictions of the dual-process model when the only change across conditions is in the value of $p$. In all cases $\mu_T = 1$, the mean and standard deviation of the normal component of the exGaussian distribution were 350 ms and 20 ms, respectively, and the mean of the exponential component was 125 ms. (b) Predictions of the mixture model when the only change across conditions is in the value of $p$. The parameters all had the same values as in panel (a), except the means of the two component target distributions were set to .5 and 2.0. (c) Predictions of the normal, unequal-variance model when the only change across conditions is in the mean of the target distribution. (c) Predictions of the normal, unequal-variance model when the only change across conditions is in the variance of the target distribution.

criterion $X_C$, RT was computed from an RT-distance model in which decision time was a power function of the distance between the memory strength and $X_C$ (see chapter 10 for a full discussion of this model). Specifically, in all cases

$$RT = T_0 + 450\,D^{-0.35}, \tag{3.8}$$

where $T_0$ was a normally distributed motor time with mean 100 ms and standard deviation 10 ms.

Consider first the dual-process SDT model with predictions shown in panel (a). On trials when the OLD response was determined by recollection, I assumed that RT had an exGaussian distribution (Hohle, 1965; Ratcliff, 1978), which is among the most popular current models of RT pdfs. The exGaussian distribution is the distribution that results when independent random samples from a normal distribution and an exponential distribution are added together (see chapter 10 for details). Figure 3.3a shows predictions of the model for four different values of $p$ (i.e., the probability of successful recollection). Note that, as the theorem requires, all four pdfs intersect at the same fixed point. Figure 3.3b shows similar predictions for the mixture SDT model. Note again that, as required by the theorem, all four distributions pass through the same fixed point.

Figures 3.3c and d show predictions of the normal, unequal-variance model for this same hypothetical experiment. In panel (c), the mean of the target distribution is assumed to change across conditions, whereas panel (d) was produced under the assumption that the variance of the target distribution changes. Note that these pdfs all intersect, but each pair intersect at a different time point, and therefore do not satisfy the fixed-point property. As a result, figures 3.3c and 3.3d show that this fixed-point property is not a feature that should be expected to hold in models that are not constructed from a binary mixture.

It is also worth stressing that this test of mixture models is parameter free. By this I mean that the test is valid for any version of the models—no matter how the models are parameterized and for any given parameterization, no matter how many free parameters they have. For example, the 4-parameter mixture model can provide good fits to a more diverse set of empirical ROC data than the 3-parameter version, but theorem 3.2 holds for any version of the mixture model, so the 3-parameter and 4-parameter versions both predict that all RT pdfs from the experiment described in theorem 3.2 must intersect at the same time point (if they intersect at all).

To my knowledge, this differential prediction of the dual-process and mixture models versus the normal, unequal-variance model has never been empirically investigated. It seems worthy of empirical test however, because it is such a strong prediction. A finding that all empirical RT pdfs cross at the same time point seems like strong evidence that OLD responses are a binary mixture of two different trial types because it seems unlikely that a model postulating a unitary process would coincidentally predict a set of pdfs that satisfy this condition. Fortunately, a statistical test of the fixed-point property has been developed, as well as an R package that implements this test (van Maanen et al., 2014). On the other hand, the fixed-point property could fail for a variety of different reasons, even if OLD responses are a binary mixture (van Maanen et al., 2016). The most likely scenario is probably that the experimental manipulations caused one or both of the component distributions to change across conditions. For these reasons, a failure of the fixed-point property should be interpreted with caution.

**3.1.4.2 ROC curves** A qualitatively different possibility is that a binary-mixture target distribution will leave an empirical signature in the ROC curves that it produces. An advantage of an ROC-curve test of mixture models is that no extra assumptions about processing time are needed. For example, there is no need to assume the RT-distance hypothesis. As this subsection shows, a large class of mixture models do make a striking ROC curve prediction. This prediction is described in the next result, which is also due to Ashby (2024).

**Theorem 3.3** Consider an experiment with a variety of different conditions. Now consider a model that assumes a single lure distribution and that the target distribution is a binary mixture—that is, in which

$$f_L(x) = g_0(x), \tag{3.9}$$

and

$$f_T(x) = p_i g_1(x) + (1 - p_i) g_2(x), \tag{3.10}$$

where $p_i$ is the mixture probability in condition $i$. Suppose this model satisfies the following conditions:

1. The mixture probability $p_i$ varies across conditions, but the lure distribution and the two component target distributions do not.
2. The cumulative distribution functions $G_1$ and $G_2$ of the two component target distributions cross—that is, there exists some value $x^*$ for which $G_1(x^*) = G_2(x^*)$.

Then all ROC curves predicted by this model must intersect at the same point in ROC space.

*Proof* See Ashby (2024). ■

Theorem 3.3 says that if the component cumulative distribution functions of a mixture model cross, then the ROC curves predicted by that model must all cross at the same point in ROC space. This result is illustrated in figure 3.4. Panels (a) and (b) show ROC curves for two different mixture models, and panels (c) and (d) show the same two curves, except plotted in Z space. Within each panel, the different curves were generated from different values of the mixture probability $p$. Because $\sigma_1 \neq \sigma_2$ in all panels, this model predicts that the cumulative distribution functions of the target

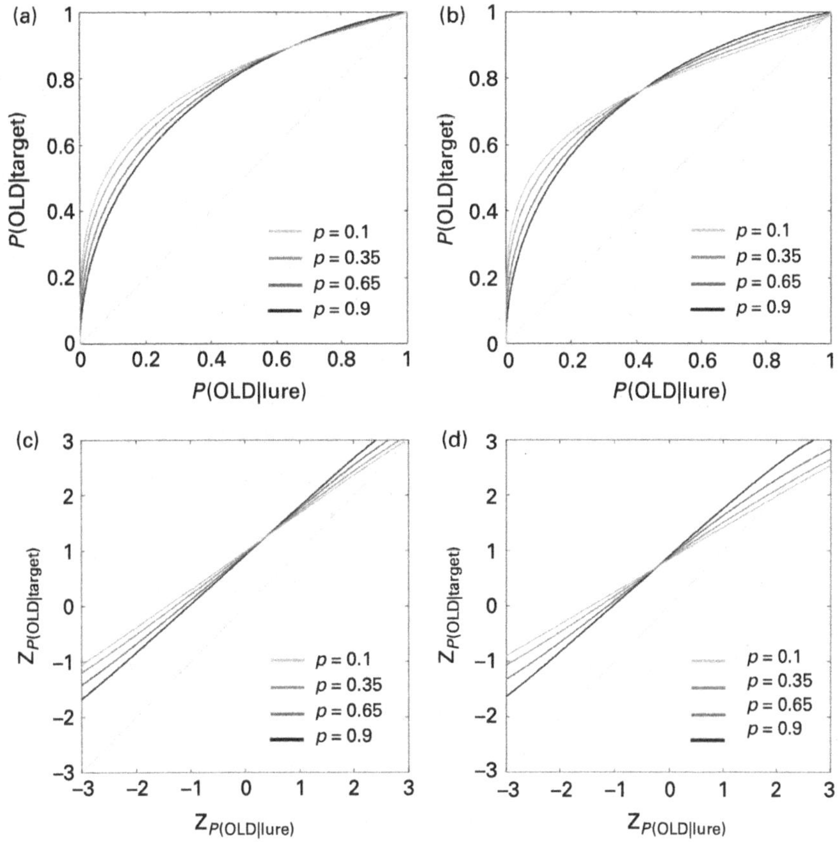


**3.4** ROC curves predicted by the unequal-variance mixture SDT model of recognition memory. In all cases, $d_1 = 1.5$, $d_2 = 1$, and $\sigma_2 = 1.1$. The curves differ only in the value of the mixture probability (i.e., $p$ = .1, .35, .65, and .9). (a) ROC curves predicted by the model when $\sigma_1 = 1.5$. (b) ROC curves predicted by the model when $\sigma_1 = 1.8$. (c) Same curves as in (a), except plotted in Z space. (d) Same curves as in (b), except plotted in Z space.

component distributions cross, and therefore the conditions of theorem 3.3 are met, and as a result, the ROCs must cross.

There are several points of note here. First, as comparison, note from figures 3.1 and 3.2, that the dual-process model and the equal-variance mixture model both predict ROC curves that are ordered by the value of the mixture probability $p$ (and therefore do not intersect). Second, the unequal variances in the two component distributions of the mixture model used to generate figure 3.4 are consistent with the expectation that in nature, variance increases with mean. In figure 3.4, the lure, target component 2, and target component 1 distributions have means equal to 0, 1, and 1.5, respectively, and standard deviations equal to 1, 1.1, and either 1.5 or 1.8, respectively. So in all cases, the variance increases with the mean. As a result, the model illustrated in figure 3.4 can be seen as a melding of the (equal-variance) mixture model (DeCarlo, 2002) and the normal, unequal-variance SDT model in which familiarity and recollection have additive effects on memory strength (Rotello et al., 2004; Wixted & Stretch, 2004).

Third, it is important to note that although theorem 3.3 guarantees that the ROC curves will all intersect at the same point if the component distributions have different variance, this prediction may have limited empirical utility if the variance difference is small. This is because the intersection will occur so many standard deviations out in Z space that it will be essentially impossible to test empirically. As a result, a failure to find that a set of ROC curves all intersect at the same point says little about whether the target distribution is a binary mixture. On the other hand though, if the ROC curves do all intersect at the same point, then a mixture model might be strongly suspected because this is not a feature of ROC curves that is predicted by non-mixture models.

Theorem 3.3 does not depend on any processing-time assumptions, but it does not hold for all mixture models. Specifically, it does not hold for any model in which the likelihood ratio $g_2(x)/g_1(x)$ is monotonic in $x$. For all such models, condition 2 of theorem 3.3 is violated. This is because a monotonic likelihood ratio guarantees that the cumulative distribution functions are ordered (Townsend & Ashby, 1983, p. 281). As a result, condition 2 can only be met if the $g_2/g_1$ likelihood ratio is nonmonotonic. Note that this implies that the mixture model described in section 3.1.3 in which $g_1$ and $g_2$ are both normal with equal variance does not predict intersecting ROC curves (e.g., see figure 3.2) because the likelihood ratio of two normal distributions with equal variance is monotonic. Even so, more general versions of the mixture model have been proposed in which the two component target distributions have different variances (Koen et al., 2017). Theorem 3.3 applies to these models.

It is also important to note that theorem 3.3 makes no assumption about the lure distribution. Specifically, it does not require nonmonotonicity of the target-to-lure likelihood ratio $f_T(x)/f_L(x)$. Therefore, it does not assume that extremely low familiarities are more likely on target trials than on lure trials, as predicted, for example, by the unequal-variance SDT model.

## 3.2 VISUAL SEARCH AND EYEWITNESS IDENTIFICATION

In visual-search tasks, observers are presented with a multielement display and their task is to respond YES or NO depending on whether the display contains a previously identified target item. The items used in these tasks can be almost anything—from simple geometric figures to complex, real-world objects. Eyewitness identification tasks are a special type of visual search task in which the items are human faces. They are patterned after the lineups that police often use to test the ability of eyewitnesses to identify suspects of some crime. Eyewitness identification tasks typically include the extra requirement that if observers respond YES, they are then also asked to point out or identify the target face.

The visual search and eyewitness-identification literatures are large and complex. SDT-based modeling approaches are common in both fields, but so are a variety of other approaches, including for example, multinomial modeling (Menne et al., 2022) and approaches rooted in information theory (Starns et al., 2023). Surveying either literature, or even the most prominent modeling approaches within each literature, is well outside the scope of this book (for a review of the visual search literature, see, e.g., Wolfe, 2020). Instead, the goal of this section is modest. The focus is exclusively on SDT-based modeling approaches, and more specifically, on describing some key mathematical identities that are widely used in SDT-based models of visual search and eyewitness-identification. This section therefore serves as a bridge from chapter 2 to the use of SDT-based models in these specialized fields.

All SDT models of these tasks assume that each item in the multielement display has its own sensory distribution. On each trial, the sensory experience elicited by each item is modeled as a single sample from the associated distribution. So if the display includes $r$ items, then the observer's experience on each trial is represented by $r$ random samples from $r$ sensory distributions. Denote these by $X_1, X_2, \ldots, X_r$. If the display items are simple stimuli that vary on a single sensory dimension then the $X_i$ samples are interpreted as perceived values on that dimension. For example, if the stimuli are spots of white light that vary in luminance, then $X_i$ is the brightness of the $i$th spot. However, the models are quite general and can be applied to tasks that include any kind of display elements, and therefore, in general, the target element will often be defined by a constellation of multiple sensory features. For example, in a typical police lineup, no single sensory feature will identify the perpetrator. In these more usual applications, the $X_i$ samples are typically assumed to describe abstract evidence that the associated item is a target, rather than any particular sensory feature of the item. In almost all applications, the $X_i$ are assumed to be mutually independent. Even so, there has been some effort to investigate models that make the more realistic assumption that in eyewitness identification at least, the sensory values associated with the various faces in the lineup may be correlated (Wixted et al., 2018).

In a typical task, the display includes only nontargets on half the trials and, on the other half, one or more targets are included. Denote the former trial type by N (for nontarget) and the latter by T (for target). A common assumption is that the sensory evidence elicited by all nontarget items is identically distributed—most often, normal with zero mean and unit variance. The evidence elicited by targets is then assumed to have a distribution with a larger mean. In the simplest models, the target distribution will also be normal with the same variance as the nontarget distribution. So this is an extension of the normal, equal-variance SDT model to visual search and eyewitness-identification tasks. But it is also common to assume that the target distribution has larger variance than the nontarget distribution, in which case the resulting model is an extension of the normal, unequal-variance SDT model. In the eyewitness-identification literature, more complex models that assume multiple nontarget and target distributions are also common (J. Lee & Penrod, 2019; Wixted & Mickes, 2014; Wogalter et al., 2004).

Almost all visual-search tasks and most eyewitness-identification tasks have studied simultaneous displays in which all items are presented at the same time and the observer's task is to respond YES or NO depending on whether the display includes a target (and in the case of most eyewitness-identification tasks, also to identify the target). The simplest and most common assumption about how a decision is reached in this scenario is that a YES or target-present response is made if the largest of the $X_i$ samples is greater than some criterion. In the eyewitness-identification literature, some more recent work has extended SDT-based modeling to sequential lineups, in which the faces are presented one at a time (Dunn et al., 2022; Kaesler et al., 2020). In this case, more complex self-terminating decision rules must also be considered. With simultaneous presentations though, the most common decision rule—that is, to respond YES if the largest of the the sensory samples is greater than some criterion—is equivalent to the rule:

$$\text{Respond YES if } \max\{X_1, X_2, \ldots, X_r\} > X_C\text{; Otherwise respond NO.}$$

where $X_C$ is some constant response criterion. Graham (1989) called SDT models that use this decision rule maximum-of-outputs models, whereas Swensson and

Judy (1981) referred to these as extreme-detector models. Shaw (1980) proposed a mathematically equivalent model that she called the independent-decisions model, which assumes the following decision rule:

Respond YES if $X_1 > X_C$ *or* if $X_2 > X_C$ *or* ... if $X_r > X_C$; Otherwise respond NO.

These two rules are equivalent because they both respond YES if any of the $X_i$ samples are greater than $X_C$ and they both respond NO only if all of the $X_i$ samples are less than $X_C$ (Palmer et al., 2000).

Note that with this max rule, there are many ways that a YES response can occur—for example, there could be 1 or 2 or 3 (or more) $X_i$ all greater than $X_C$. Computing the probability associated with all of these different ways would be tedious. Fortunately, though, there is only one way that a NO response can occur, and that is that all $X_i \leq X_C$. As a result, the probability of a YES response equals one minus the probability that all $X_i \leq X_C$. If the $X_i$ are all statistically independent of each other and identically distributed, then, following this logic, the probability of a false alarm equals

$$\begin{aligned} P(\text{YES}|\text{N}) &= 1 - P(X_1 \leq X_C, X_2 \leq X_C, \ldots, X_r \leq X_C|\text{N}) \\ &= 1 - [P(X_1 \leq X_C|\text{N})P(X_2 \leq X_C|\text{N}) \cdots P(X_r \leq X_C|\text{N})] \\ &= 1 - [F_\text{N}(X_C)]^r, \end{aligned} \tag{3.11}$$

where $F_\text{N}(x)$ is the cumulative distribution function of the sensory values elicited by a nontarget item. The second equality in this derivation holds because of our assumption that all of the $X_i$ are statistically independent of each other. The probability of a hit can be computed in a similar manner. Suppose the display includes $r_\text{T}$ targets and $r - r_\text{T}$ nontargets. A hit occurs if any of the sensory values elicited by these $r$ items exceeds $X_C$. Most models make the same predictions regardless of the spatial locations of the targets. In this case, the probability of a hit equals

$$\begin{aligned} P(\text{YES}|\text{T}) &= 1 - [P(X_1 \leq X_C|\text{T})P(X_2 \leq X_C|\text{T}) \cdots P(X_r \leq X_C|\text{T})] \\ &= 1 - [F_\text{N}(X_C)]^{r-r_\text{T}}[F_\text{T}(X_C)]^{r_\text{T}}, \end{aligned} \tag{3.12}$$

where $F_\text{T}(x)$ is the cumulative distribution function of the sensory values elicited by a target item.

Equations 3.11 and 3.12 can be used to test this SDT model against empirical visual-search ROC curves (e.g., Clark et al., 2011; Palmer et al., 2000). To construct these curves, observers would typically complete a standard visual-search task, except rather than respond YES or NO on each trial, they would respond with a rating of how confident they are that a target is present. For example, with a 7-point scale, 7 would indicate that they are certain a target is present, 1 would indicate that they are certain a target is absent, and 4 would indicate complete uncertainty. Equations 2.17–2.19 would then be used to convert the rating responses to $[\hat{P}(\text{YES}|\text{N}), \hat{P}(\text{YES}|\text{T})]$ ordered pairs, where $\hat{P}(\text{YES}|\text{N})$ plays the role of $\hat{P}(\text{FA})$ and $\hat{P}(\text{YES}|\text{T})$ plays the role of $\hat{P}(\text{H})$ in the subsequent ROC analysis.

The next step is to use equations 3.11 and 3.12 to test the model against these data. The standard approach is to assume that discriminability is the same for all $[\hat{P}(\text{YES}|\text{N}), \hat{P}(\text{YES}|\text{T})]$ pairs, but that the different points are associated with different values of $X_C$—in other words, that the different points describe an empirical isosensitivity curve. In the simplest, normal, equal-variance SDT model, only one parameter needs to be estimated to generate a predicted isosensitivity curve (i.e., from equations 3.11 and 3.12) that can be compared to our empirical curve—namely, $d'$. In this version of the model, $F_N(X_C)$ is the cumulative distribution function of the standard, normal Z distribution (i.e., a normal distribution with mean 0 and variance 1) evaluated at the point $X_C$. In other words, $F_N(X_C)$ is the area under the Z distribution to the left of $X_C$. Similarly, $F_T(X_C)$ is the cumulative distribution function of a normal distribution with mean $d'$ and variance 1 (i.e., the area under this normal distribution to the left of $X_C$). There is no closed-form expression for $F_N(x)$ or $F_T(x)$, but most software packages include a one-line command that will supply a numerical approximation. For example, in Matlab, the command "normcdf(x,mu,sigma)" returns the area to the left of x under a normal distribution with mean mu and standard deviation sigma. If a closed-form expression is needed for some reason, then the logistic distribution could be used in place of the normal distribution (e.g., as in Wixted & Mickes, 2014). The normal and logistic SDT models make almost identical predictions, but the logistic distribution has a simple, closed-form expression for its cumulative distribution function.[3]

In models of eyewitness identification, it is common to include at least two different nontarget distributions. This is because the police commonly construct their lineups by including people that appear similar to the actual perpetrator (Wogalter et al., 2004). Wixted and Mickes (2014) referred to these high-similarity nontargets as lures to distinguish them from the low-similarity nontargets, which they called foils. So all nontargets in a fair lineup should be foils. In a nontarget lineup that includes lures, at least some of the lineup members will evoke stronger evidence values than would be expected from a fair lineup. In this case, equations 3.11 and 3.12 need to be amended. For example, equation 3.11 could be revised to

$$P(\text{YES}|\text{N}) = 1 - [F_{\text{N}_{\text{lure}}}(X_C)]^{r_{\text{lure}}}[F_{\text{N}_{\text{foil}}}(X_C)]^{(r-r_{\text{lure}})}, \tag{3.13}$$

where $F_{N_{lure}}(x)$ and $F_{N_{foil}}(x)$ are the cumulative distribution functions of the sensory evidence elicited by lures and foils, respectively, and $r_{lure}$ is the number of the $r$ display items that are lures. Equation 3.12 can be amended in a similar fashion.

Eyewitness identification requires an extra decision on trials when a target is detected—in particular, the observer must pick out or identify the perpetrator among all the people in the lineup. In experiments of this type, the key dependent variable is identification accuracy. To respond correctly on perpetrator-present trials, the observer must correctly identify the single perpetrator present in the lineup. So a false alarm occurs on perpetrator-present trials if the observer identifies a lure or foil as the perpetrator, even if they correctly respond YES.

Clark (2003) distinguished between two general approaches to this problem, which he called absolute and relative (following Wells, 1984). An absolute decision strategy depends only on the single item that most resembles the target, whereas a relative strategy also depends on other display items. For example, suppose we rank order the $r$ sensory values on each trial from largest to smallest. Let $X_{max}$ denote the largest of these and $X_{next}$ denote the next largest. So if the display contains only one target and $X_{max}$ is the sensory evidence elicited by this target, then $X_{next}$ is the maximum evidence of all nontarget items. Given these definitions, an absolute strategy is one that identifies a target only if $X_{max} > X_A$, for some criterion $X_A$. In contrast, a relative strategy identifies a target only if $X_{max} - X_{next} > X_R$, for some criterion $X_R$. Note that, in general, we expect $X_A > X_R$.

One advantage of the absolute strategy is that an identification response occurs only if there is strong evidence in favor of a target. On the other hand, if the difference between $X_{max}$ and $X_{next}$ is small, then in a police lineup, a person could be identified as a perpetrator even though the evidence almost equally favors some other person in the lineup. This problem is avoided with a relative strategy, but the danger here is that despite weak evidence that anyone in the lineup is the perpetrator, a person could be identified as a perpetrator just because the other lineup members all evoke even weaker evidence.

To formalize this discussion, consider an $r$-person lineup. A common choice is $r = 6$. On perpetrator-present trials, let $X_p$ denote the sensory evidence elicited by the single perpetrator in the lineup (i.e., so $p \in \{1, 2, \ldots, r\}$). In the absolute-strategy model, the probability of a hit equals

$$P(\mathrm{H}) = P(X_p = \max\{X_1, X_2, \ldots, X_r\} \,\&\, X_p > X_A | \mathrm{T}). \tag{3.14}$$

The SDT model that assumes a relative-decision strategy predicts that the probability of a hit equals

$$P(\mathrm{H}) = P\left(X_p - \max_{i \neq p}\{X_i\} > X_{\mathrm{R}} | \mathrm{T}\right), \tag{3.15}$$

where $X_{\mathrm{R}} > 0$. Note that the requirement that $X_{\mathrm{R}} > 0$ implies that $X_p$ is the largest of all the $X_i$. Therefore, the condition that $X_p - \max_{i \neq p}\{X_i\} > X_{\mathrm{R}}$ is equivalent to the condition that $X_{\max} - X_{\mathrm{next}} > X_{\mathrm{R}}$, and $X_{\max} = X_p$. Unfortunately, there is no closed-form expression for either of these probabilities. The problems with the absolute model are that the two terms are positively correlated and the distribution of $\max_{i \neq p}\{X_i\}$ itself often has no closed-form expression. For example, this is the case if the $X_i$ are normally distributed. The problems with the relative model are that $X_p$ and $\max_{i \neq p}\{X_i\}$ are both random variables, and again the latter of these often has no closed-form expression. For these reasons, the predictions of both models are typically investigated via Monte Carlo simulation (e.g., Clark, 2003; Clark et al., 2011).

## 3.3 GROUP DECISION-MAKING

In his famous book *Politics*, Aristotle (1905) wrote that "it is possible that the many, though not individually good men, yet when they come together may be better, not individually but collectively." He called this phenomenon the "wisdom of the crowd." There are now large empirical literatures documenting that groups of individuals often outperform the best individual in the group in a variety of different decision-making tasks and theoretical literatures that try to understand and model this effect (for reviews, see, e.g., Kerr & Tindale, 2004; Tindale & Winget, 2019). SDT-based models play a large, if not dominant, role in this theoretical literature (e.g., Batchelder & Romney, 1986; Erev et al., 1995; M. D. Lee et al., 2012; Sorkin & Dai, 1994; Sorkin et al., 2001).

There are two basic approaches, depending on whether or not the group members interact before making their group decision. A standard scenario where the noninteracting model is plausible is a typical election in which each voter examines two alternative candidates and then votes for one, with the winner being the candidate who receives the most votes. In the decision-making literature, the groups in such scenarios are referred to as *Condorcet groups*, after the French mathematician Marquis de Condorcet (1785), who proved that if the accuracy of each group member is above chance, then the probability that the majority vote will be correct increases quickly to 1 with increases in group size (assuming one alternative is correct and the other is incorrect). As one might expect, Condorcet groups have been studied extensively in political science (e.g., Austen-Smith & Banks, 1996; Gehrlein & Lepelley, 2010), but within psychology, SDT models of Condorcet groups have also been proposed and tested (e.g., Batchelder & Romney, 1986; Sorkin et al., 1998).

A common example of an interacting group is a jury in a courtroom trial, where the jury members deliberate over the evidence in a group setting and then reach a consensus decision. There are many different ways that members of the group can influence each other and thereby affect the group's overall decision. One especially important focus is on optimal decision-making—specifically, in how the sensory evidence of each group member can be combined to reach the most accurate possible group decision. Once this is worked out, the performance of actual groups can then be compared to this theoretical ideal. The following two subsections consider these two alternative group decision-making modeling approaches in turn.

### 3.3.1 CONDORCET GROUPS

SDT models of Condorcet groups are conceptually simple but are analytically intractable except under the simplest possible conditions. In this simplest version, all group members are assumed to be identical, in the sense that the performance of every member is characterized by the exact same values of $d'$ and $X_C$ (Sorkin et al., 1998). Note that in this case, for any task, the predicted probability of a hit will be the same for all members, as will the predicted probability of a false alarm. Denote these probabilities by $P_I(\text{H})$ and $P_I(\text{FA})$, respectively ("I" for individual). Now consider a group with $r$ members who each complete a YES-NO detection task. Let $R_N$ and $R_{SN}$ equal the number of individuals who respond YES on N and SN trials, respectively. Then $R_N$ is binomially distributed with parameters $r$ and $P_I(\text{FA})$ and $R_{SN}$ is binomially distributed with parameters $r$ and $P_I(\text{H})$. Specifically, the probability that the majority vote of the group on SN trials is a hit, denoted by $P_G(\text{H})$, equals:

$$\begin{aligned} P_G(\text{H}) &= P(R_{SN} \geq C|\text{SN}) \\ &= \sum_{i=C}^{r} \binom{r}{i} P_I(\text{H})^i [1 - P_I(\text{H})]^{r-i}, \end{aligned} \tag{3.16}$$

where $C$ is the smallest integer greater than $r/2$. Similarly, the probability that the majority vote of the group on N trials is a false alarm, denoted by $P_G(\text{FA})$, equals:

$$\begin{aligned} P_G(\text{FA}) &= P(R_N \geq C|\text{N}) \\ &= \sum_{i=C}^{r} \binom{r}{i} P_I(\text{FA})^i [1 - P_I(\text{FA})]^{r-i}. \end{aligned} \tag{3.17}$$

Unfortunately, despite its analytic tractability, this model is not very realistic since in any real-world setting, we would expect considerable individual difference in both $d'$ and $X_C$.

A more realistic SDT model of Condorcet groups would assume that the performance of each group member is characterized by a unique value of both $d'$ and $X_C$ (Sorkin et al., 1998). Although conceptually simple, this model is analytically intractable because under these more general conditions, $R_N$ and $R_{SN}$ are no longer binomially distributed. As a result, predictions require Monte Carlo simulations. These are straightforward to run, but like all such simulations, some relatively arbitrary choices must be made about the specific parameter values to choose, and there is always the worry that the predictions derived in such simulations might not be characteristic of the model in general, but rather unique to the specific parameter values that were chosen.

### 3.3.2 INTERACTING GROUPS

In cases where the group is required to make a consensus decision, almost all SDT models assume that the group decision is based on a weighted average of the sensory evidence available to each group member. To formalize this discussion, consider a group that includes $r$ members. Let $X_i$ denote the strength of evidence perceived by the $i$th member of this group. Then almost all SDT models assume that on each trial, the group decision is based on the numerical value of

$$W = \sum_{i=1}^{r} \alpha_i X_i, \tag{3.18}$$

where $\alpha_i$ is the weight the group assigns to the sensory evidence of member $i$. If the $\alpha_i$ all sum to 1 and the $X_i$ are random samples from some population, then $W$ is an unbiased estimator of the mean of the $X_i$ population. For example, if $\alpha_i = 1/r$ for all $i$, then note that $W$ reduces to the familiar sample mean $\bar{X}$.

If the group decision depends on the numerical value of $W$, then an obvious decision rule, which is a straightforward extension of the equation 2.1 decision rule, is to

$$\text{Respond YES if } W > W_C; \text{ Otherwise respond NO,} \tag{3.19}$$

for some criterion $W_C$. In this case,

$$P(\text{H}) = P(W > W_C \mid \text{SN}) \tag{3.20}$$

and

$$P(\text{FA}) = P(W > W_C \mid \text{N}). \tag{3.21}$$

Evaluating these probabilities is straightforward, but it requires that we know the distribution of $W$ on both N and SN trials. These distributions depend on the assumptions we make about group members and about the effects of group interactions.

A first critical question to address is whether the interactions that are necessary to reach consensus are restricted to the decision-making process or whether they also extend to the sensory and perceptual levels. In other words, do the group interactions affect the distribution of each $X_i$? One could imagine scenarios in which either answer is plausible. For example, in some cases, jury members might form their sense of the overall strength of the evidence as they sit in the jury box listening to the case—without any immediate communication with other jury members. However, in other cases, it seems plausible that during deliberations, a jury member might be reminded of a piece of evidence that they had discounted or ignored at the time when it was presented, and this reconsideration could plausibly cause them to reevaluate the overall evidence and thereby change their value $X_i$. We will consider both possibilities in turn.

**3.3.2.1 Groups that interact only at the decision stage** First, consider group performance in a YES-NO detection task in which the group interactions do not affect the distributions of the $X_i$. In this class of models, the group interactions only affect the $\alpha_i$ weights of equation 3.18. In other words, the group interactions determine how much influence each group member will have on the eventual group consensus decision, but they do not affect how each group member perceives the stimulus. Suppose the N and SN distributions for each observer are normal with equal variance, and denote the $d'$ for observer $i$ by $d'_i$. Specifically, this is the $d'$ that would describe the performance of observer $i$ in this same group decision-making task if observer $i$ performed the task in isolation—that is, in a single participant experiment. Note that in this model, on N trials all $X_i$ are normally distributed with mean 0 and variance 1. As a result, $W$ is normally distributed with mean 0 and variance

$$\begin{aligned}\text{Var}(W \mid N) &= \text{Var}\Big(\sum_{i=1}^{r} \alpha_i X_i \mid N\Big)\\ &= \sum_{i=1}^{r} \text{Var}\,(\alpha_i X_i \mid N)\end{aligned}$$

$$= \sum_{i=1}^{r} \alpha_i^2 \text{Var}(X_i | N)$$

$$= \sum_{i=1}^{r} \alpha_i^2, \tag{3.22}$$

since $\text{Var}(X_i \mid \text{N}) = 1$ for all $i$. Note that if each $\alpha_i = 1/r$ and $\text{Var}(X_i \mid \text{N}) = \sigma^2$, then this equation reduces to the familiar equation

$$\text{Var}(\bar{X}) = \sum_{i=1}^{r} \left(\frac{1}{r}\right)^2 \sigma^2 = r \times \frac{\sigma^2}{r^2} = \frac{\sigma^2}{r}. \tag{3.23}$$

Things are slightly more complex on SN trials because in this case, $X_i$ has a different distribution for each group member. Specifically, on SN trials, for group member $i$, $X_i$ is normally distributed with mean $d'_i$ and variance 1. As a result, $W$ is normally distributed with mean

$$\mu_\text{G} = \sum_{i=1}^{r} \alpha_i \, d'_i, \tag{3.24}$$

and variance $\sum_{i=1}^{r} \alpha_i^2$.

Note that on both N and SN trials, $W$ is normally distributed with variance $\sum_{i=1}^{r} \alpha_i^2$. As a result, this group model reduces to the normal, equal-variance SDT model in which $d'$, defined as the standardized distance between the means, equals

$$d'_{\text{G}_\text{D}} = \frac{\mu_\text{G} - 0}{\sqrt{\sum_{i=1}^{r} \alpha_i^2}} = \frac{\sum_{i=1}^{r} \alpha_i \, d'_i}{\sqrt{\sum_{i=1}^{r} \alpha_i^2}}, \tag{3.25}$$

where the subscript D indicates that this model assumes that the only group interactions are at the decision stage. If the $\alpha_i$ all sum to 1, then note that the denominator of equation 3.25 is less than 1 (i.e., $\sum_{i=1}^{r} \alpha_i^2 < 1$). As a result, $d'_{\text{G}_\text{D}}$ will be greater than $\mu_\text{G}$. In other words, regardless of the weights that are assigned to each group member, this model displays wisdom of the crowd—that is, the performance of the group is better than the performance of the average group member.

Given these distributions, the predicted probabilities of a false alarm and hit can be computed from equations 2.2 and 2.3 to be

$$P(\text{FA}) = P(W > W_\text{C} \mid \text{N})$$

$$= 1 - P(W \leq W_\text{C} \mid \text{N})$$

$$= 1 - P\left(Z \leq \frac{W_C - 0}{\sqrt{\sum_{i=1}^{r} \alpha_i^2}} \middle| \text{N}\right)$$

$$= 1 - \Phi\left(\frac{W_C}{\sqrt{\sum_{i=1}^{r} \alpha_i^2}}\right)$$

$$= \Phi\left(\frac{-W_C}{\sqrt{\sum_{i=1}^{r} \alpha_i^2}}\right), \tag{3.26}$$

and

$$P(\text{H}) = P(W > W_C \mid \text{SN})$$

$$= 1 - \Phi\left(\frac{W_C - \sum_{i=1}^{r} \alpha_i d_i'}{\sqrt{\sum_{i=1}^{r} \alpha_i^2}}\right)$$

$$= \Phi\left(\frac{\sum_{i=1}^{r} \alpha_i d_i' - W_C}{\sqrt{\sum_{i=1}^{r} \alpha_i^2}}\right). \tag{3.27}$$

A group that uses the equation 3.19 decision rule will tend to perform better than the average group member performing alone—no matter how the contributions of each individual member are weighted. But some weightings will produce better results than others. Given this, then a natural next question is to ask how a group should set the various individual member weights (i.e., the $\alpha_i$) if the goal is to maximize accuracy. An inefficient group would likely set the weights based on factors that are irrelevant to the task, such as social influence, for example. But what about a group that wants to maximize accuracy? Is there some scheme for assigning weights that is guaranteed to maximize accuracy in the YES-NO detection task? In fact, it turns out that accuracy is maximized if

$$\alpha_i = d_i', \tag{3.28}$$

where, as before, $d_i'$ is the $d'$ of group member $i$ if that person completes the YES-NO detection task in isolation, without any assistance from other group members (Sorkin & Dai, 1994). This makes sense because this weighting scheme says that the optimal strategy is to assign weights based on sensory ability. Inputs from members who are best at discriminating between N and SN are given the greatest weights and inputs from members who are worst at this sensory discrimination are given the smallest weights. Note from equation 3.25 that the predicted $d'$ of this model equals

$$d'_{\mathrm{GD}} = \frac{\sum_{i=1}^{r}(d'_i)^2}{\sqrt{\sum_{i=1}^{r}(d'_i)^2}} = \sqrt{\sum_{i=1}^{r}(d'_i)^2}. \tag{3.29}$$

We don't expect any real group to set the weights at these exact optimal values, but comparing the performance of a real group to this ideal provides a method to assess the efficiency of the group. The best groups will come close to the performance of this ideal group, while the worst groups will fall far short. Fortunately, testing this model is straightforward. An obvious approach would be to first run each group member through a YES-NO detection task by themselves. These experiments should use exactly the same stimuli and presentation conditions as in the group experiment. The data from the single-observer conditions would then be used to estimate a $d'$ for each group member—that is, a $d'_i$. Finally, the whole group would complete the task (i.e., making a consensus decision on each trial). Given that we already have estimates of each $d'_i$ and therefore of the optimal value of each weight $\alpha_i$, note from equations 3.26 and 3.27 that there is only one free parameter left to estimate—namely, $W_{\mathrm{C}}$. The group data have two degrees of freedom [i.e., $\hat{P}(\mathrm{H})$ and $\hat{P}(\mathrm{FA})$], so, after parameter estimation, there is still one degree of freedom left to test the validity of the model.

Note that our constraint that the sum of the $\alpha_i$ should equal 1 is violated if we set each $\alpha_i$ to $d'_i$. This is fine because it turns out that none of equations 3.25, 3.26, or 3.27 depend on this constraint. In other words, all three equations are valid no matter what the sum of the $\alpha_i$ equals. To see this, suppose $\sum \alpha_i = K$, for some constant $K$. Now consider, for example, equation 3.26. Note that

$$\begin{aligned} P(\mathrm{FA}) &= \Phi\left(\frac{-\frac{1}{K}W_{\mathrm{C}}}{\frac{1}{K}\sqrt{\sum_{i=1}^{r}\alpha_i^2}}\right) \\ &= \Phi\left(\frac{-W_{\mathrm{C}}/K}{\sqrt{\sum_{i=1}^{r}(\alpha_i/K)^2}}\right) \\ &= \Phi\left(\frac{-W_{\mathrm{C}}^*}{\sqrt{\sum_{i=1}^{r}(\alpha_i^*)^2}}\right), \end{aligned} \tag{3.30}$$

for some new constants $W_{\mathrm{C}}^* = W_{\mathrm{C}}/K$ and $\alpha_i^* = \alpha_i/K$, except note now that $\sum \alpha_i^* = 1$. Similar derivations show that $d'_{\mathrm{G}}$ and $P(\mathrm{H})$ also do not change if we constrain the $\alpha_i$ to sum to 1. The flip side of this demonstration is that if $\alpha_i = d'_i$, then none of the predicted probabilities change if instead, we set $\alpha_i = d'_i/K$, where $K = \sum d'_i$. In other words, either method of setting each $\alpha_i$ is optimal.

Given that the sum of the $\alpha_i$ does not affect predictions of the model about $d'_G$, $P(\text{H})$, or $P(\text{FA})$, then why did I recommend earlier that the $\alpha_i$ should be chosen to sum to 1? Note in the equation 3.30 derivation that although dividing the numerator and denominator by $K$ does not change the predicted probability of a false alarm, it does change the numerical value of $W_C$ to $W_C^* = W_C/K$. If the sum of the weights is not standardized to 1, then the response criteria estimated from the single-observer experiments (i.e., the $X_C$) cannot be compared to the response criterion estimated from the group experiment (i.e., to $W_C$) because these values will be on scales with different units of measurement. So it would be like comparing temperatures in Fahrenheit to temperatures in Celsius. Standardizing the weights so they add to 1 places all the criteria on the same scale, and therefore allows them to be compared. For example, standardizing is the only way one could investigate whether the group decision is more or less conservative than the individual decisions of each group member.

Next, consider the special case in which all group members share the same value of $d'$—that is, in which $d'_i = d'_{YN}$ for all values of $i$, where $d'_{YN}$ is the $d'$ that describes the performance of each group member when they complete the YES-NO detection task by themselves. The group $d'$ for this model can be computed by substituting $d'_{YN}$ for each $d'_i$ in equation 3.29:

$$d'_{GD} = \sqrt{\sum_{i=1}^{r}(d'_{YN})^2} = \sqrt{r}\, d'_{YN}. \tag{3.31}$$

Note that this equation is identical to equation 2.27 from theorem 2.4. That theorem describes the YES-NO detection performance of a single optimal observer who is given $r$ independent looks at the stimulus. In contrast, equation 3.31 describes the YES-NO detection performance of a group of $r$ identical observers who optimally combine their individual perceived strengths. Equation 3.31 shows that SDT makes identical predictions for these two scenarios.

**3.3.2.2 Groups that interact at the perceptual and decision stages** Next we consider groups in which the interactions affect both the perceptual processing of each member as well as the group decision. There are many different ways to model how a group could affect the sensory representations of its members. Sorkin and Dai (1994) assumed the group would affect the variability of each individual's perceived strength, but not the mean. More specifically, they assumed that group member

$i$ would perceive the strength of evidence to equal:

$$Y_i = X_i + X_G, \tag{3.32}$$

where $X_i$ is the amount of this strength that is unique to member $i$ and $X_G$ is the amount that is common to all group members. Sorkin and Dai (1994) assumed that $X_i$ is normally distributed with mean 0 on N trials and mean $\mu_i$ on SN trials and that the variance of the $X_i$ equals $\sigma_I^2$ in both cases, where the I indicates that this is the variance that is unique to each individual observer. They also assumed that $X_G$ is normally distributed with mean 0 and variance $\sigma_G^2$. Note that $X_G$ does not depend on $i$ and so is common to all group members. In other words, on each trial of the group task, the group must decide whether to respond YES or NO by evaluating the set of perceived strengths $\{X_1 + X_G, X_2 + X_G, \ldots, X_r + X_G\}$, where $X_1, X_2, \ldots, X_r$ are all statistically independent of each other, but $X_G$ is exactly the same value for all group members on each trial. So $X_G$ models how much the group colors the perception of each member. On half the trials, $X_G$ will be positive and, therefore, the group will cause all members to increase their perceived strength of evidence, and on half the trials the opposite will occur—that is, $X_G$ will be negative, causing all members to decrease their perceived strength.

In this model, the group decision is based on the $r$ different values of the $Y_i$. On N trials, each $Y_i$ is normally distributed with mean 0 and variance $\sigma_I^2 + \sigma_G^2$ and on SN trials each $Y_i$ is normally distributed with mean $\mu_i$ and variance $\sigma_I^2 + \sigma_G^2$. So at the level of the individual observer, this is the normal, equal-variance SDT model. However, one fundamental difference between this model and the one we considered in the previous subsection that assumed no group effect at the perceptual stage is that the common group term $X_G$ will cause the various $Y_i$ to be positively correlated, even though the $X_i$ are all statistically independent. For example, on trials when $X_G$ is unusually large, all of the $Y_i$ will also tend to be large, causing them to be positively correlated.

The correlation between $Y_i$ and $Y_j$ is defined as

$$\rho_{ij} = \frac{\text{cov}(Y_i, Y_j)}{\sigma_i \sigma_j}, \tag{3.33}$$

where $\text{cov}(Y_i, Y_j)$ is the covariance between $Y_i$ and $Y_j$ and $\sigma_i$ and $\sigma_j$ are their respective standard deviations. Now

$$\begin{aligned}\text{cov}(Y_i, Y_j) &= E[(Y_i - \mu_i)(Y_j - \mu_j)] \\ &= E[(X_i + X_G - \mu_i)(X_j + X_G - \mu_j)]\end{aligned}$$

$$
\begin{aligned}
&= E\{[(X_i - \mu_i) + X_G][(X_j - \mu_j) + X_G]\} \\
&= E[(X_i - \mu_i)(X_j - \mu_j)] + E[(X_i - \mu_i)X_G] + E[(X_j - \mu_j)X_G] + E(X_G^2) \\
&= \text{cov}(X_i, X_j) + \text{cov}(X_i, X_G) + \text{cov}(X_j, X_G) + \sigma_G^2 \\
&= \sigma_G^2, \qquad (3.34)
\end{aligned}
$$

where $E$ denotes expected value. The penultimate equality holds because $\mu_G = 0$ and the last equality holds because $X_i$, $X_j$, and $X_G$ are all mutually independent, in which case all three pairwise covariances equal 0. Note that this covariance can never be negative and does not depend on either $i$ or $j$ and thus is the same for any two members of the group. The associated correlation has these same properties. Equation 3.34 says that the covariance between the sensory values of any pair of group members increases with the variance of the group's influence on the perceptual experience of each member—that is, with the variance of $X_G$. Note that $X_G$ has mean 0, so if $\sigma_G^2 = 0$, then there is no group influence (because $X_G$ is always 0) and equation 3.34 tells us that in this case all covariances (and correlations) are 0. In other words, equation 3.34 confirms that in the absence of any group influence, the sensory experiences of all group members are statistically independent.[4] If $\sigma_G^2$ is large though, equation 3.34 says that each pair of sensory values will have a large positive covariance and therefore a large positive correlation. If $\sigma_G^2$ is large, then some $X_G$ samples will have large positive values and some will have extreme negative values. In the former case, all sensory values will tend to be large because of the large $X_G$ value and the opposite will be true in the latter case (see equation 3.32). The significant contribution of the same $X_G$ value to the sensory value $Y_i$ causes the resulting positive correlation.

Similarly, the variances all equal

$$
\begin{aligned}
\sigma_i^2 &= E[(Y_i - \mu_i)^2] \\
&= E[(X_i + X_G - \mu_i)^2] \\
&= E\{[(X_i - \mu_i) + X_G]^2\} \\
&= E[(X_i - \mu_i)^2] + 2E[(X_i - \mu_i)X_G] + E(X_G^2) \\
&= \sigma_I^2 + 2\text{cov}(X_i, X_G) + \sigma_G^2 \\
&= \sigma_I^2 + \sigma_G^2. \qquad (3.35)
\end{aligned}
$$

Note that the same derivation shows that $\sigma_j^2 = \sigma_I^2 + \sigma_G^2$, so this is also a normal, equal-variance SDT model. Substituting equations 3.34 and 3.35 into equation 3.33

produces

$$\rho_{ij} = \rho = \frac{\sigma_G^2}{\sqrt{\sigma_I^2 + \sigma_G^2}}, \text{ for all } i \text{ and } j. \tag{3.36}$$

Note that this correlation can never be negative.

As in all SDT models, the sensory values—that is, the $Y_i$—lie on an interval scale, and, as a result, both the origin and unit of measurement are arbitrary. We fixed the origin when we assumed the mean of the $Y_i$ on N trials equals 0, but we haven't yet fixed the unit of measurement. The standard is to set the common variances to 1. In this case, that means that we set

$$\sigma_i^2 = \sigma_I^2 + \sigma_G^2 = 1. \tag{3.37}$$

Given this, note that the following equalities hold:

$$d_i' = \mu_i, \ \sigma_G^2 = \rho, \text{ and } \sigma_I^2 = 1 - \rho. \tag{3.38}$$

Sorkin and Dai (1994) showed that the optimal weights in this model—that is, the weights that cause the group to be as accurate as possible—are:

$$\alpha_i = [1 + \rho(r-1)]d_i' - r\rho\bar{d}_i', \tag{3.39}$$

where $\bar{d}_i'$ is the mean of the $d_i'$—that is, of the individual group member $d'$s. First, note that, as in the model in which the only group interactions are at the decision stage, this scheme assigns greater weight to individual members with higher individual $d'$s. Second, note that if there is no group influence—that is, if $\rho = \sigma_G^2 = 0$—then $\alpha_i = d_i'$, exactly as in the model in which the only group interactions are at the decision stage.

Sorkin and Dai (1994) also showed that this model predicts that the $d'$ of this group that optimally weights the sensory evidence from each group member equals

$$d'_{G_{PD}} = \sqrt{r}\left[\frac{\mathrm{Var}(d_i')}{1-\rho} + \frac{(\bar{d}_i')^2}{1+(r-1)\rho}\right]^{1/2}, \tag{3.40}$$

where the subscript PD indicates that this model assumes that group interactions affect both perceptual and decision processes, and that $\mathrm{Var}(d_i')$ is the variance of the individual $d_i'$. Note that equation 3.40 predicts that the performance of the ideal group increases with the size of the group (i.e., with $r$), with the average performance of each group member (i.e., with $\bar{d}_i'$), and with the within-group variance in performance (i.e., with $\mathrm{Var}(d_i')$). The former two predictions seem more obvious than the latter. For a given mean, note that the larger the variance, the more extreme the outliers. So for any given value of $\bar{d}_i'$, a larger $\mathrm{Var}(d_i')$ means more members with small $d_i'$, but also more members with large $d'_i$. Because the group uses optimal weighting, input from the members with small $d'_i$s will mostly be discounted, whereas input from the members with large $d'_i$s will heavily influence the group decision. So overall, a large variance improves performance. Gaining a few members with large $d'_i$s more than compensates for the addition of a few members with small $d'_i$s.

Equation 3.40 also shows that the relative contributions of $\bar{d}'_i$ and $\text{Var}(d'_i)$ to the group performance trade off as the value of $\rho$ increases. Specifically, note that when $\rho = 0$, the contribution of the $\text{Var}(d'_i)$ term is at its minimum and the contribution of $\bar{d}'_i$ is at its maximum. However, as $\rho$ increases, the contribution of $\text{Var}(d'_i)$ increases and the contribution of $\bar{d}'_i$ decreases. Recall that $\rho$ is the correlation between the sensory values of each pair of group members. When the correlation is near 1, if one $d'_i$ is below the population mean then all $d'_i$s will tend to be below the mean. Similarly, if one $d'_i$ is above the population mean then all $d'_i$s will tend to be above the mean. As a result, $\bar{d}'_i$ will be a poor predictor of group performance. $\text{Var}(d'_i)$ will be better because a large variance means large outliers, and any large outlier will dramatically improve group performance because of optimal weighting.

Although it might not be obvious, another feature of equation 3.40 is that, as it should, it reduces to equation 3.29 when $\rho = 0$. Specifically, when there is no group influence on perception—that is when $\sigma_G^2 = \rho = 0$ and all members perceive the stimulus independently—then $d'_{\text{GPD}}$ in equation 3.40 reduces to[5]

$$d'_G = \sqrt{\sum_{i=1}^{r} (d'_i)^2}. \tag{3.41}$$

Therefore, note that the model in which there is no group-level influence on perception is a special case of the model in which the group does color the perception of each of its members.

Equations 3.40 and 3.29 specify the overall group $d'$s when optimal weights are assigned to each group member and there is or is not a group influence on perception, respectively. Even so, equation 3.40 is a bit trickier to use than equation 3.29. As we discussed in the last subsection, the $d'_i$ in equation 3.29 can be estimated from control conditions in which each group member performs the task in isolation. However, in equation 3.40, each $d'_i$ includes the group influence on perception. Specifically, $d'_i$ is the mean of $Y_i = X_i + X_G$ on SN trials after the variance has been standardized to equal 1. Therefore, this model predicts that the $d'$s estimated from individual observer control conditions will differ from the $d'$s that characterize the performance of the same observers when they operate in a group.[6] Sorkin et al. (2001) overcame this problem by collecting confidence ratings from each group member during the group decision-making task and then used these ratings to estimate individual group member $d_i'$s.

The performance of every group should fall below the value of $d'_{\text{GPD}}$ predicted by this optimal model. No real group will optimally weight the input of all of its members. A more useful question is to ask how close to optimal a group can come. One way to address this question is also to compare group performance to the predictions of a model that assumes supoptimal weighting. Toward this end, Sorkin et al. (2001) also computed the predictions of a model that assumed all group members contribute equally to the group decision—in other words, that all weights are equal. They showed that this equal-weighting (EW) model predicts a group $d'$ equal to

$$d'_{\text{GEW}} = \frac{\sqrt{r}\,\bar{d}'_i}{\sqrt{1+(r-1)\rho}}. \tag{3.42}$$

So if the performance of an actual group is suboptimal but better than this value—that is, if the observed group $d'$ falls between $d'_{\text{GEW}}$ and $d'_{\text{GPD}}$, then we can conclude that the group used a differential weighting scheme that prioritized input from members with higher discriminability levels over input from members with lower levels.

Sorkin et al. (2001) used these SDT models to analyze results from two group decision-making experiments that both used a perceptual discrimination task. They reported that, as all of these models predict, group performance improved with the size of the group, but perhaps most significantly that the sensory inputs from individual group members were weighted in approximate proportion to each member's perceptual expertise. The groups were formed randomly, and all groups included members who had no prior association. Given this, it is natural to ask how such a group could estimate the expertise of its members. Although it isn't clear how such knowledge is acquired, the evidence is good that groups can estimate the ability of individual group members, even in the absence of any feedback that might be used to assess this expertise (Henry, 1993).

# 4

# GENERALIZING TO MULTIPLE DIMENSIONS: GENERAL RECOGNITION THEORY

SDT is useful for understanding behavior in any task in which the observer's decision is based on a single sensory dimension. Most real-world stimuli vary on many dimensions, however, and many perceptual decisions require attention to more than one dimension. For example, there is no single sensory dimension that allows accurate face identification. For this reason, there is obvious value in extending SDT to multiple stimulus dimensions.

At first glance, this seems like a straightforward exercise. An obvious place to begin is by replacing the unidimensional probability distributions that are used to represent the sensory effects of a stimulus in SDT with multivariate probability distributions. But complications quickly arise, even in the case of two sensory dimensions. First, some sensory dimensions interact, and the perceptual literature includes a bewildering number of terms that have been proposed to describe perceptual interactions or non-interactions. This list includes perceptual independence, separability, integrality, holism, configurality, sampling independence, dimensional orthogonality, and performance parity. How should these different types of sensory interactions be modeled? And how are they all related to each other?

Second, how should the decision process be modeled? In SDT, the sensory space is a line, and in two-alternative tasks, the observer is typically assumed to divide the line into two regions—one associated with each response alternative. Fortunately, there are only a few ways to do this. In fact, a standard SDT lecture is to show that many seemingly different decision strategies are all equivalent to the equation 2.1 decision rule, in which a single criterion is set on the sensory dimension, and a YES decision is given on all trials for which sensory magnitude exceeds this criterion value (i.e., in YES-NO detection tasks). For example, as we saw in chapter 2, if the SN-to-N likelihood ratio is monotonically increasing in sensory magnitude, then a decision to respond YES when the likelihood ratio is large is equivalent to responding YES when the sensory magnitude is large. Even in the normal, unequal-variance SDT model, in which likelihood ratio is not monotonic with sensory magnitude, the optimal likelihood-ratio decision rule (i.e., the equation 2.5 rule) is equivalent to only a slight generalization of the equation 2.1 rule, in which two criteria are placed on the sensory magnitude dimension, rather than one. In contrast, however, if there are two sensory dimensions, then the sensory space is a plane, and there are an infinite number of qualitatively different ways to divide a plane into two regions—even into two connected regions.

For example, consider a two-alternative identification task with two stimuli, A and B, that differ on two stimulus dimensions. Figure 4.1 shows hypothetical results from 400 trials of this experiment. The dots show the perceived values on the two sensory dimensions on trials when stimulus A was presented and the x's show sensory values on B trials. The four panels show four different plausible decision strategies. In all four cases, the observer divides the sensory plane into regions associated with the two response alternatives, A and B. As in SDT, the idea is that the observer notes which region the current percept is in, and then emits the response associated with that region.

The various strategies depicted in figure 4.1 will be described in detail in the next chapter. The important point here is that a variety of qualitatively different strategies seem plausible for this simple task. Panel (a) shows the optimal strategy that maximizes accuracy. Panel (b) shows the strategy of the minimum-distance classifier (Ashby & Gott, 1988), which is the strategy predicted by prototype theory in which the observer gives the response associated with whichever prototypical sensory representation is most similar to the current representation. In this case, the prototype is interpreted as the mean or mode of the sensory representations elicited by each stimulus. Panel (c) shows an independent decisions strategy. In this case, the observer makes an independent decision about the value of the current percept on each stimulus dimension. This leads to four possible outcomes. Let (H,L) denote the outcome in which the observer judges the current stimulus to have a high value on dimension 1 and a low value on dimension 2. Then the four outcomes are: (L,L), (L,H), (H,L), and (H,H). Next the observer assigns a response to each of these outcomes. In the figure

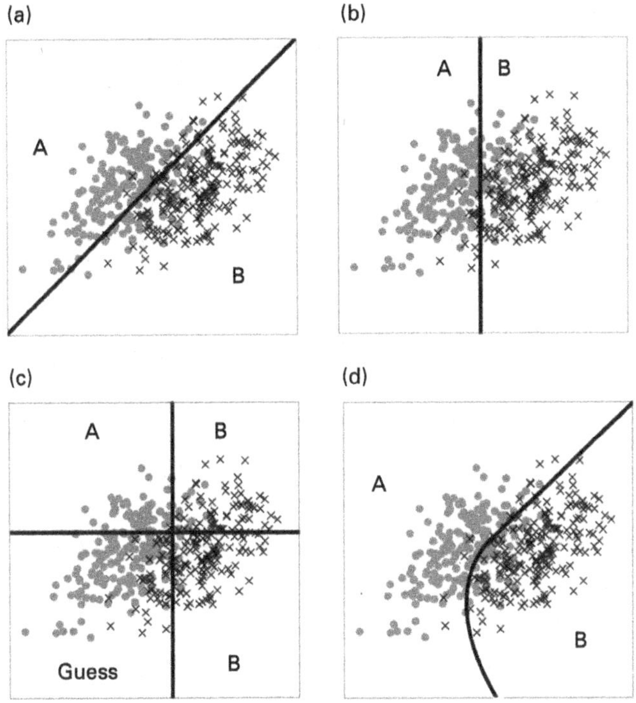


**4.1** Different possible decision strategies in a two-alternative identification task with two stimuli, A and B, that differ on two stimulus dimensions. (a) The optimal strategy; (b) the strategy predicted by prototype theory; (c) an independent decisions strategy; (d) a nonlinear strategy.

4.1 example, the observer guesses if the outcome is (L,L), responds A if the outcome is (L,H), and responds B for either of the other two outcomes. Finally, panel (d) shows one of the infinite possible strategies in which the observer divides the sensory space into two connected response regions by weighting the two dimensions nonlinearly. The four decision models illustrated in figure 4.1 make very different predictions. Which one should be used to develop a multidimensional generalization of SDT? This is not at all obvious, and to make matters worse, there are, of course, many other plausible decision strategies that an observer might use.

Not surprisingly, the first attempt to generalize SDT to multiple stimulus dimensions, by Tanner in 1956, ignored virtually all of these issues. In particular, Tanner (1956) assumed that (1) the sensory distributions are bivariate normal, (2) the variances on both dimensions and in all sensory distributions are equal, (3) all covariances are zero, and (4) the observer responds as predicted by the minimum-distance classifier, which is optimal under these distributional assumptions. Tanner did allow for one very specific type of perceptual interaction, which we will consider in some detail in section 4.3.1 below. Briefly, though, Tanner allowed the angle between the perceptual dimensions to be oblique, rather than always orthogonal, as a way to model perceptual dependencies.

Because the Tanner (1956) model assumes that variability is equal in all directions and that the observer always gives the response associated with the nearest distribution mean, all predictions of the model depend only on the locations of these means, and more specifically, on the distances between each pair of means. The predicted probability that one stimulus is confused for another depends on the overlap of the distribution associated with the first stimulus into the response region associated with the second. When variability is equal in all directions, this overlap depends on the distance between the distribution means, but for any fixed distance apart, the orientation or direction of this fixed distance is irrelevant. Therefore, the distributions could be replaced by point representations located at each distribution mean, and not much would change, so long as some probabilistic decision rule is used that converts distance to response probability. The Tanner (1956) model, then, is not qualitatively different from traditional multidimensional scaling. For example, Tucker (1972) generalized the standard Euclidean multidimensional scaling model to include oblique dimensions in the same way that Tanner assumed oblique dimensions can be used to model a perceptual interaction in his model.

Despite its simplicity, Tanner's contribution was significant because it was the first to consider multiple sensory dimensions. Even so, it was another 30 years before a more useful multidimensional version of SDT was developed. During the late 1980s, a flurry of articles significantly generalized Tanner's approach by dropping virtually all of his restrictive assumptions. The title of Tanner's (1956) article was "Theory of Recognition." To pay homage to his contributions, Ashby and Townsend (1986) called their more general approach general recognition theory (GRT). GRT quickly developed: Ashby and Townsend (1986) proposed a GRT-based theory of perceptual interactions, Ashby and Gott (1988) studied decision rules in multidimensional perceptual spaces, and Ashby and Perrin (1988) used GRT to develop a unified theory of similarity and identification. Since then, more than 1,000 articles on GRT have been published.

## 4.1 IDENTIFICATION VERSUS CATEGORIZATION

GRT has been applied to a wide variety of different tasks. But two tasks—identification and categorization—have emerged as the most popular, and which one is used depends on the goals of the research. In particular, identification tasks are better if the primary goal is to study perceptual representations, whereas categorization tasks are better if the primary goal is to study decision processes.

In identification tasks, there are $N_S$ stimuli and $N_S$ unique identifying responses. On each trial, one of the stimuli is presented, and the observer's task is to identify the stimulus by emitting the appropriate response. The data are collected in an $N_S \times N_S$ confusion matrix, in which the entry in row $i$ and column $j$ is the frequency with which the observer gave response $j$ on trials when stimulus $i$ was presented. Because the number of stimulus presentations is known, there is one constraint on each row of the confusion matrix. Specifically, each row sum must equal the total number of trials on which that stimulus was presented. As a result, every identification confusion matrix has $N_S \times (N_S - 1)$ degrees of freedom. Note that the YES-NO detection task is a special case of this identification task in which $N_S = 2$ and the two stimuli to be identified are N and SN.

As an example identification experiment, Townsend (1971) studied the confusions that people make when they identify letters of the English alphabet. In this experiment, one of the 26 letters was briefly presented on each trial and the observer's task was to name the presented letter. The resulting confusion matrix had order $26 \times 26$, and the entry in row $i$ and column $j$ was the number of trials when letter $i$ was presented on which the observer responded that the stimulus was letter $j$. This matrix has 650 degrees of freedom ($26 \times 25$) that can be used to estimate the free parameters of an alphabetic confusion model and test its validity.

The most useful information in identification tasks is in the confusions that observers make, so experimental conditions are selected to guarantee errors. This is usually accomplished by using highly similar stimuli, but sometimes brief exposure durations or noise masks are used instead. For example, Townsend (1971) guaranteed errors by using brief stimulus presentations and a post-stimulus noise mask. Perceptual confusions occur when the distribution of percepts induced by different stimuli overlap in perceptual space. The overlap causes identification errors and by carefully noting the frequency with which a particular stimulus is confused with each of the other stimuli, the experimenter is able to make inferences about the underlying perceptual distribution induced by that stimulus. Of course, an error can also occur because observers fail to remember which response button is associated with which stimulus. In identification experiments, these are typically nuisance errors because the ability of observers to learn these stimulus-response associations is almost never the focus of the experiment. Therefore, to eliminate these response confusions, feedback is usually provided to help observers learn the stimulus-response associations, and some training trials are included that are then excluded from the data analysis. Obviously, the amount of training that is necessary will tend to increase with the number of stimuli that are used—at least, if the stimuli and/or responses are novel. Because of these training trials, after performance asymptotes, most misidentifications are made because of errors in perception, rather than because of a suboptimal decision strategy, and it is for this reason that identification experiments are a good choice if the goal is to study perceptual representations.

In the most widely used identification tasks, the stimuli are constructed by factorially combining a small number of discrete values on two sensory dimensions. The most common choice is to factorially combine two values on two dimensions to create a total of four stimuli. Each $4 \times 4$ confusion matrix collected from such a $2 \times 2$ factorial design includes 12 degrees of freedom (i.e., $4 \times 3$) for parameter estimation and model testing. If we call the two stimulus components A and B, then we can denote the stimulus in which dimension or component A is at level $i$ and component B is at level $j$ by $A_iB_j$, and the corresponding response by $a_ib_j$. Figure 4.2 shows examples of stimuli that might be used in two different possible $2 \times 2$ factorial designs. The top panel shows four sine-wave gratings (Gabor patches) that differ in bar width (spatial frequency) and bar orientation. And the bottom panel shows four faces that differ in emotional expression (neutral and sad) and facial identity.

Categorization experiments are identical to identification experiments, except they include fewer response alternatives than stimuli. In a categorization experiment, one of $N_S$ stimuli is presented on each trial and the observer's task is to assign it to one of $N_R$ categories, where $N_R < N_S$. The confusion matrix is therefore $N_S \times N_R$, and it contains $N_S \times (N_R - 1)$ degrees of freedom. The most common choice is $N_R = 2$. Note that in this case, the data include $N_S$ degrees of freedom. In most cases the categories are novel, in the sense that they were created specifically to use in the experiment. As a result, accurate responding requires the observer to learn the structure of these categories, most commonly via trial-by-trial feedback provided by the experimenter. Errors are most likely to occur because the observer is using a suboptimal strategy to assign stimuli to categories. Misperceptions are just as likely as in identification experiments, but they tend to have little effect on accuracy. For example, confusing

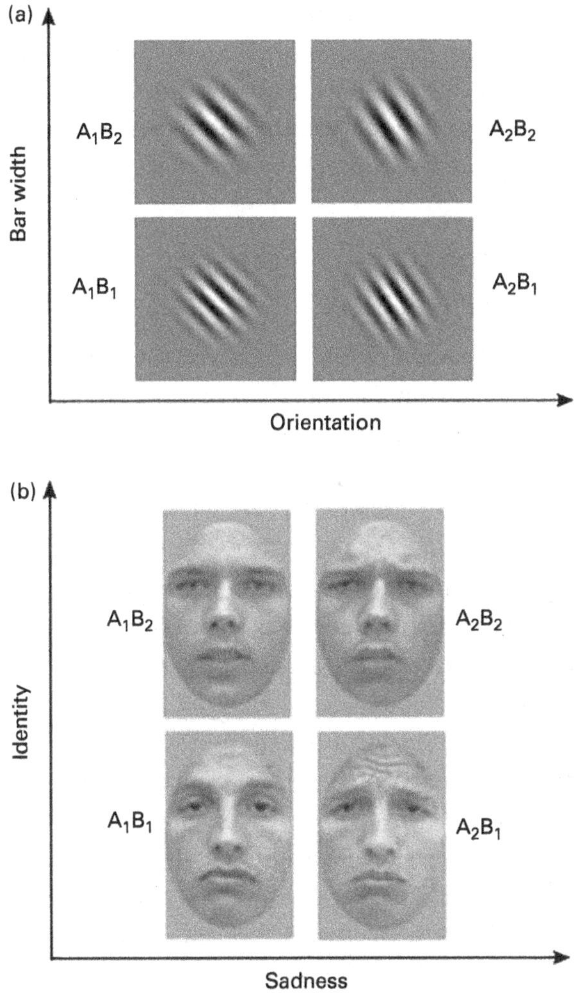


**4.2** Stimuli for two different 2 × 2 factorial identification experiments.

one stimulus with another in the same category does not change the response, and therefore has no observable effect on behavior. Errors are common in categorization experiments even if the stimuli are distinct and non-confusable because of the arbitrary nature of the experimenter-defined categories. For these reasons, categorization experiments are a good choice if the goal is to study decision processes.

Because errors in categorization experiments are caused mostly by suboptimal decision strategies, when modeling the data it is common to assign most of the free parameters to the decision process. For example, a common approach is to model the perceptual representations in the same way as Tanner (1956)—that is, to assume equal variance in all perceptual distributions on every stimulus dimension and that all covariances equal 0. GRT models that are applied to categorization experiments and make these strong distributional assumptions are often called *decision-bound models*.

## 4.2 THE GRT MODEL FOR THE 2 × 2 FACTORIAL IDENTIFICATION EXPERIMENT

Consider a $2 \times 2$ factorial identification experiment with stimuli $A_1B_1$, $A_1B_2$, $A_2B_1$, and $A_2B_2$ that are created by factorially combining two levels of stimulus component A with two levels of some other stimulus component B. Example stimuli that might be used in this task are shown in figure 4.2. On each trial, one of these four stimuli is randomly selected and presented to the observer, whose task is to identify it uniquely. Denote the four respective response alternatives by $a_1b_1$, $a_1b_2$, $a_2b_1$, and $a_2b_2$.

GRT assumes the percepts elicited by each stimulus vary in an unpredictable way across trials. Variability should occur on all stimulus dimensions, but for modeling purposes, it is typical to assume that the four stimuli are all identical except for their values on the two manipulated dimensions A and B. As a result, variability on any other irrelevant dimension should not affect performance, so we only need to account for variability in the perceived values of the two relevant components A and B. Denote the perceived value of component A as $X_1$ and the perceived value of component B as $X_2$. Then GRT assumes that each of the four stimuli elicit a different set of possible $(X_1, X_2)$ values. Let $f_{ij}(x_1, x_2)$ denote the likelihood that presentation of stimulus $A_iB_j$ elicits the specific percept $(x_1, x_2)$—that is, $f_{ij}(x_1, x_2)$ is the joint pdf that describes the likelihoods that the observer experiences any possible specific percept on stimulus $A_iB_j$ trials. Although we will see in chapter 6 that some GRT approaches make no assumptions about the form of this distribution, when distributional assumptions are made, the most common choice is to assume that $f_{ij}(x_1, x_2)$ is a bivariate normal distribution. Appendix A provides a brief tutorial on the bivariate normal distribution.

A GRT model of the perceptual effects elicited by the four stimuli in the $2 \times 2$ factorial identification experiment is shown in the top panel of figure 4.3. This model assumes bivariate normal distributions. The height of each distribution at the point $(x_1, x_2)$ equals the likelihood that the specific percept $(x_1, x_2)$ is elicited on trials when that stimulus is presented. In bivariate normal distributions the most likely percept is the distribution mean (i.e., in normal distributions, the mean equals the mode).

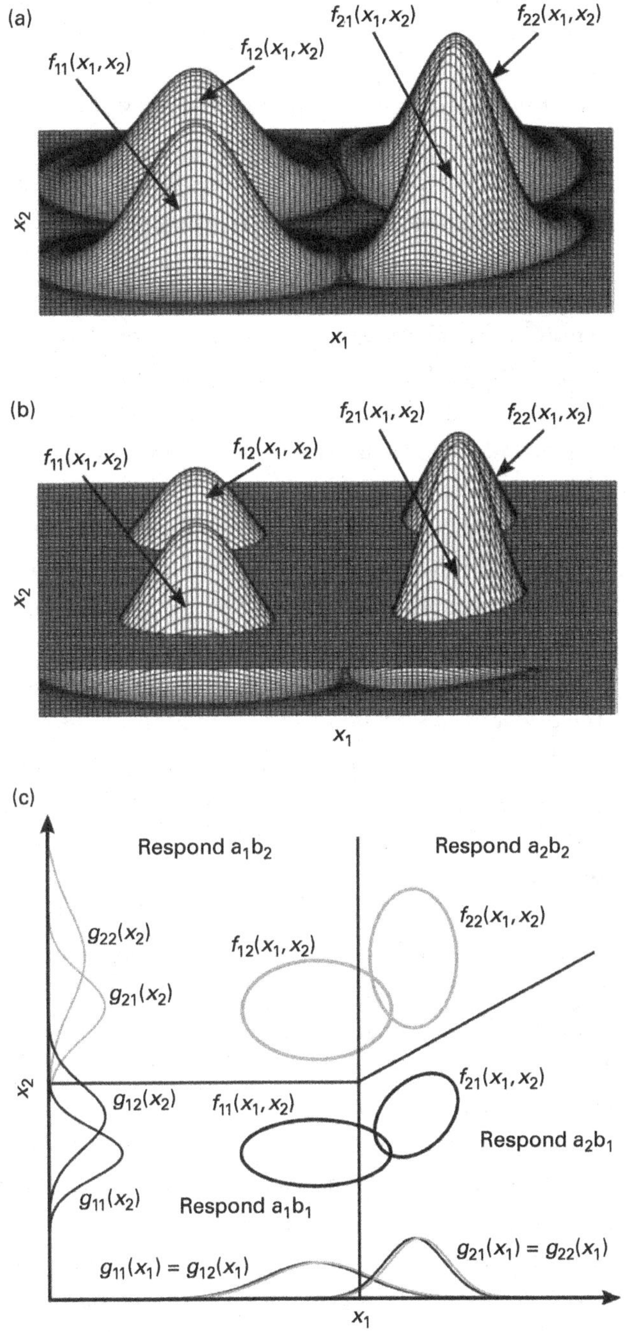


**4.3** A GRT model of the $2 \times 2$ factorial identification experiment. (a) The bivariate normal distributions elicited by the four stimuli. (b) Same as panel (a) except with a plane associated with a constant, fixed likelihood. (c) The ellipses denote the contours of equal likelihood for the four bivariate perceptual distributions, the bell-shaped curves denote the marginal distributions, and the line segments mark the decision boundaries.

The three-dimensional nature of figure 4.3a makes it difficult to interpret. As a result, the convention is to reduce it to two dimensions (see appendix A for details). This is done, as in panel (b), by passing a plane through the four distributions that is parallel to the $(x_1, x_2)$ perceptual plane. Note that every point on this plane has the same height above the perceptual plane and, therefore, has the same fixed value of likelihood. If we remove the parts of each distribution that rise above the plane, and look down at the result from above, we see the ellipses shown in panel (c). These are called contours of equal likelihood because each point on these four ellipses come from the same height above the $(x_1, x_2)$ perceptual plane and, therefore, denote percepts that are all equally likely to occur. Figures like figure 4.3c are the most common graphical method for describing GRT models.

The ellipses that define each contour are always centered at the distribution means. As a result, figure 4.3c indicates that the means do not fall in a rectangle, despite the factorial nature of the stimuli. And we can also tell that the perceived values of component A are more variable on trials when stimuli $A_1B_1$ or $A_1B_2$ are presented than on trials when stimuli $A_2B_1$ or $A_2B_2$ are presented. This is because the ellipses associated with the distributions elicited by the former two stimuli are wider on the $x_1$ dimension than the ellipses associated with the latter two stimuli. Also, note that by examining the orientation of each ellipse, we can tell that the perceived values of components A and B are statistically independent in all stimuli except $A_2B_1$, where they are positively correlated.

The bell-shaped curves drawn on the two axes of figure 4.3c are the marginal distributions associated with the four stimuli. These are created by collapsing each bivariate distribution onto either the $x_1$ or $x_2$ axis. In the bivariate normal distribution, the marginal distributions are always (univariate) normal distributions. Denote the marginal distributions elicited by presentation of stimulus $A_iB_j$ by $g_{ij}(x_1)$ and $g_{ij}(x_2)$ for dimensions $x_1$ and $x_2$, respectively. These univariate distributions have exactly the same interpretation as the distributions in standard SDT. For example, $g_{21}(x_1)$ describes the likelihood that stimulus component A has a perceived value of $x_1$ on trials when stimulus $A_2B_1$ is presented.

The line segments in figure 4.3c denote the decision boundaries that divide the $(x_1, x_2)$ plane into four response regions. So, for example, according to this model, any percept that falls in the upper-left quadrant elicits an $a_1b_2$ response. GRT is flexible enough to model a wide variety of different decision strategies and, therefore, a wide variety of different types of decision bounds. For example, alternative models of this same identification task could be based on any of the decision strategies depicted in figure 4.1.

## 4.3 PERCEPTUAL INDEPENDENCE, PERCEPTUAL SEPARABILITY, AND DECISIONAL SEPARABILITY

GRT was first proposed in its modern form by Ashby and Townsend in 1986. The primary goal of that article was to model perceptual interactions in a theoretically rigorous way. For much of the middle portion of the twentieth century, this issue was addressed almost completely in terms of operational definitions, and, as a result, the literature at that time included a bewildering variety of independence-related terms. To complicate matters further, in the absence of any underlying theory, it was unknown how or if any of these terms were related to each other. As part of their explication of GRT, Ashby and Townsend (1986) defined the psychological primitives of perceptual independence, perceptual separability, and decisional separability, and they then showed how many of the then-popular independence-related terms are related through these primitives. The concepts of perceptual independence, perceptual separability, and decisional separability are now standard in the field. Because they describe relations between perceptual dimensions, they necessarily do not exist in SDT and are not even defined in one-dimensional approaches.

In summary, extending SDT to multiple stimulus dimensions requires a theory of how perceptual and decisional interactions between dimensions are mediated. In GRT, the most important interaction-related concepts are *perceptual independence, perceptual separability*, and *decisional separability*. This section formally describes each of these terms.

### 4.3.1 PERCEPTUAL INDEPENDENCE

It seems clear that the perceptual independence of components A and B should imply that their perceived values are statistically independent of each other. In other words, if components A and B satisfy perceptual independence, then knowledge of whether the perceived value of component A is above or below its mean value on a trial should convey no information about whether the perceived value of component B is above or below its mean value on that same trial. The question is how to formalize this concept within the language of GRT.

The first such attempt was by Tanner (1956), who was interested in determining whether a pair of tones $S_A$ and $S_B$ that differ in frequency are perceived independently when embedded in white noise N. His experiment involved three conditions: a YES-NO detection task in which the two stimuli were noise N and tone $S_A$ embedded in noise ($N + S_A$), a YES-NO detection task in which the two stimuli were noise N and

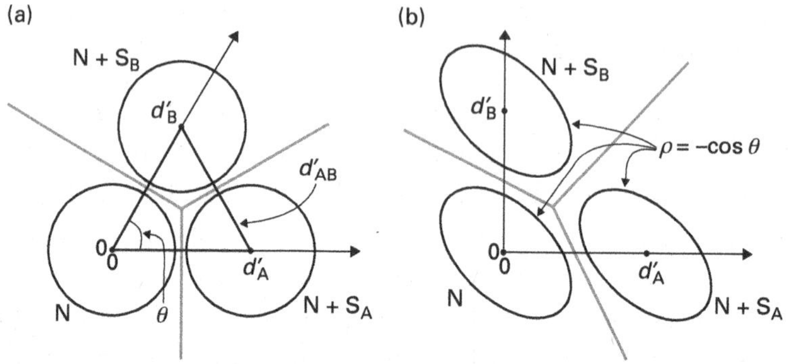


**4.4** (a) Tanner's (1956) GRT model of his three-stimulus experiment. (b) The GRT model with orthogonal dimensions that makes identical accuracy predictions. In both panels, the gray lines are the optimal decision bounds.

tone $S_B$ embedded in noise ($N + S_B$), and a two-stimulus identification experiment with stimuli $N + S_A$ and $N + S_B$.

With respect to testing for perceptual independence, note that this design suffers from an obvious liability—namely, that the two tones are never simultaneously presented (i.e., there is no stimulus $N + S_A + S_B$). As a result, at best, this design offers only indirect information about whether the perceived values of $S_A$ and $S_A$ will interact. Therefore, this experimental design can be used to test for perceptual independence only if some extra assumptions are made.

As described earlier in this chapter, Tanner's (1956) version of GRT assumed that each perceptual distribution is bivariate normal, that all variances are equal, that all covariances are zero, and that the observer gives the response associated with the nearest perceptual mean. This model of his three-stimulus experiment is described in figure 4.4a. These distributional assumptions are restrictive, but nevertheless still sometimes appear in modern versions of GRT, especially in applications to categorization experiments (e.g., see section 8.2). The one extra assumption that Tanner made though, which no longer appears in GRT applications, is that the angle between the perceptual dimensions might not be 90° and that the amount by which the angle differs from 90° is a measure of perceptual dependence. More specifically, he assumed that orthogonal dimensions imply that perception of the two signals satisfies perceptual independence, whereas oblique dimensions signal a perceptual dependence, and the more oblique the dimensions, the stronger the dependence. The idea that perceptual independence is related to the angle between the perceptual dimensions has its roots in linear algebra, where orthogonal vectors are linearly independent.

In figure 4.4a, the dimensions are oblique because of the extra assumption that Tanner made that the means of the N + $S_A$ and N + $S_B$ perceptual distributions are fixed on the two perceptual dimensions—that is, he assumed that the mean of the N + $S_A$ distribution has coordinates $(d'_A, 0)$ and that the mean of the N + $S_B$ distribution has coordinates $(0, d'_B)$. Given this model, the angle between the dimensions in figure 4.4a, denoted $\theta$, can be estimated by solving for $\theta$ in the law of cosines:

$$(d'_{AB})^2 = (d'_A)^2 + (d'_A)^2 - 2d'_A d'_B \cos\theta, \tag{4.1}$$

where each of these $d'$s is estimated in one of Tanner's three experimental conditions. Using this approach, Tanner (1956) reported that when $S_A$ and $S_B$ are tones that are widely separated in frequency, then perceptual independence is approximately satisfied.

Modern versions of GRT no longer define perceptual independence in this way because Ashby and Townsend (1986) showed that Tanner's (1956) GRT model, with oblique dimensions, is a special case of the modern version of GRT in which the dimensions are constrained to be orthogonal, and that more general measures of perceptual dependence are possible in the modern form. The key is to note that the angle $\theta$ between dimensions in Tanner's model is closely related to the correlation coefficient in the modern version of GRT. As described in appendix A, every bivariate normal distribution has an associated variance-covariance matrix (see equation A.4). Given these parameters, the associated correlation coefficient is defined as the standardized covariance:

$$\rho = \frac{\text{cov}_{12}}{\sigma_1 \sigma_2}. \tag{4.2}$$

The Tanner (1956) model constrains all covariances to equal 0, whereas in the modern version of GRT, the covariances can take any values and can even differ from one stimulus to the next. So the Tanner model is more flexible than the current GRT model with respect to the angle between dimensions, whereas the current GRT model is much more flexible than the Tanner model with respect to the covariance (and variance) parameters.

The basic idea of the Ashby and Townsend (1986) result is that there is a linear transformation that obliquely rotates the figure 4.4a model to the one shown in figure 4.4b that has orthogonal dimensions, and that this transformed model is mathematically equivalent to the Tanner model, in the sense that the two models make exactly the same accuracy predictions for all experiments. The key is that although the transformation rotates the decision bounds, like all linear transformations, it preserves all order relations. More specifically, no points cross any decision bounds during the rotation process. So, for example, all points in the upper response region of figure 4.4a are transformed to the upper response region of figure 4.4b. As a result, the predicted probability of a stimulus $S_B$ response is the same in both models. Note also that in the rotated model, the means of the new $N + S_A$ and $N + S_B$ distributions are still on the two axes and still the same distance from the origin, but now those axes are orthogonal. In figure 4.4a, the abscissa is already horizontal, so the oblique rotation does not change this dimension. Instead, it just rotates the ordinate to vertical. Note that this rotation deforms the circular contours of equal likelihood into ellipses. In the figure 4.4 example, the major axes of the ellipses that result from the rotation have a negative slope, and therefore the contours of equal likelihood in the rotated model are characterized by a negative covariance, and therefore a negative correlation coefficient. These intuitive arguments are formalized in the following result.

**Theorem 4.1** Tanner's (1956) GRT model is a special case of the modern version of GRT (i.e., as described, for example, in figure 4.3) in which all covariance matrices are equal, with correlation coefficient equal to

$$\rho = -\cos\theta, \tag{4.3}$$

where $\theta$ is the angle between dimensions in the Tanner model.

*Proof* See section 4.6. ■

If two random variables $X_1$ and $X_2$ have a bivariate normal distribution, then $X_1$ and $X_2$ are statistically independent if and only if they are uncorrelated. So Tanner's (1956) idea to link perceptual independence to the orthogonality of the perceptual dimensions agrees with our intuitive notion that a pair of stimulus components are perceived independently if and only if their perceived values are statistically independent. In Tanner's model, orthogonal dimensions implies that all covariances in the equivalent modern version of GRT equal 0, and therefore if the percepts have a bivariate normal distribution, the perceived values of the two components are statistically independent in all perceptual representations. This is all good, except in the Tanner model, this measure of perceptual independence/dependence is operationalized in an unnecessarily constrained way. In particular, note that the 2D perceptual space in the Tanner model is characterized by one angle $\theta$ between the dimensions—no matter how many stimuli are represented in that space. Therefore, the Tanner model assumes that components A and B share exactly the same degree of perceptual dependence in all stimuli. If components A and B are perceived independently in stimulus $S_i$, then they also must be perceived independently in any other stimulus $S_j$. Unfortunately, there is strong empirical evidence against this prediction. For example, in psychoacoustics, the critical band is a narrow range of frequencies within which tones strongly interact (Fletcher, 1940). So violations of perceptual independence should be expected for tones that are within the same critical band, whereas perceptual independence might be expected for tones that differ widely in frequency (as reported, e.g., by Tanner, 1956). Unfortunately, there is no way to account for such stimulus-specific differences in perceptual independence in the Tanner model.

Ashby and Townsend (1986) conceived of perceptual independence as a stimulus-specific effect. In particular, they wanted a definition general enough that would allow perceptual independence to hold for some stimuli and be violated in others. As a result, they proposed that perceptual independence of components A and B occurs for stimulus $A_iB_j$ if the perceived value of component $A_i$ is statistically independent of the perceived value of component $B_j$. This definition allows perceptual independence to hold, for example, for stimulus $A_1B_1$ and be violated for stimulus $A_1B_2$. The formal definition is as follows.

**Definition of perceptual independence** *Stimulus components A and B satisfy perceptual independence in stimulus $A_iB_j$ if and only if the perceived value of component A is statistically independent of the perceived value of component B on trials when stimulus $A_iB_j$ is presented. More specifically, perceptual independence of components A and B holds in stimulus $A_iB_j$ if and only if*

$$f_{ij}(x_1, x_2) = g_{ij}(x_1)g_{ij}(x_2), \tag{4.4}$$

*for all values of $x_1$ and $x_2$. If perceptual independence is violated, then components A and B are perceived dependently.*

So, in agreement with our intuition, if components A and B are perceived independently, then knowledge of the perceived value of component A on some trial provides no information about the perceived value of component B.

If two variables are bivariate normally distributed, then statistical independence holds if and only if the variables are uncorrelated. However, we must be careful here because, in general, uncorrelated variables are not necessarily independent. So Tanner's idea to equate perceptual independence with dimensional orthogonality is valid only if the extra assumption is added that the distributions are all bivariate normal. The Pearson correlation coefficient that models statistical dependencies in the multivariate normal distribution only measures the strength of the linear relationship between two variables. Statistical independence requires that there is no relationship of any kind between the variables. In the multivariate normal distribution, the only relationships possible are linear. So if variables are multivariate normally distributed, then either they have a linear relationship or no relationship. With other distributions though, nonlinear relationships are possible. For example, suppose a scatterplot of two variables has a perfect U shape. Note that the correlation between the variables is zero (i.e., the best-fitting regression line is horizontal), even though there is a perfect relationship between the variables—in the sense that knowledge of the value of the first variable allows perfect prediction of the second. These variables are uncorrelated but statistically dependent (and they do not have a bivariate normal distribution).

The stimulus-specific nature of this definition is illustrated in figure 4.3c. In this example, the distributions are all bivariate normal, so independence is equivalent to zero correlation. Note that perceptual independence appears to be satisfied in all stimuli except $A_2B_1$, which displays a positive correlation between perceived values of the A and B stimulus components.

### 4.3.2 PERCEPTUAL SEPARABILITY

Tanner (1956) did not mention separability. Instead, the GRT concepts of perceptual and decisional separability evolved from the operationally defined notions of separable versus integral stimulus dimensions that were popularized during the 1960s and 1970s (Garner, 1974; Lockhead, 1966; Shepard, 1964). The idea was that two stimulus dimensions are separable if it is possible to attend to one and ignore variation in the other. If this is impossible, then the dimensions were said to be integral. Classic separable dimensions are color and shape. The judgments that observers make about the color of an object are unaffected if the shape of the object changes from circle to square. Classic integral dimensions are the brightness and saturation of a color patch. Brightness judgments change when the saturation of the color patch changes.

The classic concepts of separable and integral dimensions did not differentiate between perceptual and decisional processing. In effect, these definitions implicitly assumed the classical notion of an absolute threshold in which responding in perceptual tasks depends only on perception and does not require a decision. Of course, like SDT, a fundamental assumption of GRT is that every behavior requires a decision (i.e., the axiom of universal decisions). As a result, GRT distinguishes between whether the separability holds at the perceptual or decisional levels. In particular, Ashby and

Townsend (1986) proposed that perceptual separability holds if the perception of a stimulus component is unaffected by changes in the level of some other component, whereas decisional separability holds if the decision about the level of one component does not depend on the perceived value of the other component. These ideas are formalized as follows.

**Definition of perceptual separability** *Component A is perceptually separable from component B if the observer's perception of A does not change when the level of B is varied. Formally, component A is perceptually separable from component B if and only if*

$$g_{11}(x_1) = g_{12}(x_1), \text{ and } g_{21}(x_1) = g_{22}(x_1), \text{ for all values of } x_1. \tag{4.5}$$

*If perceptual separability fails, then component A is perceptually integral with component B.*

In GRT, all information about the perception of component A on trials when stimulus $A_iB_j$ is presented is contained in the marginal distribution $g_{ij}(x_1)$. Therefore, if component A is perceptually separable from component B then, for example, the perception of $A_1$ when stimulus $A_1B_1$ is presented must be the same as when stimulus $A_1B_2$ is presented. In other words, $g_{11}(x_1)$ must equal $g_{12}(x_1)$, for all values of $x_1$.

In the figure 4.3c example, note that component A is perceptually separable from component B, but component B is not perceptually separable from A. In particular, changing the level of B does not change the perception of A. The $g_{11}(x_1)$ and $g_{12}(x_1)$ marginal distributions are superimposed, and so are the $g_{21}(x_1)$ and $g_{22}(x_1)$ marginal distributions. However, note that increasing the level of A from $A_1$ to $A_2$ increases the perceived value of component B. In particular, the mean of the $g_{12}(x_2)$ marginal distribution is greater than the mean of the $g_{11}(x_2)$ marginal distribution, and the same change in mean occurs for the $g_{22}(x_2)$ and $g_{21}(x_2)$ marginal distributions.

Two stimulus components are perceptually integral if they are not perceptually separable. Therefore, because there are an infinite number of ways that perceptual separability could fail, there are an infinite number of ways that a perceptual integrality could manifest itself. One of the simplest of these, which is illustrated in figure 4.5, is called *mean-shift integrality* (Ashby & Maddox, 1994). In this case, increasing the level of component B, from $B_1$ to $B_2$, increases the mean percept of component A but has no effect on the variance or any other distributional statistics.

Note that perceptual separability is a property of multiple stimuli (i.e., all that share a common value on one stimulus dimension) because it requires examining how the perception of a component is affected by changing the level of some other component. In contrast, perceptual independence is a property of a single stimulus

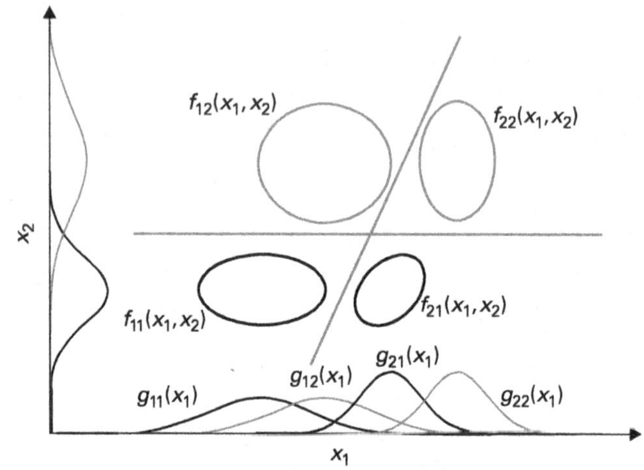


**4.5** A GRT model of the 2 × 2 factorial identification experiment. Component B is perceptually separable from component A, but component A is mean-shift integral from component B.

because it requires examining how the perception of one component is affected by the perceived value of another component on that same trial.

### 4.3.3 DECISIONAL SEPARABILITY

Perceptual separability focuses on whether the perception of one component is affected by the level of another component. In contrast, decisional separability focuses on whether the decision about the level of a component is affected by changes in the perceived value of the other component. More formally, decisional separability is defined as follows.

**Definition of decisional separability** *Decisional separability holds on dimension $x_1$ if the decision about whether component A is at level 1 or level 2 does not depend on the perceived value of component B. Mathematically, this condition holds if and only if the observer uses the following decision rule to determine the level of component A:*

$$\text{The level of component A is 1 if } X_1 \leq X_{C_1}\text{; Otherwise the level is 2,} \tag{4.6}$$

*for some constant criterion $X_{C_1}$. Simlarly, decisional separability holds on dimension $x_2$ if the decision about whether component B is at level 1 or level 2 does not depend on the perceived value of component A.*

Decisional separability is equivalent to using a decision bound on dimension $x_1$ that is parallel to the $x_2$ axis (and therefore orthogonal to the $x_1$ axis)—in other words, a vertical line. In the figure 4.3 example, decisional separability holds on dimension $x_1$, but not on dimension $x_2$. In contrast, in figure 4.5, decisional separability holds on dimension $x_2$, but not on dimension $x_1$.

Note that decisional and perceptual separability are logically unrelated. It is logically possible that either one, both, or neither could hold. It is true that decisional separability is often a sensible decision strategy if perceptual separability holds. For example, note that in both figures 4.3 and 4.5, decisional separability holds on the dimension where perceptual separability is satisfied but not on the dimension where perceptual separability fails. Even so, it is important to note that decisional separability is usually not the optimal decision strategy—even if perceptual separability holds. For example, consider the model illustrated in figure 4.6. This model assumes an optimal decision strategy, so the bounds shown there maximize response accuracy. Note that perceptual separability holds on both dimensions, but because of the positive perceptual dependence, the optimal bounds strongly violate decisional separability.

Decisional separability is a fundamental assumption of independent-decisions models of decision-making (e.g., Ashby & Gott, 1988). This class of models assumes

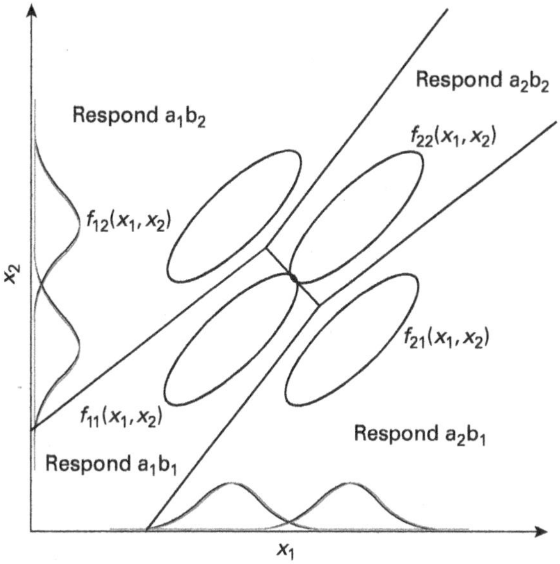


**4.6** A GRT model of the 2 × 2 factorial identification experiment that assumes optimal responding. Note that perceptual separability holds on both dimensions, but decisional separability is strongly violated.

a two-stage decision process. In the first stage, a separate decision is made about the level or presence of each stimulus component. For example, in the $2 \times 2$ factorial identification experiment, the observer is assumed to make separate decisions about whether component A in the presented stimulus is at level 1 or 2 and about whether component B is at level 1 or 2. Note that these decisions require decisional separability on both dimensions. For example, a decision about whether component A is at level 1 or 2 requires a decision rule like the one specified in equation 4.6.

The second stage of independent-decisions models uses the results of these separate decisions to select a response. For example, in the $2 \times 2$ factorial identification experiment an independent-decisions model might assume the following decision rule:

Respond $a_1b_1$ if the levels of both components of the stimulus are small;
Respond $a_1b_2$ if the level of component A is small and B is large;
Respond $a_2b_1$ if the level of component A is large and B is small;
Respond $a_2b_2$ if the levels of both components of the stimulus are large.

More complex decision rules are possible, but the decision rules of all independent-decisions models share one common property—they can all be described using Boolean algebra, where the "truth" values are established via decisions like the one described in equation 4.6. Furthermore, the decision bounds of any independent-decisions model must include piece-wise horizontal and vertical line segments (e.g., as in figure 4.1c), and therefore must satisfy decisional separability, at least locally.

## 4.4 HOLISTIC PERCEPTION

GRT also provides a powerful platform from which to formalize and study holistic or configural perception. Holistic perception, which offers a sharp contrast to the feature-analytic approach that was described in chapter 1, refers to phenomena in which the observer seemingly processes the stimulus as a whole or "gestalt." The most widely studied class of stimuli that seem to evoke holistic processing are human faces (e.g., Farah et al., 1998; Piepers & Robbins, 2012; Richler & Gauthier, 2014). While this literature almost universally acknowledges that face perception is holistic, there is widespread disagreement about what this means at the level of perceptual and/or decisional processing. For example, there are competing proposals that faces are encoded as a single object, rather than as a collection of features (e.g., Farah et al.,

1998), that faces are encoded as a set of independent features but attentional training causes failures of selective attention (e.g., Richler et al., 2012), and that holistic perception includes a significant decisional component (Richler et al., 2008). GRT offers the potential to resolve these debates.

GRT was first used to model holistic or configural perception by O'Toole et al. (2001), and it was first applied to holistic face perception by Wenger and Ingvalson (2002, 2003). More recently, Townsend and Wenger (2015) used GRT to propose a set of working axioms for holistic or configural perception.

One widely replicated phenomenon of face perception, which is thought to provide strong evidence of holistic perception, is the composite-face illusion (Murphy et al., 2017; A. W. Young et al., 1987). This illusion occurs in tasks where observers are presented with an image of a face, divided into top and bottom portions roughly at the nose. Observers are asked to identify either the top or bottom half while ignoring the other half. The top and bottom portions can be created either from the same or from different faces, the faces can be either familiar (e.g., famous) or unfamiliar, and the two halves can be either aligned or misaligned. When the top and bottom halves are from different faces, each half is often easy to identify when they are misaligned, but when aligned, they seem to fuse together to form a novel face, and observers are now impaired at identifying either half. The accuracy advantage of misaligned faces is referred to as the composite-face illusion.

As an illustration of how GRT might be used to study holistic perception, consider various GRT models of the composite-face illusion. For simplicity, assume the simplest case in which the top and bottom halves are always aligned. Consider a $2 \times 2$ factorial identification experiment in which the top and bottom halves of two different faces are factorially combined. Suppose the two faces are from person 1 and person 2. Let $T_i$ denote the top half-face of person $i$ ($i = 1$ or 2) and let $B_i$ denote the bottom half-face of person $i$. Then the four stimuli are $T_1B_1$, $T_2B_1$, $T_1B_2$, and $T_2B_2$. Note that the top and bottom halves of stimuli $T_1B_1$ and $T_2B_2$ are from the same person, whereas the top and bottom halves of stimuli $T_2B_1$ and $T_1B_2$ are from different people. Since this is an identification experiment, the observer's task is to identify the stimulus uniquely, which requires them to make separate identification responses for the top and bottom half-faces. Denote the four identifying responses as $t_1b_1$, $t_2b_1$, $t_1b_2$, and $t_2b_2$.

Because the top and bottom halves are always aligned in this experiment, the composite-face illusion predicts that the two halves of stimuli $T_2B_1$ and $T_1B_2$ will fuse to form a novel face, which will make it more difficult for observers to identify each half-face. As a result, we expect performance to be better on $T_1B_1$ and $T_2B_2$ trials than on $T_2B_1$ and $T_1B_2$ trials. Any such accuracy difference is an example of a composite-face effect. The question that our GRT analysis will try to answer is: what causes this performance difference? GRT offers a number of logically different alternative accounts of why $T_1B_1$ and $T_2B_2$ performance might be better than $T_2B_1$ and $T_1B_2$ performance. Specifically, the composite-face illusion might be primarily due to decision processes, it could be due primarily to a violation of perceptual separability or to a violation of perceptual independence, or it could be caused by some combination of two or more of these.

Figure 4.7 shows five alternative GRT models of this task. Figure 4.7a shows a null model that assumes perceptual independence, perceptual separability, and decisional separability, and therefore does not display any type of holism or configurality. This

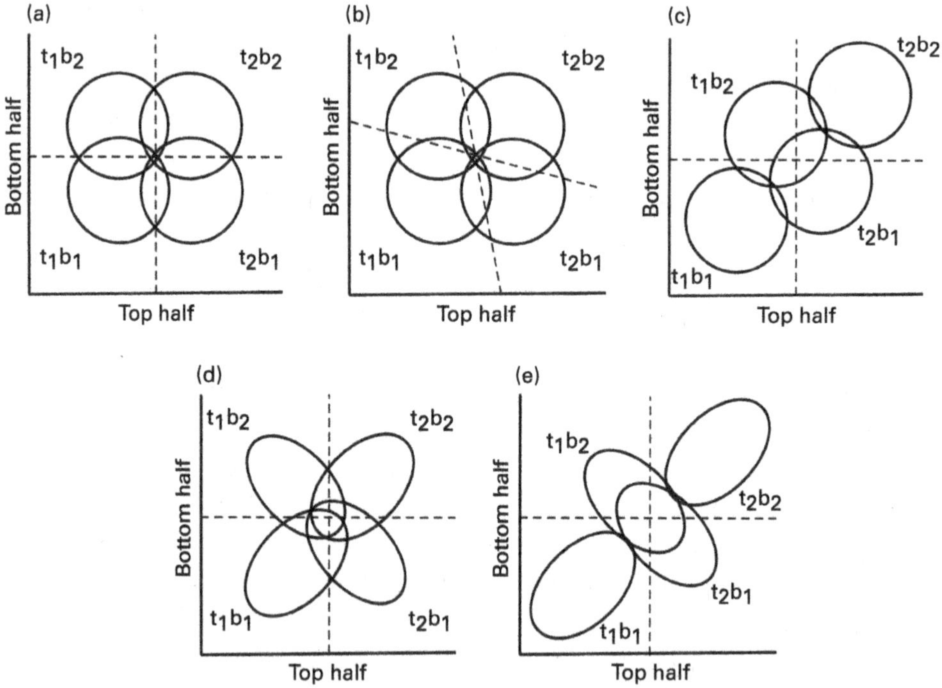


**4.7** Alternative GRT models of holism or configurality in the composite-face effect task: (a) null model that lacks holism or configurality; (b) configurality as a decisional effect; (c) configurality as a type of perceptual integrality; (d) configurality as a perceptual dependence; (e) configurality as perceptual integrality and perceptual dependence.

model predicts equal accuracy on all trials and would be used primarily as a baseline model against which the other GRT models could be compared.

Figure 4.7b models the composite-face illusion as a purely decisional effect—that is, as a violation of decisional separability. Note that with the decision bounds shown in figure 4.7b, the $t_1b_1$ and $t_2b_2$ response regions are larger than the $t_1b_2$ and $t_2b_1$ regions, and as a result, this model correctly predicts that accuracy will be higher on $T_1B_1$ and $T_2B_2$ trials than on $T_1B_2$ and $T_2B_1$ trials. On the other hand, because GRT assumes that decision bounds can be modified with training (as does SDT), this account of holism makes the strong, and seemingly dubious prediction that holism would disappear in this task if observers were rewarded more for correct $T_1B_2$ and $T_2B_1$ responses than for correct $T_1B_1$ and $T_2B_2$ responses. Because we tend to think of configurality as a perceptual effect, this prediction seems unlikely. This is not to say that there is no decisional component to the composite-face effect, only that the phenomenon is likely not purely decisional.

Figure 4.7c models the composite-face illusion as purely a violation of perceptual separability. This model makes the appealing prediction that on trials when the two halves are from the same face, perception of both halves is altered to improve accuracy, even when decision bounds are used that satisfy decisional separability. Similarly, the model predicts that when the two halves are from different faces, the perception of both halves is altered to decrease accuracy. So this is an attractive model of configurality. Even so, note that perceptual independence is satisfied in this model. As a result, it predicts that when stimulus $T_2B_2$ is presented, for example, knowledge that the top-half looks especially like person 2 on a trial conveys no information about the perception of the bottom-half on that trial. This seems like a flaw because our common notion of configurality is that the face should be seen as a Gestalt and not as independent top-half and bottom-half percepts.

Figure 4.7d models the composite-face illusion as a type of perceptual dependence and therefore overcomes the weakness we identified with model 4.7c. In other words, model 4.7d predicts that when stimulus $T_2B_2$ is presented, for example, trials when the top-half looks especially like person 2 co-occur with trials when the bottom-half looks especially like person 2, and trials when the top-half looks especially like person 1 co-occur with trials when the bottom-half looks especially like person 1. Therefore, unlike any of the preceding models, model 4.7d treats holism as a within-stimulus effect. As a result, as a model of configurality, this version of GRT also has some attractive properties. On the other hand, figure 4.7d is fatally flawed because it predicts equal accuracy on all trials, and therefore fails to predict the composite-face effect.

Figure 4.7e combines the figure 4.7c and 4.7d models of configurality. It includes the same perceptual integrality as model 4.7c and the same perceptual dependencies as model 4.7d, and therefore overcomes the weaknesses of both models. A sixth model might incorporate violations of decisional separability into model 4.7e.

More work is needed to test among these alternative accounts of the composite-face effect, but even so, the existing applications of GRT to the study of holistic perception have already led to two important conclusions. The first is that, just as there are varieties of independence in perception (Ashby & Townsend, 1986), there are a variety of ways to obtain patterns of data that seem to suggest holism or configurality. The second is that some tasks might be too impoverished to test certain hypotheses about holistic processing. For example, GRT simulations reported by Richler and Gauthier (2013) demonstrated that the standard method of testing the composite-face effect does not provide data that would allow for testing the strong hypothesis that holism is a within-stimulus effect (see discussions in Richler and Gauthier, 2013; Rossion, 2013).

## 4.5 MODEL IDENTIFIABILITY

GRT has been used to analyze data from identification confusion matrices in two different ways. One approach is to compute certain summary statistics from the empirical confusion matrix and then to check whether these satisfy conditions that are characteristic of perceptual independence or perceptual separability. This is the topic of chapter 6. The other approach, which is the focus of chapter 7, is to fit GRT models to the entire confusion matrix. This requires distributional assumptions, and the standard approach is to assume that the perceptual distributions are all bivariate normals. To test various assumptions about perceptual and decisional processing—for example, to test whether perceptual independence holds—a version of the model that assumes perceptual independence is fit to the data as well as a version that makes no assumptions about independence. This latter version contains the former version as a special case (i.e., in which all covariance parameters are set to zero), so it can never fit worse. After fitting these two models, we conclude that perceptual independence is violated if the more general model fits significantly better than the more restricted model that assumes perceptual independence (R. D. Thomas, 2001).

It is important to note, however, that regardless of which method is used, there are certain nonidentifiabilities in GRT models that could limit the conclusions that are possible to draw from any such analyses (e.g., Menneer et al., 2010; Silbert & Thomas, 2013). The problems are most severe when GRT is applied to identification data from $2 \times 2$ factorial designs like those modeled in figures 4.3, 4.5, and 4.6 (i.e., with stimuli $A_1B_1$, $A_1B_2$, $A_2B_1$, and $A_2B_2$). For example, Silbert and Thomas (2013) showed that in $2 \times 2$ applications where there are two intersecting linear decision bounds that do not satisfy decisional separability, there always exists an alternative model that makes the exact same empirical predictions and satisfies decisional separability (and these two models are related by a linear transformation).

This nonidentifiability is illustrated in figure 4.8. Panel (a) shows a GRT model for the $2 \times 2$ factorial identification experiment. The black lines denote the decision bounds. Note that decisional separability is strongly violated, as is perceptual independence, whereas perceptual separability is satisfied on both dimensions. Panel (b) shows an alternative model that makes identical accuracy predictions—that is, it predicts exactly the same $4 \times 4$ confusion matrix as the model in panel (a). However, in this alternative model, decisional separability is satisfied. Therefore, if the empirical confusion matrix happened to equal the matrix predicted by these models, it would be impossible to tell from the matrix alone, which of these models was valid, and, as a result, it would be impossible to know whether the observer was using a decision strategy that satisfied decisional separability. To make matters worse, note that perceptual separability is satisfied on both dimensions in panel (a) and violated on both

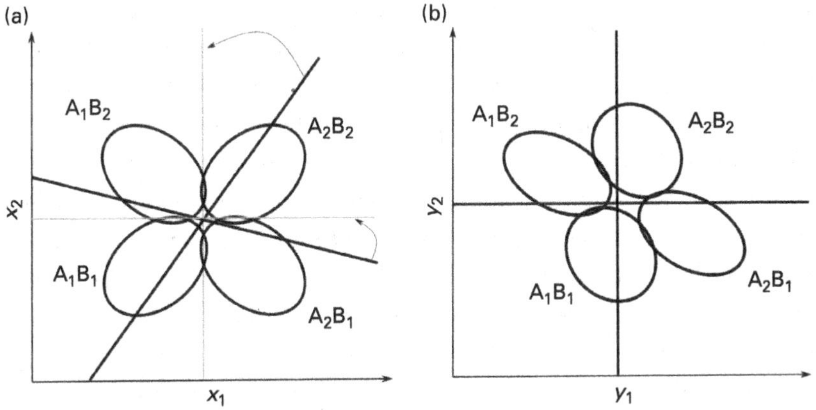


**4.8** Nonidentifiability of decisional separability in the $2 \times 2$ factorial identification experiment. (a) A GRT model in which decisional separability is violated. (b) An alternative GRT model that makes identical accuracy predictions, but in which decisional separability is satisfied. This figure was adapted from Silbert and Thomas (2013).

dimensions in panel (b), so given these data, we would also be unable to draw any inferences about perceptual separability.

Given any GRT model for the $2 \times 2$ factorial identification task in which the four response regions are created from two intersecting lines—like the model shown in figure 4.8a—there always exists an alternative GRT model that satisfies decisional separability and makes identical accuracy predictions (Silbert & Thomas, 2013). The equivalent model that satisfies decisional separability can be found by performing an oblique rotation on the space in which the original model is defined. The oblique rotation is a linear transformation that simultaneously rotates one decision bound to be vertical and the other to be horizontal. This is exactly the type of transformation we used to transform the Tanner (1956) model shown in figure 4.4a to the equivalent GRT model shown in figure 4.4b (i.e., and described in the proof of theorem 4.1). As in that case, because the transformation is linear, the rotation preserves the location of all points relative to the decision bounds—that is, no points cross a decision bound during the rotation process. As a result, all predicted response probabilities are unaffected by the transformation, and therefore the two models make identical accuracy predictions. Thus, in standard applications of GRT to identification experiments that use a $2 \times 2$ factorial design, decisional separability is not testable, nor are the slopes of the decision bounds uniquely estimable. It turns out however, that for a variety of reasons, these nonidentifiabilities are not catastrophic.

First, there are no identifiability problems if the perceptual dimensions are known. The linear transformation that rotates intersecting linear bounds so that one is vertical and one is horizontal rotates the whole space, including the perceptual dimensions. This is why the axes in the figure 4.8b model are labeled $y_1$ and $y_2$, whereas the axes in the figure 4.8a model are labeled $x_1$ and $x_2$. So although decisional separability holds in the new model, the separability is with respect to novel dimensions. In other words, one interpretation of the identifiability problem is that if the GRT model that best fits some single confusion matrix from a $2 \times 2$ factorial design assumes intersecting linear bounds that violate decisional separability, then there is always an alternative GRT model that fits equally well and assumes that the observer made decisions by selectively attending to some different perceptual dimensions. With complex stimuli, such as faces, this will often be difficult to rule out. However, with many simple stimuli, this possibility is straightforward to reject. For example, consider sine-wave gratings (e.g., such as Gabor patches), like those shown in figure 4.2a, that are created by factorially combining two spatial frequencies (bar widths) and two (bar) orientations. An enormous visual perception literature tells us that humans treat these two dimensions as primary. So any conclusions about decisional separability drawn from a GRT analysis should be immune to identifiability problems because the mathematically equivalent model that makes different assumptions about decisional separability must assume that the observer perceived the stimuli in a way that is incompatible with the visual perception literature.

Second, if different observers use different linear decision bounds that each violate decisional separability, then different oblique rotations will be required for each observer to create models in which decisional separability is satisfied for all observers. As a result, the equivalent GRT models that assume decisional separability would assume that all observers are selectively attending to different perceptual dimensions. Although this is logically possible with complex stimuli, such as human faces, it seems highly unlikely. Instead, it is commonly assumed that individual differences in perception are rare, even when individual differences in decision strategy are common.

Third, the problems do not generally exist with $3 \times 3$ or larger factorial designs (as used, e.g., by Ashby et al., 2001). In the $3 \times 3$ case, there are nine stimuli and therefore nine response regions. As a result, the GRT model with linear bounds requires at least four decision bounds to divide the perceptual space into these nine regions (e.g., in a tic-tac-toe configuration). Typically, two will have a generally vertical orientation with respect to one perceptual dimension and two will have a generally horizontal orientation. In this case, there is no linear transformation that guarantees decisional separability except in the special case where the two vertical-tending bounds are parallel and the two horizontal-tending bounds are also parallel (because parallel lines remain parallel after linear transformations). Thus, in $3 \times 3$ (or higher) designs, decisional separability is typically identifiable and testable.

This same rationale also explains why the identifiability problem associated with decisional separability is not usually extended to perceptual independence or perceptual separability. For example, consider perceptual independence, which is violated in figure 4.8a. Note that two distributions seem to share a common positive correlation (the distributions associated with stimuli $A_1B_1$ and $A_2B_2$) and two seem to share a common negative correlation (the distributions associated with stimuli $A_1B_2$ and $A_2B_1$). However, of course, in general, all four distributions could each express a unique correlation. For any one perceptual distribution characterized by a nonzero correlation, there always exists an oblique rotation that creates a new model in which that same distribution satisfies perceptual independence. Even so, to produce the new model, this same oblique rotation is applied to the whole space, and if any two distributions are characterized by different amounts of correlation, then the rotation that erases one perceptual dependence will not erase the other. In fact, if one distribution is characterized by a perceptual dependence and a second satisfies perceptual independence then the rotation that eliminates the dependence in the first distribution will create a new dependence in the second. As a result, the only conditions under which an oblique rotation can eliminate all perceptual dependencies is if all four distributions are characterized by the exact same level of dependence (i.e., they all have the exact same correlation coefficient, as in the Tanner model).

Fourth, there are simple experimental manipulations that can be added to the basic $2 \times 2$ identification experiment to test for decisional separability. These will be discussed in detail in the next chapter, but as just one example, switching the locations of the response buttons interferes with performance much more if decisional separability fails than if decisional separability holds (Ashby et al., 2003; Maddox et al., 2004). Thus, one could add one hundred extra trials to the end of a $2 \times 2$ identification experiment where the locations of the response buttons are shuffled. The observers should be instructed about this change and the new identities should be displayed on the screen above each button. If accuracy drops significantly during this period, then decisional separability is likely violated, whereas if accuracy is relatively unaffected then decisional separability is supported.

Fifth, one could analyze the $2 \times 2$ data using GRT-wIND (GRT with INDividual differences; Soto et al., 2015), which was inspired by the INDSCAL model of multidimensional scaling (Carroll & Chang, 1970) and is described in detail in chapter 7 (in section 7.7). Like INDSCAL, GRT-wIND is fit to the data from all individuals simultaneously. All observers are assumed to share the same group perceptual distributions, but different observers are allowed different linear bounds and they are assumed to allocate different amounts of attention to each perceptual dimension. The model does not suffer from the identifiability problems identified by Silbert and Thomas (2013), even in the $2 \times 2$ case, because with different linear bounds for each observer, there is no linear transformation that simultaneously makes all these bounds satisfy decisional separability.

In summary, the identifiability problems that have been attributed to decisional separability are only problematic under some extreme conditions—namely, when (1) the task is a four-stimulus identification experiment; (2) the data from individual observers are all analyzed separately (rather than simultaneously by GRT-wIND); (3) all observers use two linear decision bounds to select a response; (4) the true perceptual dimensions are unknown; and (5) all observers use exactly the same two linear bounds. If any of these conditions fail, then decisional separability should be testable. Furthermore, even if all these conditions hold, it should be possible to test for decisional separability by adding some extra experimental manipulations (e.g., such as a button switch).

## 4.6 PROOF OF THEOREM 4.1

For a complete proof, see Ashby and Townsend (1986). The focus here is on how the figure 4.4b model was derived from the figure 4.4a model. Let the vector $\underline{x}$ denote points in the figure 4.4a space and $\underline{y}$ denote points in the figure 4.4b space. Then Ashby and Townsend (1986) showed that the linear transformation that rotates the figure 4.4a space to the figure 4.4b space is:

$$\underline{y} = \begin{bmatrix} 1 & -\cos(\theta)/\sin(\theta) \\ 0 & 1/\sin(\theta) \end{bmatrix} \underline{x} = \mathrm{R}\underline{x}, \tag{4.7}$$

where, for obvious reasons, the matrix R is called a rotation matrix. Because equation 4.7 is a linear transformation, if $\underline{x}$ is bivariate normal with mean vector $\underline{\mu}$ and variance-covariance matrix $\Sigma$, then $\underline{y}$ will be bivariate normal with mean vector $\mathrm{R}\underline{\mu}$ and variance-covariance matrix $\mathrm{R}\Sigma\mathrm{R}'$ (see equations A.13 and A.14 in appendix A). When the figure 4.4a distributions are embedded in the figure 4.4b space, they have means (0,0), $(d'_{\mathrm{A}}, 0)$, and $(d'_{\mathrm{B}}\cos(\theta), d'_{\mathrm{B}}\sin(\theta))$, and they all have variance-covariance matrix equal to the identity. Applying the equation 4.7 transformation converts the means to (0,0), $(d'_{\mathrm{A}}, 0)$, and $(0, d'_{\mathrm{B}})$, respectively, and all variance-covariance matrices to

$$\Sigma_{\underline{y}} = \mathrm{RR}' = \begin{bmatrix} 1/\sin^2(\theta) & -\cos(\theta)/\sin^2(\theta) \\ -\cos(\theta)/\sin^2(\theta) & 1/\sin^2(\theta) \end{bmatrix}. \tag{4.8}$$

Note that the first variance (in row 1 and column 1) is due to the trig identity: $\sin^2(\theta) + \cos^2(\theta) = 1$. Given this variance-covariance matrix, note that the correlation coefficient equals

$$\rho = \frac{\mathrm{cov}}{\sigma_1\sigma_2} = \frac{\frac{-\cos(\theta)}{\sin^2(\theta)}}{\sqrt{\left(\frac{1}{\sin^2(\theta)}\right)^2}} = -\cos(\theta). \tag{4.9}$$

■

# 5

# DECISION PROCESSES WITH MULTIDIMENSIONAL STIMULI

GRT assumes that the sensory experience elicited by a stimulus can be represented by a point in a multidimensional sensory space that includes one dimension for every feature or component that varies across the stimulus ensemble. Because of stimulus, perceptual, and memory noise, this point varies probabilistically over repeated presentations of the same stimulus. To model the observer's decision, GRT assumes that people divide the sensory space into response regions and then emit the response on each trial associated with the region into which the percept falls. The theory makes no strong assumptions, though, about how the sensory space is carved into response regions. Logically, this could be done in an infinite number of ways. For example, GRT does not even require that the region associated with each response is connected.

Modeling decision processes in GRT is therefore more complex than in SDT. With the single sensory dimension assumed by SDT, two response regions could also be created in an infinite number of ways, but all but one of these disappear if connectedness is assumed. The only way to divide a line into two connected regions is via a single criterion that separates the two regions. Furthermore, if the ROC curve is convex, then the likelihood ratio of the two sensory distributions is monotonic (see theorem 2.2), in which case response regions separated by a single criterion are optimal (maximizing either accuracy or payoffs). This is why the SDT literature allocates so little attention to alternative decision strategies.[1]

## 5.1 ALTERNATIVE DECISION STRATEGIES WITH MULTIDIMENSIONAL STIMULI

Although there are an infinite number of ways to divide a plane into response regions, almost all of these can be ignored because they lack any psychological justification. When modeling human behavior, we should only consider decision strategies that a human might employ. The problem though, is that this criterion still leaves an uncomfortably large number of alternative strategies. This section provides a conceptual review of some of the more prominent strategies that have appeared in the psychological literature. Chapters 7 and 8 describe the computational details required to fit any of these models to identification or categorization data, respectively.

### 5.1.1 INDEPENDENT-DECISIONS MODELS

Independent-decisions models assume that observers make independent decisions about the level of each stimulus component and then these separate decisions are combined to select a response in a way that is consistent with Boolean algebra (Ashby & Gott, 1988; Feldman, 2000; Shaw, 1982; Townsend & Ashby, 1982). An example is shown in figure 5.1a for a hypothetical identification task with four stimuli, A, B, C, and D. The circles denote the contours of equal likelihood of the perceptual distributions elicited by the presentation of each stimulus and the solid lines show the decision bounds. The bounds in figure 5.1a are compatible with the independent-decisions rule:

> Respond D if the stimulus is small on dimension 2, and respond B if the stimulus is large on dimension 2. If the dimension 2 value is intermediate, respond A if the stimulus is small on dimension 1 and C if the dimension 1 value is large.

Note that this rule, like all independent-decisions strategies, is compatible with a two-stage decision process. During stage 1, the observer makes independent decisions about whether the stimulus is small or large on dimension 1, and small, medium, or large on dimension 2. Denote the outcomes of the decision about the level of dimension 1 as $S_1$ and $L_1$, respectively, and the outcomes of the decision about the level of dimension 2 as $S_2$, $M_2$, and $L_2$, respectively. Next, during stage 2, these separate decisions are combined via the rules of Boolean algebra to select an identification decision. The final decision strategy can be expressed using the notation of Boolean algebra as follows:

$$
\begin{aligned}
M_2 \wedge S_1 &= A \\
L_2 &= B
\end{aligned}
$$

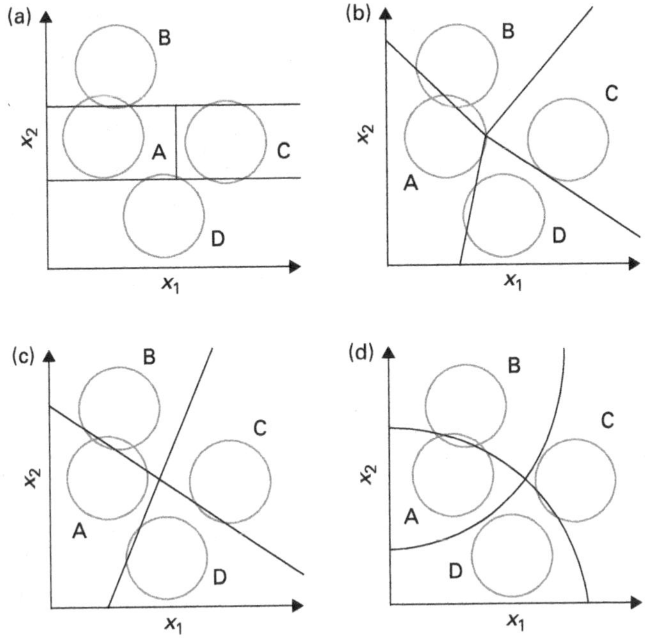


**5.1** Alternative decision strategies for a hypothetical four-stimulus (i.e., A, B, C, and D) identification task. The circles identify contours of equal likelihood associated with the perceptual distributions, and the solid lines/curves identify the decision bounds. (a) An independent-decisions strategy. (b) A minimum-distance or nearest-prototype strategy. (c) The general-linear classifier. (d) The general-quadratic classifier.

$$\begin{aligned} &M_2 \wedge L_1 = C \\ &\quad S_2 = D. \end{aligned} \tag{5.1}$$

From the perspective of GRT, the key to modeling an independent-decisions strategy is that no matter how complex the rule—that is, no matter how complex the Boolean expression that describes the rule—the resulting decision bounds are always piecewise combinations of horizontal and vertical line segments. This is because of the stage 1 independent decisions. By definition, an independent decision about the level of one stimulus component can only depend on the perceived value of that component, and therefore the criterion about whether the value is small or large (or intermediate) can not depend on the perceived value of any other component. In other words, decisional separability holds, at least locally, and decisional separability always produces a vertical or horizontal decision bound.

Independent-decisions strategies play a critical role in many GRT applications for two primary reasons. First, as we will see in the next chapter, virtually all current summary-statistics tests of perceptual independence and perceptual separability assume that decisional separability holds. In other words, these tests are invalid if decisional separability is violated, and unfortunately, there is no way to know from the tests alone, whether or not the assumption of decisional separability is valid. Therefore, when using such tests, it is important to arrange the experimental conditions so that an independent-decisions strategy is as likely as possible. Later in this chapter, we will consider what conditions might encourage observers to adopt an independent-decisions strategy (see section 5.5).

Second, as described in chapter 7, GRT models are easier to fit to identification confusion data if decisional separability holds. One advantage is that only a few parameters are needed to describe such strategies and another is that decisional separability often greatly simplifies the numerical integrations required to fit the model. On the other hand, because fast computers and efficient numerical integration algorithms are now readily available, this benefit is now much less important than during the early years of GRT.

### 5.1.2 THE MINIMUM-DISTANCE CLASSIFIER

The minimum-distance classifier, illustrated in figure 5.1b, gives the response associated with the perceived mean that is closest to the percept elicited by the stimulus (Ashby & Gott, 1988). In other words, the observer is assumed to compute the distance from the current perceptual sample to the mean of each perceptual distribution and to give the response associated with the smallest of these distances.

In the multivariate normal distribution, the mean is also the mode, and the mode can be interpreted as the prototypical percept elicited by that stimulus. Therefore, an alternative interpretation of the minimum-distance classifier is that the observer gives the response associated with the most similar prototypical percept. In the categorization literature, models postulating this classification scheme are known as prototype models (e.g., Homa et al., 1981; Posner & Keele, 1968; J. D. Smith & Minda, 1998).

A third interpretation of the minimum-distance classifier is as a matched filter (Fukunaga, 2013). The idea here is that the observer stores a template image of each stimulus, correlates the current percept with each of these templates, and then gives the response associated with the highest correlation. This scheme is equivalent to minimum-distance or most-similar-prototype classification because an obvious choice

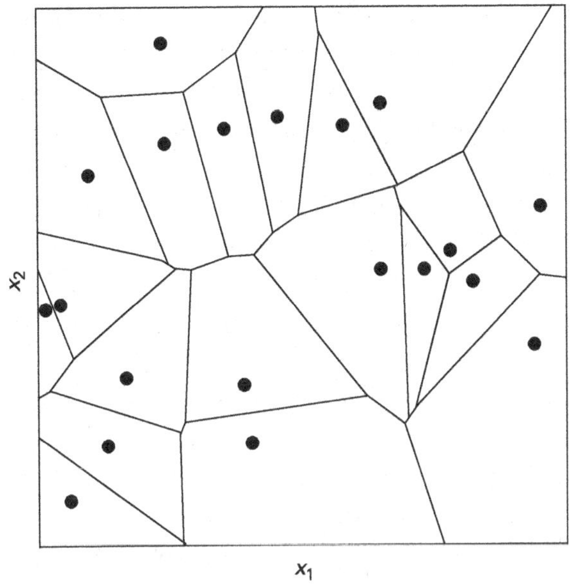


**5.2** Decision bounds predicted by the minimum-distance classifier in an identification task with 20 stimuli. The dots denote the perceived means, prototypes, or templates.

for a template is the image associated with the mean or mode of the perceptual distribution, and correlation is sometimes used as a measure of similarity (Ashby, 2019).

The minimum-distance classifier always produces piece-wise linear bounds. For example, every point on the line segment separating the A and B means in figure 5.1b is equidistant to those means. So all points above the segment are closer to the B mean and therefore elicit a B response, whereas all points below the segment are nearer the A mean. A more complex example is shown in figure 5.2. This figure shows the bounds that result from minimum-distance classification in an identification task with 20 stimuli, where the 20 mean percepts are denoted by the dots. In mathematics, the tiling patterns produced by this classification scheme are called *Voronoi tessellations*.

Despite the complexity of the bounds shown in figure 5.2, the minimum-distance classifier that produced those bounds has no free parameters. The location of the bounds is determined completely by the location of the perceptual means. Therefore, a GRT model that assumes responses are selected by the minimum-distance classifier can allocate all available degrees of freedom to estimating the perceptual distributions.

A more general version of the minimum-distance classifier assumes that the prototypes or templates used in the decision process might not equal the true distribution means. The idea here is that the decision process might not have access to the perceptual distributions, and so might have to estimate the prototypes/templates separately. If so, then there is no reason to expect these estimates to equal the perceptual means exactly. This more general model includes free parameters in the decision process that represent the coordinates of the estimated prototypes/templates. Ashby et al. (2001) reported that this model accounted for 99.9% of the variance in the confusion matrices from a $3 \times 3$ factorial identification experiment (which therefore included nine different stimuli).

The minimum-distance classifier ignores all information about the perceptual distributions, except the means. Even so, it is optimal in the sense that it maximizes accuracy if the following conditions all hold: (1) the perceptual distributions are multivariate normal, (2) all perceptual variances are equal, (3) all covariances equal 0, and (4) all base rates are equal (e.g., Ashby & Alfonso-Reese, 1995). If any of these conditions is violated, however, then in general, the accuracy of the minimum-distance classifier is less than optimal. This is because it predicts exactly the same bounds regardless of the distribution variances and covariances, whereas the optimal decision strategy is highly sensitive to these parameters. For example, look back at figure 4.6. Note that in that model, the means of the four perceptual distributions lie on the corners of a rectangle. As a result, the minimum-distance classifier predicts that the decision bounds will include a vertical and horizontal line—in other words, that decisional separability will hold on both dimensions. The $X_1$ and $X_2$ variances are equal in all four distributions, but note that in each distribution there is a strong positive perceptual dependence. This strong dependence causes the optimal bounds, shown in figure 4.6, to be very different from the decisional separability bounds predicted by the minimum-distance classifier.

A large literature shows that people are highly sensitive to correlations between feature values in categorization tasks (e.g., Anderson & Fincham, 1996; Ashby & Gott, 1988; Medin et al., 1982). However, in these experiments the values of each feature are correlated across stimuli rather than across percepts, there are generally only two contrasting categories, and the correlations are extreme enough that the predicted accuracy difference between the minimum-distance classifier and the optimal classifier is large. The scenario illustrated in figure 4.6 meets this last criterion because all four distributions display the same large perceptual dependence. In cases in which the correlation and variance differences are less extreme, and perhaps vary from one stimulus to the next, the accuracy difference between the minimum-distance classifier and the optimal classifier will be much smaller. As a result, minimum-distance classification might be a reasonable identification strategy, even though it is rarely optimal.

### 5.1.3 THE OPTIMAL CLASSIFIER

The optimal classifier assumes the participant adopts the decision strategy that maximizes accuracy (Ashby & Gott, 1988). If the perceptual distributions are multivariate normal, then the decision bounds that maximize accuracy are always piecewise linear or quadratic (e.g., Fukunaga, 2013). As the next result shows, for any two multivariate normal distributions, the optimal bound is linear if the two variance-covariance matrices are equal and quadratic if they are unequal.

**Theorem 5.1** Consider two categories of stimuli or percepts, A and B, that vary on $r$ dimensions. Consider a task where on each trial a random sample, denoted by the $r \times 1$ random vector $\underline{X}$, is drawn from a randomly selected category and presented to the observer, whose task is to identify whether the sample belongs to category A or B. Suppose the two categories are equally likely and are each defined by a multivariate normal distribution. Let $\underline{\mu}_J$ and $\Sigma_J$ denote the mean vector and variance-covariance matrix, respectively, for category $J$ (where $J$ = A or B).

(1) If $\Sigma_A = \Sigma_B = \Sigma$, then accuracy is maximized by partitioning the stimulus space into A and B response regions where the partition (i.e., the decision bound) is the line (or plane) that satisfies

$$\underline{a}'\underline{x} + b = 0, \tag{5.2}$$

where

$$\underline{a}' = 2(\underline{\mu}_A - \underline{\mu}_B)'\Sigma^{-1}$$

and

$$b = \underline{\mu}_A'\Sigma^{-1}\underline{\mu}_A - \underline{\mu}_B'\Sigma^{-1}\underline{\mu}_B.$$

(2) If $\Sigma_A \neq \Sigma_B$, then accuracy is maximized by partitioning the stimulus space into A and B response regions where the partition (i.e., the decision bound) is the quadratic curve (or surface) that satisfies

$$(\underline{x} - \underline{\mu}_A)'\Sigma_A^{-1}(\underline{x} - \underline{\mu}_A) - (\underline{x} - \underline{\mu}_B)'\Sigma_B^{-1}(\underline{x} - \underline{\mu}_B) + b = 0, \tag{5.3}$$

where

$b = \ln|\Sigma_A| - \ln|\Sigma_B|.$

Note that $|\Sigma|$ denotes the determinant of $\Sigma$.

*Proof* See section 5.7. ■

Equation 5.3 describes a quadratic curve (or surface) because after the specified multiplications, the equation will include squared term for each entry in $\underline{x}$. Furthermore, any type of quadratic curve is possible—a parabola, hyperbola, ellipse, or circle. The key result of this theorem however, is that given any two multivariate normal distributions, so matter what their mean vectors or variance-covariance matrices, the optimal decision bound is always linear or quadratic. No other bounds are possible.

### 5.1.4 THE GENERAL-LINEAR CLASSIFIER

The general-linear classifier (GLC) assumes that the perceptual plane is carved up into response regions by a set of lines of arbitrary slope and intercept (Ashby & Gott, 1988; Medin & Schwanenflugel, 1981). This classification scheme is illustrated in figure 5.1c. Note that, in this case, two lines divide the plane into four response regions. On the one hand, the GLC is more general than the minimum-distance classifier because the two lines can have any slopes and intercepts. On the other hand however, the GLC bounds are constructed from two lines, whereas the minimum-distance classifier bounds shown in figure 5.1b are constructed from four line segments. If there are only two response regions, as in a two-stimulus identification task or a two-category categorization task, then both classifiers predict single-line decision bounds. In this case, the minimum-distance classifier is a special case of the GLC, in which the slope and intercept are fixed to the minimum-distance values. However, the more general version of the minimum-distance classifier in which the estimated prototypes/templates might not equal the distribution means, is equivalent to the GLC, at least if no constraints are placed on the locations of the estimated prototypes.

The key to fitting the general linear classifier to either identification or categorization data is to leverage the discriminant function associated with each linear bound (e.g., Ashby & Gott, 1988; Townsend & Landon, 1983). This strategy is illustrated in figure 5.3. Suppose as before, that the stimuli vary on two perceptual dimensions and denote the perceived value of the presented stimulus on these two dimensions

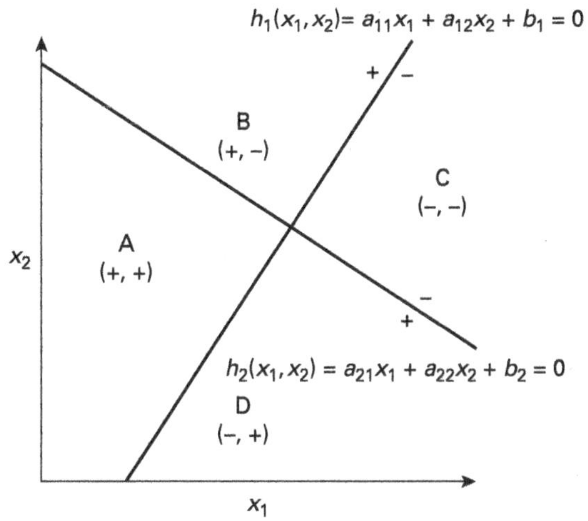


**5.3** Decision bounds predicted by the GLC in a task with four response regions.

by $(X_1, X_2)$. Then for any decision bound, we can always define a discriminant function $h(X_1, X_2)$ with the property that $h(x_1, x_2) > 0$ for any specific percept—that is, for any numerical values $(x_1, x_2)$—that fall on one side of the bound [e.g., the side that includes the point (0,0)], $h(x_1, x_2) = 0$ for any specific percept that falls exactly on the bound, and $h(x_1, x_2) < 0$ for any percept that falls on the other side of the bound. The GLC assumes linear decision bounds, and therefore linear discriminant functions. With two stimulus dimensions $x_1$ and $x_2$, all linear discriminant functions can be written as

$$h(x_1, x_2) = a_1x_1 + a_2x_1 + b, \tag{5.4}$$

for some constants $a_1$, $a_2$, and $b$. For example, consider a diagonal bound that has slope 1 and passes through the origin. The equation of this line is $x_2 = x_1$, and therefore we could create the discriminant function $h(X_1, X_2) = X_2 - X_1$ (so $a_1 = -1$, $a_2 = 1$, and $b = 0$). Note that $h(x_1, x_2) = 0$ for any point that falls on this bound, $h(x_1, x_2) > 0$ for any point above and to the left of the bound, and $h(x_1, x_2) < 0$ for any point below and to the right of the bound.

Once the decision bounds are described by a set of discriminant functions, each response region then can be identified by the unique pattern of pluses and minuses that result when any point in that region is inserted into all of the discriminant functions. These are identified in figure 5.3 directly below the response labels associated with each region. So if the values of the current percept $(X_1, X_2) = (x_1, x_2)$ cause

$h_1(x_1, x_2) > 0$ and $h_2(x_1, x_2) > 0$, then the percept must lie in response region A. As a result, the probability of responding A on any given trial equals

$$P(\mathrm{R_A}) = P[h_1(X_1, X_2) > 0, h_2(X_1, X_2) > 0]. \tag{5.5}$$

As we will see in Chapters 7 and 8, this equation can be used to fit the GLC to identification or categorization data.

### 5.1.5 THE GENERAL-QUADRATIC CLASSIFIER

The decision bounds of the general-quadratic classifier (GQC) are illustrated in figure 5.1d. This model is identical to the GLC, except each decision bound is a quadratic curve, rather than a line (Ashby & Maddox, 1993). This means that the discriminant functions are quadratic, rather than linear. As such, the GLC is a special case of the GQC in which the coefficients of all quadratic terms in every discriminant function equal 0. Both models use equation 5.5 for model fitting. They just use qualitatively different functional forms for the $h_i$ (i.e., linear versus quadratic).

## 5.2 THE GENERAL RECOGNITION RANDOMIZATION TECHNIQUE

In contrast to SDT, an enormous amount of research has investigated decision strategies in tasks where the stimuli vary on multiple stimulus dimensions. If the percept was observable on each trial, then this problem would be straightforward. For example, consider the hypothetical case, depicted in figure 5.4, in which two stimuli, A and B, vary on two stimulus dimensions. Suppose the observer's task is to identify the presented stimulus. So this experiment is a two-stimulus identification task. Figure 5.4a shows the hypothetical percepts elicited by 75 presentations of stimulus A and 75 presentations of stimulus B. In panel (a), each percept is coded by its stimulus. So the circles denote all percepts elicited on stimulus A trials and the stars show all percepts from B trials. Note that the percepts overlap and as a result, perfect accuracy is impossible in this task.

Panels (b) and (c) of figure 5.4 show responses of two hypothetical observers in this task. Now the same percepts from figure 5.4a are coded by response. So circles denote percepts for which the observer responded A and stars denote percepts that caused a B response. In both cases, the line was drawn to separate the A and B responses. Note that in panel (b) the observer responds A if the value on dimension 2 exceeds the value on dimension 1, and otherwise responds B. Theorem 5.1 specifies that this is the

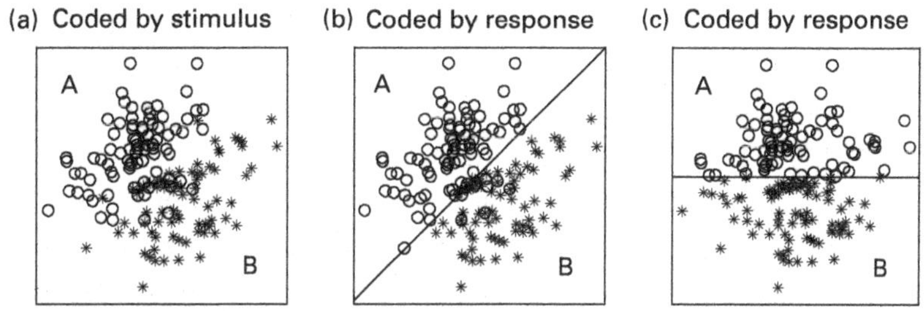


**5.4** Hypothetical percepts and responses from an experiment in which stimuli A and B are each presented 75 times. (a) The percepts are coded by stimulus. Each circle denotes a percept elicited by stimulus A and each star denotes a percept elicited by stimulus B. (b) Percepts are coded by response. Each circle denotes a percept that elicited an A response and each star denotes a percept that elicited a B response. (c) Same as panel (b), except for a different hypothetical observer.

optimal strategy with these perceptual distributions (i.e., the strategy that maximizes accuracy). In contrast, the panel (c) observer adopts a strategy that satisfies decisional separability by ignoring dimension 1 and responding A if the value on dimension 2 is large and B if the dimension 2 value is small. Therefore, if the percepts in this task were observable, it would be a simple matter to see that the panel (b) observer is using the optimal classifier and the panel (c) observer is using an independent decisions strategy that satisfies decisional separability. The problem, of course, is that in a real two-stimulus identification task, the percepts are not observable, so partitioning the percepts that elicited A versus B responses, as in figure 5.4b and 5.4c, is not possible.

Ashby and Gott (1988) proposed an experimental paradigm that they called the *general recognition randomization technique* that simulates the observer's decision task in this experiment, but in a way that makes these decisions observable. This paradigm has been a major source of data about GRT decision strategies. In the figure 5.4 example, the idea is to create a stimulus with the same coordinate values as each percept illustrated in figure 5.4a. So each circle in figure 5.4a is a different stimulus that requires an A response and each star is a different stimulus requiring a B response. Technically, this is a categorization task with two categories of stimuli, A and B, rather than a two-stimulus identification task, because many different stimuli are associated with each response. Observers are told that there are two categories and that their task is to use the trial-by-trial feedback to learn to assign each stimulus to its correct category. Just as in figure 5.4, we can code each stimulus by its response and then examine the resulting scatterplot to see how observers separated their A and B responses.

In the randomization technique, the stimuli that comprise each category are selected by random sampling from some multivariate probability distribution in which each dimension specifies numerical values of a stimulus component. The most common choice is bivariate normal distributions, but uniform distributions are also used. In fact, the two sets of stimuli depicted in figure 5.4a were generated by drawing random samples from two different bivariate normal distributions.

Whatever distributions are chosen, a key feature is to select those distributions, and therefore the resulting categories, so that they overlap or nearly overlap. If the categories are widely separated and the observer eventually learns to respond with perfect accuracy, then it is impossible to know what decision strategy the observer used because an infinite number of boundaries will separate their A and B responses. In contrast, with overlapping categories, as in figure 5.4a, note that only one boundary separates the A and B responses of the hypothetical observers who produced the responses shown in figures 5.4b and c.

The multivariate normal distribution is popular for a variety of reasons. First, like many natural categories, it contains an unlimited number of exemplars that vary on continuous-valued dimensions, and it includes a well-defined prototype (e.g., the distribution mode; Ashby, 1992). Second, the normal distribution is the most popular choice for modeling perceptual distributions in both SDT and GRT. Therefore, it is especially important to understand how people create response regions when the stimuli and/or the percepts are normally distributed. Third, the optimal decision strategy is well known. In fact, as we saw in theorem 5.1, with two multivariate normal distributions, the decision bound that maximizes accuracy is always linear or quadratic.

Section 5.6 below describes step-by-step computational methods for running an experiment using the randomization technique. The next section summarizes empirical results from many such experiments. Briefly though, the first important results established that humans are not constrained to any one type of decision bound. Instead they can learn to respond as if they are using any type of strategy available to an ideal observer—that is, a linear bound, any type of quadratic bound, or a strategy that obeys decisional separability (Ashby & Maddox, 1992). The next important realization was that humans have multiple learning systems that for the most part are neuroanatomically and functionally distinct, that evolved at different times for different purposes, that learn in qualitatively different ways, and that are ideally suited for learning different types of information (e.g., Ashby & Maddox, 2005; Eichenbaum & Cohen, 2001; Squire, 2004). This insight was critical because the different learning systems rely on qualitatively different types of decision strategies, which require different types of decision bounds to model.

## 5.3 RULE-BASED AND INFORMATION-INTEGRATION CATEGORY-LEARNING TASKS

Much of the evidence for multiple learning systems comes from experiments that contrasted learning and performance in rule-based (RB) and information-integration (II) category-learning tasks (Ashby & Ell, 2001). In RB tasks, the optimal strategy requires independent decisions about the level of the stimulus on one or more stimulus dimensions. In contrast, accuracy is maximized in II tasks only if information from two or more incommensurable stimulus dimensions is integrated perceptually at a pre-decisional stage. The most compelling evidence for multiple systems comes from carefully matched RB and II tasks in which the category structures are created using the randomization technique. These studies were designed to test predictions of the COVIS theory of category learning, which was among the first theories to propose that humans have multiple learning systems. Specifically, COVIS assumes that humans have separate explicit-reasoning and procedural-learning systems that compete for access to response production (Ashby & Valentin, 2017; Ashby et al., 1998). The explicit-reasoning system uses executive attention and working memory to learn explicit rules, whereas the procedural system uses dopamine-mediated reinforcement learning to establish associations between specific stimuli and their appropriate response.

Figure 5.5 shows examples of matched RB and II tasks. Each task includes two categories of stimuli that vary across trials on two stimulus dimensions. In the case of figure 5.5, these stimuli are circular disks in which luminance varies sinusoidally across each disk. The disks vary across trials in bar width and bar orientation. The categories in the two tasks are carefully matched because the stimuli in the II task were created simply by rotating the RB stimuli 45° clockwise in stimulus space. As a result, the RB and II categories are identical except for the orientation of their optimal category bound, and therefore they are exactly matched on most category-separation statistics. Section 5.6 describes a step-by-step algorithm that creates the stimuli in these two conditions, and appendix B.1 gives the Matlab code that automates this process. In standard applications, stimuli in each condition are presented one at a

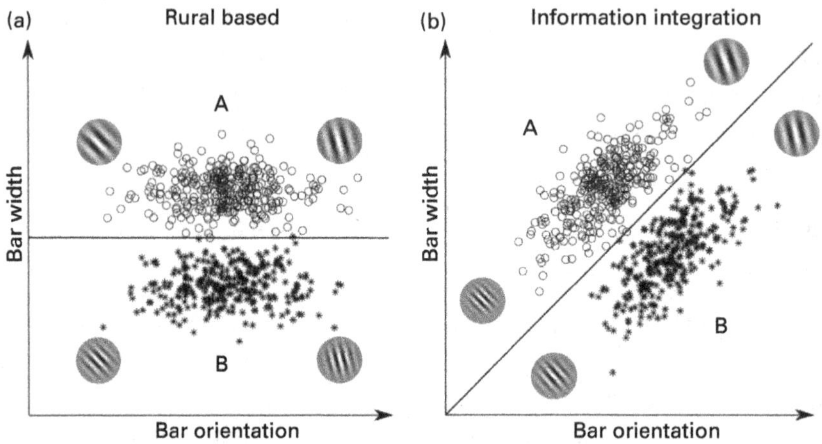


**5.5** Examples of matched (a) rule-based (RB) and (b) information-integration (II) category-learning tasks. Each stimulus is a sine-wave disk that varies across trials in bar width and bar orientation. For each task, two illustrative Category A and B stimuli are shown. The circles denote the specific values of all stimuli in each Category A, whereas the stars denote stimuli in each Category B. In the RB task, only bar width carries diagnostic category information, so the optimal strategy is to respond with a one-dimensional bar-width rule (thick versus thin), while ignoring orientation. In the II task, both bar width and orientation carry useful but insufficient category information. The optimal strategy requires integrating information from both dimensions in a way that is impossible to describe verbally.

time, participants assign each stimulus to a category by pressing a response key, and feedback is given after each response (i.e., correct versus incorrect). Participants are told that there are two categories of stimuli and that their task is to use the feedback to learn how to assign each stimulus to its correct category. Critically, they are given no prior information about the structure of the categories.

Figure 5.5a shows the simplest possible RB task, in which only one stimulus dimension is relevant (bar width), and the observer's task is to discover this dimension and then map the different dimensional values to the relevant categories. Therefore, note that in this RB task, the optimal strategy satisfies decisional separability. As might be expected, healthy young adults quickly discover this optimal strategy and accuracy in this task increases rapidly with practice (J. D. Smith and Ell, 2015). The independent decisions about single stimulus dimensions that are necessary to optimize performance in RB tasks require selective attention to single stimulus dimensions. A large literature associates this ability with stimulus dimensions that are separable (Ashby & Townsend, 1986; Garner, 1974; Lockhead, 1966; Maddox, 1992; Shepard, 1964). In fact, the classic definition of separable dimensions is that it is easy to attend to one and ignore the others, whereas with integral dimensions, this type of selective attention is difficult or impossible (Garner & Morton, 1969; Lockhead, 1966). Prototypical separable dimensions are color and shape, whereas prototypical integral dimensions are hue and saturation. As a result, RB tasks require that the stimuli are constructed from separable stimulus dimensions.

Despite the fact that the figure 5.5b II categories are identical to the 5.5a RB categories, except for their orientation in stimulus space, note that the 45° rotation profoundly changes the verbalizability of the optimal strategy. Whereas the optimal strategy in the figure 5.5a RB task has a simple verbal description, the optimal strategy in the figure 5.5b II task is impossible to describe verbally. It is mathematically simple—"respond A if $X_2 > X_1$; otherwise respond B"—and this mathematical rule is easy to translate into words—"respond A if bar width is greater than bar orientation; otherwise respond B"—but these words are nonsensical. What does it mean for a bar width to be greater than an orientation? This is literally like comparing apples and oranges. My lab has run thousands of participants through tasks that are essentially identical to the figure 5.5b II task, and during debriefing after all the data have been collected, we typically ask participants what strategy they used to categorize each stimulus. No participant has ever said that they responded A if bar width was greater than orientation. In fact, even if they are told explicitly at the outset of the experiment that this is the optimal strategy they should try to adopt, these instructions are of no benefit (Rosedahl et al., 2021). This contrasts sharply with RB tasks, where prior instruction about the optimal strategy significantly improves performance (Rosedahl et al., 2021). Some participants do use a strategy that has a simple verbal description in II tasks, but these strategies always lead to poor performance. Although the optimal strategy is impossible to describe, most participants do learn II categories if they are given enough time and reliable, immediate feedback. Even so, learning is slow and incremental and qualitatively different from the all-or-none learning that is often observed in the RB task (J. D. Smith and Ell, 2015).

A critical requirement of II tasks is that the stimulus dimensions are in incommensurable units. If not, then the figure 5.5b II task could transform into an RB task. For example, suppose the stimuli are rectangles that vary across trials in height and width. Note that in this case, the dimensions are commensurable, since they are both in units of length. The commensurabity of the stimulus dimensions makes it easy to describe the optimal strategy, which is simply to "respond A if the rectangle is taller than it is wide; otherwise respond B". As might be expected, performance in such tasks is much more similar to traditional RB tasks than to II tasks (Ashby & Gott, 1988).

Despite the structural similarity of the figure 5.5 RB and II categories, many studies have shown that during early learning, human accuracy is much higher in the RB task than in the II task (i.e., given the same amount of training). The RB advantage lasts for at least a few thousand trials, but after that, performance asymptotes at the same high levels of accuracy in both tasks (e.g., Hélie et al., 2010). COVIS naturally predicts this human RB advantage because it predicts that in the RB task, people quickly discover the optimal rule (i.e., respond A if the bars are wide and B if they are thin). In the II task however, because there is no verbal description of the optimal strategy, people eventually abandon the search for an explicit rule. At this point, control is transferred to the procedural system, which learns in a slow incremental fashion.

Experiments that used RB and II tasks that are almost identical to those shown in figure 5.5 have been run in at least five different species: rats (Broschard et al., 2019), pigeons (J. D. Smith et al., 2011), capuchin monkeys (J. D. Smith et al., 2012), macaque monkeys (J. D. Smith et al., 2010), and of course, many studies have used human participants. All species successfully learn both types of category structures. Humans and both species of monkeys learn the RB categories considerably faster than the II categories, whereas rats and pigeons learn both tasks at exactly the same rate. COVIS accounts for this across-species difference by predicting that pigeons and rats lack the robust logical reasoning system of humans and monkeys. As a result, these phylogentically simpler species must learn both tasks with the same associative-learning system. Since the categories are equally separated in the two tasks, pigeons and rats therefore learn both at exactly the same rate.

Many human studies have compared performance in RB and II tasks across a variety of different training and testing conditions. These studies varied the methods and order of stimulus presentation, the nature and timing of feedback, the quality of the training conditions, and the generalizability of the knowledge acquired during training. This line of research has produced scores of articles that have reported more than 30 qualitatively different dissociations between training and performance in RB and II tasks (for a review of many of these, see Ashby & Valentin, 2017). In other words, the experimental manipulation affected one task more than the other. In some studies, the greater effect was on the RB task and in others, it was on the II task. Virtually all of these dissociations were predicted a priori by COVIS—that is, for all values of its free parameters, COVIS predicted that the experimental manipulation would impair the task that was more affected at least as much as the task that was less affected, and it always predicted that a dissociation in the opposite direction was impossible. As a result, collectively, all these dissociations provide strong evidence that humans have, at least, the two functionally separate learning systems postulated by COVIS.

Reviewing all of these reported dissociations is well beyond the scope of this chapter. Even so, it is instructive to consider a few. One set of reported dissociations tested predictions that follow from the COVIS assumption that the explicit rule-learning system presumed to dominate in RB tasks uses executive attention and working memory, whereas the procedural-learning system does not. One obvious prediction that follows from these assumptions is that a simultaneous dual task that requires working memory and executive attention should interfere more with RB tasks than with II tasks. This prediction has been supported in several studies (Crossley, Paul, et al., 2016; Waldron & Ashby, 2001; Zeithamova & Maddox, 2006). Another dissociation, reported by Maddox, Ashby, et al. (2004), confirmed the COVIS prediction that follows from these same assumptions that reducing the amount of time available to interpret the trial-by-trial feedback should impair learning more in RB tasks than in II tasks.

A different set of dissociations tested predictions that follow from the COVIS assumptions about the nature of trial-by-trial learning in the two systems. COVIS assumes that learning in the procedural system is of stimulus-response associations that depend on a dopamine-mediated reinforcement signal that must arrive in the relevant synapses within a couple of seconds of the rewarded behavior (see section 11.2.3 of chapter 11 for details). This dopamine signal is presumed to be driven by the feedback. Positive feedback briefly elevates dopamine levels in the relevant synapses, whereas negative feedback briefly depresses dopamine levels. In contrast, COVIS assumes that learning in the explicit system is a matter of selecting and testing alternative candidate decision rules while they are maintained in working memory. As a result, COVIS predicts that the absence or delay of trial-by-trial feedback should interfere with II category learning much more than with RB learning. Many studies have supported these predictions. First, when training is unsupervised and there is no trial-by-trial feedback, people can learn some RB categories but there is no evidence that they can learn II categories (Ashby et al., 1999; Ell et al., 2012). Second, II learning is impaired with feedback delays as short as 2.5 s, whereas RB learning is unaffected by feedback delays as long as 10 s (Maddox et al., 2003; Maddox & Ing, 2005). Third, II learning is impaired if the feedback is batch delivered after every six trials rather than given after each response, whereas RB learning is not (J. D. Smith et al., 2014). Fourth, the same dissociation is observed with one-back feedback in which the feedback about the accuracy of the response on trial $n$ is given immediately after the response on trial $n+1$ (J. D. Smith et al., 2018).

There are no alternative theoretical accounts of these dissociations. As a result, there is no theory that assumes a single learning system, which could be viewed as a viable alternative to COVIS. Even so, a number of studies singled out one dissociation in which II performance was worse than RB performance and hypothesized that the results of this one experiment were because the II task is inherently more difficult than the RB task (Edmunds et al., 2015; Le Pelley et al., 2019; Nosofsky et al., 2005; Zaki & Kleinschmidt, 2014). In each of these cases, the authors postulated that there is a single (unspecified) learning system used in all tasks, but that the II task shown in figure 5.5 is more difficult than the RB task, and that this difficulty difference was the cause of the one dissociation that the study examined. One obstacle to testing this hypothesis is that none of these studies defined what they meant by difficulty. Even so, at first glance, this seems like a sensible hypothesis. The figure 5.5 RB task certainly seems easier. After all, learning the RB categories is simply a matter of figuring out that Category A disks have thick bars and Category B disks have thin bars, whereas learning the II categories requires . . . well, something else. But difficulty is a tricky construct. It seems so familiar that one is tempted to apply the criterion that Supreme Court Justice Potter Stewart used to identify obscenity in Jacobellis v. Ohio—namely, "I know it when I see it" (Lattman, 2007, September 27). However, difficulty is a useful scientific construct only if it can be defined rigorously. Otherwise, arguments flirt with circularity. One can always examine any set of results, note which condition has the worst performance, and then assert by *fiat* that this is the most difficult condition. But this approach has no predictive power—the results are used as an explanation of themselves.

Ashby et al. (2020) tested and rejected the difficulty account of performance differences in the figure 5.5 RB and II tasks. First, if the II task is more difficult than the RB task, then there must be some single measure of task difficulty that simultaneously accounts for performance differences in all RB and II tasks. In contrast, if performance in RB and II tasks is mediated by separate systems, then no such difficulty measure should exist because all learning systems struggle under certain experimental conditions and flourish under others. As a result, if RB and II learning are mediated by separate systems, then the conditions that slow learning will be different in RB and II tasks (else the systems would not be different), and so, instead of one difficulty measure, two would be required—one that is sensitive to the conditions that make RB learning difficult and one that is sensitive to the conditions that make II learning difficult.

In fact, the best current predictors of RB and II task difficulty are qualitatively distinct. Furthermore, the best predictor of RB difficulty—the Boolean complexity of the optimal strategy (Feldman, 2000)—is not even defined in II tasks, whereas the best predictor of II difficulty—the striatal difficulty measure (Rosedahl & Ashby, 2019)—is a poor predictor of RB task difficulty (e.g., it predicts that the figure 5.5 RB and II tasks should be equally difficult). However, to test whether some other single difficulty measure might have been missed, Ashby et al. (2020) examined the predictions for the figure 5.5 RB and II tasks of 14 different quantitative measures of classification difficulty that are popular in the machine-learning, cognitive-psychology, and neural-network literatures. Ten of the 14 measures predicted that the RB and II tasks are equally difficult. This is not surprising since the categories used in the two tasks are identical except for their orientation in stimulus space. Two measures actually predicted that the RB task is more difficult than the II task. Finally, two other measures—collective feature efficiency and volume of overlapping regions—correctly predicted that the II task is more difficult. They made this prediction because both measures privilege the stimulus dimensions over other directions through stimulus space. So these measures might seem like good candidates for a single metric that predicts difficulty in all category-learning tasks. To test this hypothesis, Ashby et al. (2020) examined the ability of both measures to predict the difficulty of 17 previously published II datasets that used different stimuli and different II category structures and were collected in multiple laboratories. Unfortunately, the two measures accounted for only 2% and 18% of the variance in the final-block accuracies of these 17 datasets, respectively. This compares poorly with the 87% of variance accounted for by the best current measure of difficulty in II tasks (i.e., the striatal difficulty measure; Rosedahl & Ashby, 2019). So collective feature efficiency and volume of overlapping regions are both poor predictors of human category-learning difficulty, even though they both correctly predict a human RB performance advantage in the figure 5.5 tasks. In summary, there is one current difficulty measure that accurately predicts the relative difficulty of RB tasks and a qualitatively different measure that accurately predicts the relative difficulty of II tasks, but all currently known difficulty measures fail miserably at predicting the relative difficulty of all RB and II tasks.

Second, Ashby et al. (2020) pointed out that if the figure 5.5 II task is inherently more difficult than the figure 5.5 RB task then the RB versus II performance difference should be even greater in a cognitively simpler species. Tasks that are difficult for humans should be even more difficult for a simpler species. In contrast to this prediction, pigeons learn rotated RB and II categories at exactly the same rate (Qadri et al., 2019; J. D. Smith et al., 2011) and so do rats (Broschard et al., 2019). Therefore, both of these species perform as if they learn both types of category structures in the same way. In contrast, the more cognitively sophisticated monkeys, like humans, learn II categories slower than RB categories (J. D. Smith et al., 2010, 2012).

Third, Ashby et al. (2020) noted that if the II task is inherently more difficult than the RB task, then any manipulation that slows learning should have a greater effect on II performance than on RB performance. However, many different types of dissociations have been reported that are exactly opposite to this prediction. For example, as noted earlier, several studies have reported that a simultaneous (working-memory) dual task interferes more with one-dimensional RB learning than with II learning (Crossley, Paul, et al., 2016; Waldron & Ashby, 2001; Zeithamova & Maddox, 2006). This is opposite to what the difficulty hypothesis predicts because trying to do two things at once should be harder if the two tasks are difficult than if they are easy.[2]

In summary, the available evidence overwhelmingly rejects the difficulty hypothesis and instead strongly favors the multiple-systems account of the human RB advantage and of the many RB versus II dissociations that have been reported. These results are important for GRT because they suggest that the type of decision bounds that best describe observer performance will depend on which learning system observers are relying on in the task under study. On the other hand, it is natural to ask whether the results summarized in this section generalize from the categorization tasks used in these studies to the identification tasks that are common with GRT. Note that identification is similar to categorization because in both tasks, observers are asked to respond to the single stimulus presented on each trial with its assigned label. The only difference is that in identification, each stimulus has a unique label, whereas in categorization, at least two stimuli have the same label. Even so, if there is stimulus or perceptual noise, then from the perspective of the decision process, the two tasks are identical (Ashby & Lee, 1993). At the level of the percept, both are categorization tasks. Not surprisingly, there have been multiple proposals that exactly the same decision processes mediate both identification and categorization (e.g., Ashby & Lee, 1991; Nosofsky, 1986; Shepard et al., 1961), and there is even evidence that exactly the same two explicit-rule and procedural-learning systems thought to dominate during category learning also mediate stimulus identification (Ashby et al., 2001).

## 5.4 DECISION STRATEGIES USED DURING EXPLICIT RULE DISCOVERY AND PROCEDURAL LEARNING

The good news for GRT users is that there is strong evidence that the default decision strategy of the explicit system can be modeled by decision bounds that satisfy decisional separability, at least locally. The bad news though, is that evidence suggests the procedural system does not use decision bounds of any kind. Even so, decision bounds that are of the same type as the optimal classifier commonly provide good fits to data thought to be mediated by procedural learning.

### 5.4.1 DECISION BOUNDS COMPATIBLE WITH EXPLICIT RULE USE

The COVIS explicit system uses simple verbalizable rules to make classification judgments. For example, in the figure 5.5 RB task, the optimal decision bound is compatible with the rule: "Respond A if the bars are thick and B if the bars are thin." Note that if we denote the perceived value of bar orientation on a randomly selected trial as $X_1$ and the perceived value of bar width as $X_2$, then this rule is equivalent to "Respond A if $X_2 > X_W$ and B if $X_2 \leq X_W$," where $X_W$ is the criterial value of bar width that separates thick from thin. As a result, the decision bound for this strategy is the set of $x_2$ equal to $X_W$, or in other words, the set of all $x_2$ that satisfy $x_2 = X_W$, which is the horizontal bound shown in panel (a) of figure 5.5. Similarly, a simple rule that assigned disks with steep orientations to A and disks with shallow orientations to B is equivalent to a vertical-line boundary. If the stimulus dimensions are separable and in incommensurable units—as they are with the disks used to create the figure 5.5 RB categories—then these are the only one-dimensional rules that can be easily verbalized. As such, they are the base strategies of the explicit system. More complex explicit rules are created by combining these base rules using the operators of Boolean algebra. In fact, an RB task is defined as one in which the optimal strategy is some Boolean combination of these base rules. Not surprisingly, the best available predictor of RB task difficulty is the Boolean complexity of the optimal strategy (Feldman, 2000). In other words, the decision strategies used by the explicit system are all compatible with some independent-decisions model (see section 5.1.1).

Figure 5.6 shows four different strategies that could be learned by the explicit system. Panel (a) shows bounds compatible with the conjunction rule: "Respond A if the orientation is shallow *and* the bars are wide; otherwise respond B." Panel (b) is compatible with the exclusive-or rule: "Respond A if the orientation is shallow and the bars are wide, *or* if the orientation is steep and the bars are narrow; otherwise respond B."

Panel (c) shows the disjunction rule: "Respond A if the orientation is very shallow *or* very steep; respond B if the orientation has an intermediate value." Finally, panel (d) depicts bounds that are consistent with the hybrid rule: "If the bar width is narrow, respond A if the orientation is very shallow; otherwise respond B. If the bar width is wide, respond B if the orientation is very steep; otherwise respond A."

Of course, even more complex rules are possible (e.g., see figure 5.1a), but the critical point is that no matter how complex, any strategy used by the explicit system, or equivalently, any independent-decisions strategy formed by some Boolean combination of the two base rules is compatible with decision bounds that are constructed from piecewise vertical and horizontal line segments. Therefore, any decision strategy implemented by the explicit system with these stimuli will satisfy decisional separability—at least locally. For example, note that decisional separability holds for the decision strategies illustrated in figures 5.6(a), (b), and (c). In panel (d), decisional separability holds on dimension 2, but technically it fails on dimension 1. For decisional separability to hold on dimension 1, the decision about the dimension 1 level of the stimulus can not depend on the perceived vale of dimension 2. This condition

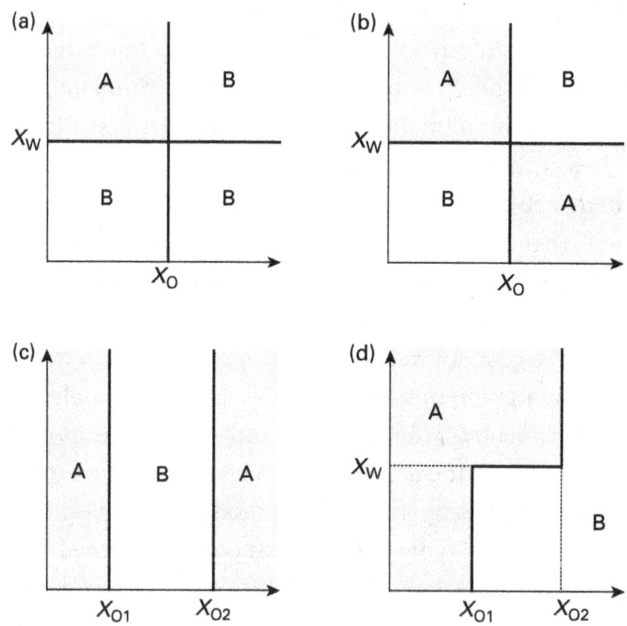


**5.6** Decision bounds consistent with different explicit rules. (a) A conjunction rule. (b) An exclusive-or rule. (c) A type of logical disjunction. (d) A hybrid rule.

is not met in panel (d). Even so, note that the panel (d) decision strategy would be easy to model in GRT using two vertical bounds and one horizontal bound, and three free parameters would be needed to model this decision process (i.e., $X_W$, $X_{O1}$, and $X_{O2}$).

### 5.4.2 DECISION MODELS COMPATIBLE WITH A PROCEDURAL STRATEGY

Human procedural learning is highly flexible and can respond optimally or nearly optimally when the optimal boundary is any type of quadratic curve (Ashby & Maddox, 1992), even when the two categories have identical prototypes (Ashby & Perrin, 1988), and even when multiple piecewise quadratic curves optimally separate four categories (Ashby et al., 2001).

Ashby and Waldron (1999) tested whether or not the procedural system learns decision bounds that have a specific functional form. For example, if the categories (or percepts) are multivariate normally distributed, then from theorem 5.1 we know the optimal bound is either linear or quadratic. In this case, rather than learn the optimal response to each stimulus, a more efficient strategy is to estimate the coefficients of the optimal linear or quadratic boundary. Participants in the Ashby and Waldron (1999) experiments learned two categories of stimuli that varied on two stimulus dimensions. The experiment used the general recognition randomization technique and sampled stimuli randomly from nonnormal distributions. In the first experiment, the variance-covariance matrices from the two categories were identical, but the optimal category bound was quadratic, whereas in the second experiment the opposite was true—that is, the two variance-covariance matrices were different, but the optimal bound was linear. Previous research had shown that when the optimal bound was linear, people responded as if using linear bounds, and when the optimal bound was quadratic, people responded as if using quadratic bounds. Ashby and Waldron (1999) argued that the only information signaling which type of bound to use that could be estimated reliably within a few hundred learning trials was in the category variances and covariances. Thus, if participants learned decision bounds, they should have responded suboptimally in both experiments—with a linear bound in the first experiment and a quadratic bound in the second. However, in both experiments, participants responded optimally, suggesting that they were not estimating coefficients of some decision bound.

Rather than learning some type of decision bound, the evidence suggests that the human procedural system learns simple stimulus-response associations. For example,

Casale et al. (2012) reported results from a series of experiments in which participants were first trained on RB and II categories like those shown in figure 5.5. After participants learned the categories, they were transferred to novel stimuli in which perfect performance was possible with the original training bound. If participants had learned decision bounds during the original training, then transfer performance should be good because there was nothing new to learn. In fact, transfer was almost perfect in the RB conditions, but there was no evidence of any transfer in the II conditions. These results suggest that learning in the RB conditions was mediated by the explicit system. The one-dimensional rule that was learned during training generalized easily to the new stimuli encountered during transfer, leading to nearly perfect transfer performance. In contrast, the II results were consistent with the hypothesis that performance was mediated by a procedural system that learns stimulus-response associations. During transfer, novel stimuli were encountered for which no prior stimulus-response associations had been formed. As a result, transfer accuracy was at chance.

The Casale et al. (2012) results suggest that procedural learning does not generalize to novel stimuli. Hélie and Ashby (2012) presented evidence that procedural learning also does not generalize to novel responses. In this experiment, participants first learned either RB or II categories that were essentially identical to those shown in figure 5.5. Next, they were shown pairs of stimuli and asked whether the two stimuli belonged to the same or different categories. In the RB condition, same-different transfer performance was good. However, in the II condition, despite the fact that categorization accuracy was about 80% correct at the end of training, same-different accuracy was near chance. Therefore, even though II participants could categorize each stimulus accurately, they could not tell whether the two stimuli were in the same or different categories.

Much evidence suggests that procedural learning is mediated largely within the striatum, and is facilitated by a dopamine-mediated reinforcement learning signal (Ashby & Ennis, 2006; Badgaiyan et al., 2007; Grafton et al., 1995; Knopman & Nissen, 1991). Chapter 11 provides much more detail, but briefly, the well-accepted theory is that positive feedback that follows successful behaviors increases phasic dopamine levels in the striatum, which has the effect of strengthening recently active synapses, whereas negative feedback causes dopamine levels to fall below baseline, which has the effect of weakening recently active synapses (Houk et al., 1995). The critical site of learning within the striatum is thought to be at synapses between pyramidal neurons projecting from visual association areas of cortex (in the case of experiments with visual stimuli) and striatal medium spiny neurons. The idea is that each medium spiny neuron becomes associated with a motor goal (e.g., press the button on the left), so the procedural system learns associations between each presented stimulus (via activation of visual cortical neurons) and the relevant motor response (Ashby and Waldron, 1999; Ashby et al., 2007). These cortical-striatal projections are characterized by massive convergence—the ratio is somewhere around 10,000-to-1 (Wilson, 1995).

Based on these anatomical considerations, Ashby and Waldron (1999) proposed the striatal pattern classifier (SPC) as a model of decision processes in the procedural-learning system. The SPC assumes a high-resolution perceptual space that maps to a low-resolution decision space. Each perceptual unit is assumed to represent a different neuron (or group of neurons) in the visual cortex, and each decision unit is assumed to represent a different medium spiny neuron in the striatum. Typical applications might include 10,000 perceptual units and only a few decision units. Each perceptual unit has a different preferred stimulus, but presentation of a stimulus causes a graded pattern of perceptual activation because the tuning of each perceptual unit is modeled via a radial basis function. Each decision unit is associated with a response. Initially, the perceptual units are fully connected to the decision units, but reinforcement learning changes the strengths of these connections. Computationally, the SPC is similar to the covering-map version of ALCOVE (Kruschke, 1992) and to the rational model of Anderson (1991).

Furthermore, Ashby and Rosedahl (2017) showed that the standard exemplar model of categorization—namely the generalized context model (Nosofsky, 1986)—is a special case of the SPC, although the SPC offers a very different psychological interpretation of the same computational model. Exemplar theory assumes that people categorize a novel object by comparing its similarity to the memory representations of all previous exemplars from each relevant category. In contrast, the SPC assumes that no memory representations are ever retrieved. For example, the SPC assumes that in tasks with two categories, stimulus presentation always activates two striatal medium spiny neurons, regardless of the number of exemplars in each category and regardless of the number of training trials. Instead of adding a new exemplar memory representation, the SPC assumes that presentation of a new stimulus either modifies the strength of an existing cortical-striatal synapse or creates a new synapse (e.g., by adding a new spine to an existing medium spiny neuron). Thus, the analogue of an exemplar representation in the SPC is either the strength of a cortical-striatal synapse or perhaps it could be interpreted as a dedicated spine on a medium spiny neuron.

The more important point, however, is that stimulation of that medium spiny neuron (or that dendritic spine) would not retrieve an exemplar memory, but rather create an urge to respond either A or B. The models are mathematically equivalent because the summed similarity of the presented stimulus to all previous exemplars of some category A turns out to equal exactly the strength of the synapse between the visual cortical unit that is maximally tuned to the presented stimulus and the striatal medium spiny neuron that is associated with an A response. The synaptic strength equals this sum because it is modified by all previous presentations of category A exemplars.

The SPC does not learn decision bounds. Instead it learns to associate a response with each region of stimulus space. Stimulus space is modeled by the high-resolution set of visual cortical units, and the decision space is modeled by the low-resolution set of striatal units. Learning is the process of strengthening the association between each visual cortical unit and whichever striatal unit is associated with the optimal response to the stimulus that maximally excites that visual unit. The decision bound has no psychological meaning in this model. Rather it is simply the set of points in stimulus space that separate the visual units associated with contrasting responses. Even so, the responses produced by the SPC in a randomization experiment can typically be modeled accurately via some type of decision bound model.

As mentioned, Ashby and Rosedahl (2017) showed that the generalized context model (Nosofsky, 1986) is a special case of the SPC. Ashby and Alfonso-Reese (1995) showed that the generalized context model is equivalent to estimating the category distributions using a Parzen (1962) kernel density estimator. This is a popular and powerful nonparametric statistical estimator of a probability density function. The estimator is consistent under very weak conditions. As a result, the generalized context model, and therefore also the SPC, can approximate almost any optimal decision bound. For these reasons, a decision bound model that assumes a bound of the same general type as the optimal bound can typically fit any data produced by the SPC.

## 5.5 EXPERIMENTAL CONDITIONS THAT ENCOURAGE EXPLICIT RULE USE

Like SDT, GRT is particularly good at separating sensory and decision processes. As described in chapter 4, a categorization experiment is better than identification if the primary interest is in decision processes, whereas identification is better than categorization if the primary interest is in perceptual representations. Therefore, in identification experiments, the decision strategies used by participants are often a nuisance variable. They are typically not the researcher's primary interest, but they must be characterized accurately, or else any inferences about perceptual processes will be biased. In chapters 6 and 7 we will see that GRT applications are greatly facilitated if decisional separability holds. First, virtually all current summary statistics tests require decisional separability, so if decisional separability is violated these tests are invalid. Second, fitting GRT models to identification confusion matrices is greatly simplified if decisional separability holds. For these reasons, when the primary interest is in sensory processes, there is typically a great advantage if experimental conditions are selected that make decisional separability as likely as possible. As described in the previous section, this means that the experiment should be designed to encourage explicit rule use, rather than procedural learning.

The independent decisions about single stimulus dimensions that form the base rules of the explicit system require selective attention in order to isolate each relevant stimulus dimension. As mentioned earlier, a large literature associates this ability with separable stimulus dimensions (Ashby & Townsend, 1986; Garner, 1974; Lockhead, 1966; Maddox, 1992; Shepard, 1964). As a result, decisional separability is more likely with stimuli constructed from separable dimensions. For example, as this hypothesis predicts, the figure 5.5 RB learning advantage disappears when the stimulus dimensions are integral (i.e., hue and saturation; Ell et al., 2012).

Second, the stimuli should be selected so that decisional separability is optimal or nearly optimal. Humans are remarkably adept at stimulus identification and category learning. Even so, when the stimulus dimensions are separable, considerable evidence suggests that decisional separability is the default human decision strategy (e.g., Ashby et al., 1999; Canini et al., 2014). More complex strategies are adopted only when decisional separability yields unacceptable results. Many experiments with the figure 5.5 RB and II category structures strongly support this hypothesis. Initial performance is far better with the RB categories, in which decisional separability yields optimal performance, and the responses of almost all practiced participants in the figure 5.5 II task are better described by a diagonal decision bound than by one that satisfies decisional separability.

One way to design an identification experiment so that decisional separability is optimal or nearly optimal is to use the popular factorial design, illustrated in figure 4.2, in which the stimuli are created by factorially combining two or more levels of each stimulus dimension. This does not guarantee that decisional separability is optimal. For example, violations of perceptual separability and/or perceptual independence will usually mean that decisional separability is not optimal (e.g., see figure 4.6). Even so, the tic-tac-toe arrangement of stimuli that characterize factorial designs will typically guarantee that the accuracy of a strategy that satisfies decisional separability is not too much worse than optimal.

Third, the instructions to the participants should encourage explicit rule use. This is most important in tasks with more than two response alternatives. If two real-world objects are chosen at random, then it is guaranteed that they can be identified perfectly by focusing all attention on some single stimulus dimension. For example, any two distinct points in the plane can be separated by either a vertical or horizontal line (or both). As a result, participants will bring to a two-stimulus identification experiment the expectation that decisional separability is a reasonable strategy. However, if more than two objects are chosen at random, then decisional separability is typically less reasonable. For example, in many cases, four random points in a plane will not be separated perfectly by single vertical and horizontal lines. And as the number of stimuli increases, decisional separability seems less likely to be a reasonable strategy. As a result, in identification experiments with four or more stimuli, participants might need encouragement to adopt decisional separability. This can be done via instruction. For example, in an experiment that uses a factorial design, participants could be shown the factorial nature of the stimuli before the experiment begins, and a keypad could be used for responding that maps onto this factorial structure in a natural way (e.g., a $2 \times 2$ keypad in the case of a $2 \times 2$ factorial design).

An example of the effects that instructions can have on the decision strategies adopted by observers in identification experiments was reported by Ashby et al. (2001). This study described the results of two identification experiments that both used nine stimuli arranged in a $3 \times 3$ factorial structure (half circles that varied in size with a radial line that varied in orientation; see figure 7.4a). The primary difference between the experiments was in the instructions to the participants and the layout of the response devices. In experiment 1, participants were told about the $3 \times 3$ factorial structure of the stimuli and they were instructed to identify each stimulus by responding first with its level of orientation (i.e., 1, 2, or 3) and then its level of size (1, 2, or 3). In experiment 2, each stimulus was randomly assigned (without replacement) one of the first nine letters of the alphabet, participants were not told about the factorial structure, and they were asked to identify the presented stimulus with its assigned letter.

A wide variety of different GRT models were fit to the identification confusion matrices collected in the two experiments. Section 7.6 of chapter 7 describes these models in some detail. Briefly though, in experiment 1, the best-fitting model assumed decisional separability on the size dimension and local decisional separability on orientation (i.e., see figure 7.4b). In experiment 2, however, the best-fitting model was the SPC (see figure 7.4c). In other words, these experiments suggested that when the instructions suggested to participants that simple explicit rules might succeed, participants used explicit strategies, whereas in the absence of such instructions, they resorted to procedural learning.

Finally, one could leverage the many empirical dissociations that have been reported between performance in the figure 5.5 RB and II tasks to encourage explicit rule use (for a review of many of these, see Ashby & Valentin, 2017). For example, many studies have reported conditions that impair RB learning more than II learning. Obviously, these should be avoided if the goal is to encourage explicit rule use. For example, a simultaneous dual task that loads on working memory impairs RB learning more than II learning (e.g., Waldron & Ashby, 2001; Zeithamova & Maddox, 2006). As a result, care should be taken to ensure that participants are not distracted by any other attention-demanding task during the identification experiment. Similarly, reducing the amount of time available to process feedback impairs RB learning more than II learning (Maddox, Ashby, et al., 2004), so the intertrial interval in identification experiments should not be too short—especially during early trials when participants are settling on a decision strategy.

Many other variables have been identified that impair II learning more than RB learning. Including some of these could increase the likelihood that participants adopt decisional separability. Perhaps most useful would be to leverage the fact that II learning requires immediate feedback, whereas RB learning is much more flexible with respect to the nature and timing of feedback. There are a variety of options here. For example, II learning is impaired more than RB learning if the feedback is delayed by a few seconds (Crossley & Ashby, 2015; Dunn et al., 2012; Maddox et al., 2003; Maddox & Ing, 2005), if observational training is used in which the correct response label is presented before stimulus presentation (e.g., as in "here is stimulus A"; Ashby, Maddox, & Bohil, 2002), or if one-back feedback is used in which the accuracy of the trial $n$ response is given immediately after the response on the ensuing trial (i.e., on trial $n+1$; J. D. Smith et al., 2018). So a feedback delay, observational training, or one-back feedback could be used during learning to encourage explicit rule use. A particularly attractive choice though is to train with batch feedback, rather than with traditional trial-by-trial feedback. J. D. Smith et al. (2014) examined the effects of batch feedback on RB and II learning. In the batch feedback conditions, each participant completed six categorization trials with no feedback of any kind. At the end of each block, they were then told how many of the six responses were correct and how many were incorrect, but they were given no information about which specific responses were correct or incorrect. J. D. Smith et al. (2014) reported that batch feedback eliminated procedural learning. Specifically, with batch feedback all participants resorted to explicit strategies, even those in the II condition. So using batch feedback during the learning period in an identification experiment should strongly encourage strategies that satisfy decisional separability, without slowing training in any significant way.

## 5.6 THE RANDOMIZATION TECHNIQUE: STEP BY STEP

This section describes the methods required to construct RB and II categories by random sampling from bivariate normal distributions. One nice consequence of defining categories as bivariate normal distributions, which was established in theorem 5.1, is that in the two-category case, the optimal decision boundary (i.e., that maximizes categorization accuracy) is always linear or quadratic. See appendix A for a more detailed description of bivariate normal distributions.

This section describes the steps (adapted from Ashby & Valentin, 2018) required to generate random samples from two bivariate normal distributions for which the optimal boundary is linear and the optimal strategy is equivalent to assigning each stimulus to the category with the most similar prototype (i.e., with the nearest mean). This is the most common application of the randomization technique. Even so, although the prototype rule always produces a linear bound, not all linear bounds are equivalent to a prototype strategy (Ashby & Gott, 1988). Thus, the methods described here are valid for only a subset of all possible linear bounds. Constructing categories that have other types of optimal bounds follows similar, but slightly more complex steps.

To make the discussion concrete, we will construct the pair of matching RB and II categories shown in figure 5.7, in which optimal accuracy is 0.99. This process is subdivided into nine sequential steps that are described in detail below. The first four of these steps must be performed by hand, but steps 5–8 are typically computerized. To facilitate this process, section B.1 of appendix B contains the Matlab code for these steps that I used to create the stimuli shown in figure 5.7.

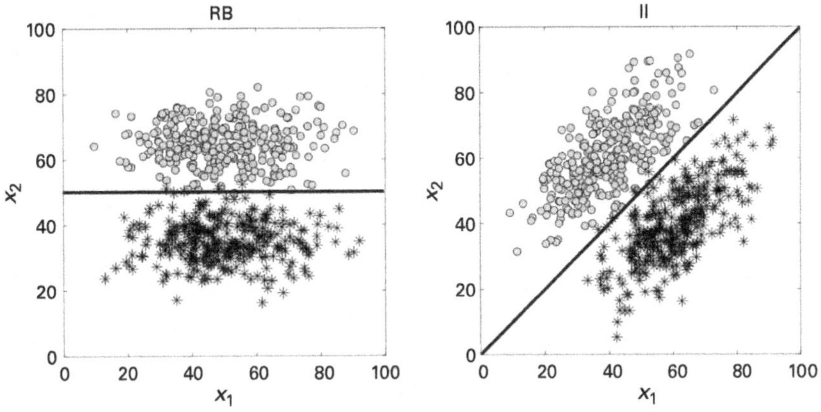


**5.7** Matching RB and II categories created from the Matlab code listed in section B.1 of appendix B. The code follows Steps 5–8 of this section.

## STEP 1. DEFINE THE STIMULUS SPACE

Because each category is defined by a bivariate normal distribution, this example assumes two stimulus dimensions. The first step is to identify those dimensions as well as the minimum and maximum values on each dimension that will be used in the experiment. For example, with many stimulus dimensions, extreme values are physically unrealizable. In the case of the disks shown in figure 5.5, there is both an upper and lower limit on the bar widths that can be shown on a computer screen. If one stimulus dimension is physical size, then there will be an upper limit that ensures the stimulus is foveal. So the first problem in step 1 is to define a range on each stimulus dimension that is measured in physical units—for example, in the case of sine-wave gratings, degrees of counterclockwise rotation from horizontal for orientation and cycles per disk for bar width. Denote this range on each dimension by [min, max].

The physical units though, should not be used to create the contrasting categories. This is because, in the space where the categories are defined, it is important that a change of say 10 units on one dimension is equally salient perceptually to a change of 10 units on the other dimension. This is necessary to ensure that the perceptual representation of the categories roughly matches the stimulus representation. For example, if we run an experiment with the RB and II categories shown in figure 5.5, we would want the perceptual representations of these categories to roughly match the stimulus representations shown in that figure.

A standard solution to this problem is to create the categories in an abstract stimulus space in which the values on each dimension are constrained to the interval [0,100] and in which a change of $n$ units on dimension 1 is equally perceptually salient to a change of $n$ units on dimension 2. So for example, one approach might be to equate a difference of 10 units on each [0,100] dimension with one *just noticeable difference* (jnd) (Wichmann et al., 2018). Then both dimensions would span 10 jnds. To determine a jnd on each dimension, one could either consult the literature or run a quick psychophysical pilot experiment that uses a staircase procedure to estimate the jnd. The final problem in this step, therefore, is to find two transformations that convert each [0,100] range in abstract space to a [min, max] range in physical space, with the provision that a change of $n$ units on each [0,100] dimension is equally perceptually salient. For example, in the case of the disks shown in figure 5.5, these transformations were identified by Treutwein et al. (1989). Steps 2–8 operate in this abstract $[0, 100] \times [0, 100]$ space. The final step 9 is to create physical stimuli by converting the coordinates of each stimulus from abstract [0,100] space to physical [min, max] space.

## STEP 2. SELECT THE OPTIMAL BOUNDS AND THE CATEGORY MEANS

Step 2 is to select the desired optimal bounds and category means in abstract [0,100] space. For the RB categories shown in figure 5.7, note that the optimal bound is a horizontal line (with slope 0), whereas the optimal bound for the II categories shown in figure 5.7 has a slope of +1 and an intercept of 0.

Selecting the category means includes two constraints. First, the means must be selected to be consistent with our optimal bounds. In the present example, this requires that both means lie on a line orthogonal to the category bound, which in our case means they must fall on a vertical line for the RB categories and on a line with slope −1 for the II categories. Second, if our experiment follows standard procedure and sets the two category base rates equal, then the two means of each pair of categories must be equidistant from the optimal bound. Denote the distance between the means of the two contrasting categories by $D$ (i.e., in [0,100] space). To create matched RB and II conditions, $D$ should be the same for both category structures, although the specific value of $D$ can be set to any value we choose. In other words, it is possible to follow all of the remaining steps in this procedure for any numerical value of $D > 0$. In practice, though, $D$ should be chosen large enough so that the two stimuli that correspond to the means are easily discriminable. Otherwise learning may be impossible. On the other hand, two problems arise if $D$ is chosen to be too large. First, it is likely that a one-dimensional rule will achieve high accuracy in the II condition. An enormous literature shows that people have a strong preference for one-dimensional rules, so if the goal is to study some aspect of procedural learning, it is imperative that the best one-dimensional rule performs poorly in the II task. Second, a large $D$ makes extreme samples more likely and as previously noted, extreme samples are often physically unrealizable. For these reasons, the best choice for $D$ is some intermediate value.

Suppose we have defined our [0,100] abstract space so that the jnd on each dimension equals 10, and we want our category means to differ by 3 jnds in each condition. In other words, we want the distance $D$ between category means to equal $D = 30$ in both conditions. If the horizontal bound in the RB condition intercepts the ordinate at $x_2 = 50$ and we want the distributions to be centered in [0,100] space, then the category A and B means will have coordinates (50, 65) and (50, 35), respectively. In the case of the II categories, some straightforward geometry locates the category A and B means at coordinates (39.4, 60.6) and (60.6, 39.4), respectively.

## STEP 3. DETERMINE THE ENTRIES IN THE VARIANCE-COVARIANCE MATRICES

In our example problem, both categories in each condition have identical variance-covariance matrices. Our next step is to determine the numerical values of the covariance and two variances that define these matrices. The key is to write the variance-covariance matrix $\Sigma$ that characterizes each condition in its diagonal form (see appendix A for details):

$$\Sigma = \begin{bmatrix} \sigma_1^2 & \text{cov} \\ \text{cov} & \sigma_2^2 \end{bmatrix} = \begin{bmatrix} r_{11} & r_{12} \\ r_{21} & r_{22} \end{bmatrix} \begin{bmatrix} w_1^2 & 0 \\ 0 & w_2^2 \end{bmatrix} \begin{bmatrix} r_{11} & r_{12} \\ r_{21} & r_{22} \end{bmatrix}'$$

$$= \begin{bmatrix} \underline{r}_1 & \underline{r}_2 \end{bmatrix} \begin{bmatrix} w_1^2 & 0 \\ 0 & w_2^2 \end{bmatrix} \begin{bmatrix} \underline{r}_1 & \underline{r}_2 \end{bmatrix}' \tag{5.6}$$

The $2 \times 1$ column vectors $\underline{r}_1$ and $\underline{r}_2$ are the eigenvectors of $\Sigma$ and $w_1^2$ and $w_2^2$ are the corresponding eigenvalues. Our approach will be to determine the necessary numerical values of $\underline{r}_1$, $\underline{r}_2$, $w_1^2$, and $w_2^2$ and then insert these values into the right side of equation (5.6) to compute $\Sigma$.

As described in appendix A, the eigenvectors of $\Sigma$ are parallel to the major and minor axes of the ellipses that define the distribution's contours of equal likelihood, and the eigenvalues of $\Sigma$ are equal to the distribution's variances along each of these directions. This property is illustrated in figure A.1d. Note that in both of our conditions, every point on the optimal decision bound is equidistant from the two category means. Ashby and Alfonso-Reese (1995) showed that under these conditions, one of the eigenvectors of $\Sigma$ must be orthogonal to the categorization decision bound. Because the eigenvectors of $\Sigma$ are always orthogonal to each other, this means that the other eigenvector must be parallel to the decision bound.

This is enough information to identify $\underline{r}_1$ and $\underline{r}_2$ for both of our conditions. The entries in any vector can be considered the endpoints of a directed line segment that begins at the origin. The diagonal representation shown in equation A.10 requires that $\underline{r}_1$ and $\underline{r}_2$ each must have a length of 1. Therefore, in the RB condition the eigenvectors should be horizontal and vertical—that is,

$$\underline{r}_1 = \begin{bmatrix} 1 \\ 0 \end{bmatrix} \text{ and } \underline{r}_2 = \begin{bmatrix} 0 \\ 1 \end{bmatrix}. \tag{5.7}$$

In contrast, in the II condition, the eigenvectors should have slopes of $\pm 1$, and therefore

$$\underline{r}_1 = \begin{bmatrix} \frac{1}{\sqrt{2}} \\ \frac{1}{\sqrt{2}} \end{bmatrix} \text{ and } \underline{r}_2 = \begin{bmatrix} -\frac{1}{\sqrt{2}} \\ \frac{1}{\sqrt{2}} \end{bmatrix}. \tag{5.8}$$

Looking back at figure 5.7 shows that in both cases, $\underline{r}_1$ is the major axis and $\underline{r}_2$ is the minor axis. The next two values to determine are the eigenvalues $w_1^2$ and $w_2^2$. As described in appendix A, the eigenvalues of $\Sigma$ equal the variances in the distribution along the directions specified by the eigenvectors. So as in figure A.1d, $w_1^2$ is the variance along the major axis (i.e., the $\underline{r}_1$ direction) and $w_2^2$ is the variance along the minor axis (i.e., the $\underline{r}_2$ direction). In our current example, the RB and II categories have different eigenvectors, but they must have identical eigenvalues in order to guarantee that they are simple 45° rotations of each other (together with our constraint that the means are 30 units apart in both conditions).

A look back at figure 5.7 shows that in both conditions, categorization accuracy depends solely on the distance from the stimulus to the optimal decision bound—that is, on the relative position of the stimulus with respect to the minor axis of the ellipse that characterizes the scatterplot of stimuli that define each category. Note that the position of the stimulus on the major axis is irrelevant to categorization accuracy. Because $\underline{r}_1$ is the major axis and $\underline{r}_2$ is the minor axis in both conditions, this means that $w_2^2$ completely determines optimal accuracy. It turns out that the eigenvalue $w_1^2$ is also important though, because it determines the difference between optimal accuracy and the accuracy of the most accurate one-dimensional rule in the II condition.

First we compute $w_2^2$. Figures 5.8a and 5.8b show conceptually how this is done for both the RB and II conditions, respectively. Specifically, since optimal accuracy depends only on variability in the direction orthogonal to the category bound (i.e., in the $\underline{r}_2$ direction), we can project the distributions onto the dimension orthogonal to the bound. This produces two univariate normal distributions, both with variance $w_2^2$. The distance between the means is $D$, which is the same as the distance between the means of the two categories—that is, between $\underline{\mu}_A$ and $\underline{\mu}_B$. Note that this projection collapses the decision bound to a criterion that is distance $D/2$ from the mean of each of these marginal distributions. A correct response occurs if a sample from these marginal distributions falls below this criterion value on A trials and above the criterion on B trails. Therefore, if the two category base rates are equal, then the optimal accuracy $A$ equals

$$A = \frac{1}{2}P\left(Z \leq \frac{D/2-0}{w_2}\right) + \frac{1}{2}P\left(Z > \frac{D/2-D}{w_2}\right)$$
$$= P\left(Z \leq \frac{D/2}{w_2}\right), \tag{5.9}$$

where $Z$ has a standard normal distribution (i.e., mean = 0, variance = 1). The first probability equals the probability correct on Category A trials and the second probability equals the probability correct on Category B trials. So for example, if we want

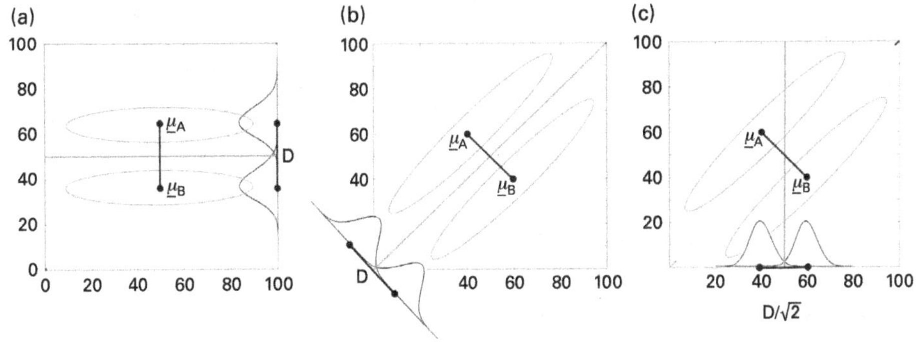


**5.8** (a): Computing optimal accuracy in the RB condition. (b): Computing optimal accuracy in the II condition. (c) Computing the accuracy of the most accurate one-dimensional rule in the II condition.

optimal accuracy to be 99% (i.e., $A = 0.99$) we simply use a Z-table to note that $P(Z \leq 2.326) = 0.99$ and therefore

$$\frac{15}{w_2} = 2.326, \tag{5.10}$$

which implies that $w_2 = 6.45$ (i.e., since $D = 30$).

The next task is to determine a numerical value of $w_1^2$—that is, the variance along the irrelevant direction. Generally, this value is selected to be as large as possible because the larger this value the greater the difference in the II condition between optimal accuracy and the accuracy of the most accurate one-dimensional rule. Even so, there are almost always upper limits on $w_1^2$ because if this variance is too large then some random samples will be physically unrealizable. So generally $w_1^2$ is set to near the physical upper limit. With any normal distribution almost all samples fall within three standard deviations from the mean (samples outside this range can be discarded). Therefore, it is important to ensure that an interval of width $6w_1$ (i.e., $\pm 3w_1$) along the major axis of each ellipse and centered on the category mean includes only stimulus values that are physically realizable—that is, only stimulus values that fall inside our $100 \times 100$ abstract stimulus space. If we leave a buffer of five units above and below the largest and smallest stimulus values on the irrelevant dimension, then in the RB condition we are left with 90 available units (and more than this in the II condition). If we divide these 90 units into six standard deviations, then we need to set $w_1 = 15$.

So in this hypothetical experiment, if we set $w_1 = 15$ and $w_2 = 6.45$, then optimal accuracy will equal 99% correct in both conditions and we will have as much variability on the irrelevant dimension as possible. We have now determined numerical values for all terms on the right side of equation 5.6. The final task in this step is to multiply the three matrices in that equation to determine the variance-covariance matrices (i.e., the values of $\Sigma$) that we will use to generate the stimuli in the RB and II conditions. In the RB condition, this multiplication produces

$$\Sigma = \begin{bmatrix} 1 & 0 \\ 0 & 1 \end{bmatrix} \begin{bmatrix} 15^2 & 0 \\ 0 & 6.45^2 \end{bmatrix} \begin{bmatrix} 1 & 0 \\ 0 & 1 \end{bmatrix}' = \begin{bmatrix} 225 & 0 \\ 0 & 41.6 \end{bmatrix}.$$

In the II condition, this multiplication produces

$$\Sigma = \begin{bmatrix} \frac{1}{\sqrt{2}} & -\frac{1}{\sqrt{2}} \\ \frac{1}{\sqrt{2}} & \frac{1}{\sqrt{2}} \end{bmatrix} \begin{bmatrix} 15^2 & 0 \\ 0 & 6.45^2 \end{bmatrix} \begin{bmatrix} \frac{1}{\sqrt{2}} & -\frac{1}{\sqrt{2}} \\ \frac{1}{\sqrt{2}} & \frac{1}{\sqrt{2}} \end{bmatrix}' = \begin{bmatrix} 133.3 & 91.7 \\ 91.7 & 133.3 \end{bmatrix}.$$

## STEP 4. COMPUTE THE ACCURACY OF THE MOST ACCURATE ONE-DIMENSIONAL RULE

In II tasks, it is always important to compute the accuracy of the most accurate one-dimensional rule. As mentioned earlier, people have a strong preference for one-dimensional rules, so if the goal is to study procedural learning, the category distributions should be constructed so that the best one-dimensional rule performs poorly in the II task.

The calculations required to compute the accuracy of the most accurate one-dimensional rule are illustrated in figure 5.8c. The most accurate possible one-dimensional rule is illustrated by the vertical bound.[3] The accuracy of this rule only depends on variability along the horizontal dimension. Thus, to compute the accuracy of this rule, we can project the bivariate normal distributions that define the II categories onto the abscissa. This produces two (univariate) normal distributions, which are just the marginal distributions of the bivariate normals on the first dimension. Note that the distance between the means of these marginal distributions is now $D/\sqrt{2}$ and the criterion is again half this distance from each marginal mean. Therefore, by a calculation almost identical to equation 5.9 we can compute the best one-dimensional accuracy, which we denote by $A_{1D}$, to be:

$$A_{1D} = \frac{1}{2}P\left(Z \leq \frac{\frac{D}{2\sqrt{2}} - 0}{\sigma_1}\right) + \frac{1}{2}P\left(Z > \frac{\frac{D}{2\sqrt{2}} - \frac{D}{\sqrt{2}}}{\sigma_1}\right)$$

$$= P\left(Z \leq \frac{\frac{D}{2\sqrt{2}}}{\sigma_1}\right). \tag{5.11}$$

Increasing $w_1^2$ will decrease this value (because it will increase $\sigma_1$).

Applying these equations to our II condition produces

$$A_{1D} = P\left(Z \leq \frac{\frac{30}{2\sqrt{2}}}{\sqrt{133.3}}\right) = P(Z \leq 0.919) = 0.82.$$

So the best a participant could do in our II task with a simple one-dimensional rule is an accuracy of 82% correct, whereas the optimal procedural strategy achieves an accuracy of 99% correct. As a result, any participant who achieves an accuracy that is higer than 82% correct must be using some strategy that is more sophisticated than a simple one-dimensional rule.

## STEP 5. GENERATE THE RANDOM SAMPLES THAT DEFINE EACH CATEGORY

Steps 1–4 are typically completed by hand, whereas the remaining steps in the process can be computerized. To facilitate this process, section B.1 of appendix B lists Matlab code that performs each of these steps.

The first of these automated steps is to generate random samples from these distributions. Many software packages have routines that will generate samples from multivariate normal distributions given numerical values for the mean vector and variance-covariance matrix. For example, in the Matlab code given in section B.1, the command

```
RB_A = mvnrnd(mu_RB_A,Sig_RB,N);
```

draws N random samples from a multivariate normal distribution that has mean vector mu_RB_A and variance-covariance matrix Sig_RB. The sample size N should be somewhat larger than the number of stimuli desired for each category. This is to compensate for the samples lost when we discard outliers in step 6. For example, in the section B.1 code, N is set to 330, but the program is designed to end up with 300 stimuli in each category.

Some software packages might only be able to generate samples from a standard (univariate) normal distribution (i.e., a "Z" distribution with mean 0 and variance 1). In this case, the first step is to generate two random (and independent) samples and load them into a vector we can call $\underline{z}$. These values can then be transformed into random samples $\underline{x}$ from a bivariate normal distribution with mean $\underline{\mu}$ and variance-covariance matrix $\Sigma$ by the linear transformation

$$\underline{x} = \mathrm{P}\underline{z} + \underline{\mu}, \tag{5.12}$$

where

$$\mathrm{P} = \begin{bmatrix} \sigma_1 & 0 \\ \frac{\text{cov}}{\sigma_1} & \sqrt{\sigma_2^2 - \frac{\text{cov}^2}{\sigma_1^2}} \end{bmatrix}. \tag{5.13}$$

The matrix P is known as the Cholesky matrix (e.g., Ashby & Soto, 2015). If the only available random number generator produces samples from a uniform [0,1] distribution, then several different methods can be used to convert these samples to samples that have an approximate Z distribution (e.g., Ashby, 1992) and then equation 5.12 can be applied.

### STEP 6. DISCARD OUTLIERS

Step 6 is to discard any sample more than 3 standard deviations from the mean. Strictly speaking, this step is not necessary. However, given the methods described above, outliers can be physically unrealizable, whereas these methods should ensure that any stimulus within 3 standard deviations from the mean can be physically constructed. Discarding outliers, however, is complicated by the fact that the numerical value of the standard deviation will typically depend on the direction from the sample to the mean. For example, with the figure 5.5 categories, the standard deviation along the minor axis of the ellipse that characterizes each distribution (i.e., $w_2$) is much less than the standard deviation along the major axis (i.e., $w_1$). Fortunately, the distance metric known as Mahalanobis distance (e.g., Fukunaga, 2013) corrects for these changes. Given any point $\underline{x}$, the Mahalanobis distance $D_M$ from $\underline{x}$ to the distribution mean $\underline{\mu}$ equals

$$D_M = \sqrt{(\underline{x} - \underline{\mu})' \Sigma^{-1} (\underline{x} - \underline{\mu})}, \tag{5.14}$$

where $\Sigma$ is the distribution variance-covariance matrix. This value equals the number of (direction-dependent) standard deviations from $\underline{x}$ to the mean. Therefore, we will remove as an outlier any sample $\underline{x}$ for which $D_M > 3$. In the Matlab program listed in section B.1, the following code discards outliers.

```
j=1; k=1;
for i=1:N
    dA=sqrt((RB_A(i,:)'-mu_RB_A)'*inv(Sig_RB)*(RB_A(i,:)'-mu_RB_A));
    if dA>3
Out_A(j)=i; j=j+1;
    end
    dB=sqrt((RB_B(i,:)'-mu_RB_B)'*inv(Sig_RB)*(RB_B(i,:)'-mu_RB_B));
    if dB>3
Out_B(k)=i; k=k+1;
    end
end
if length(Out_A)>.5                    % make sure there is at least 1 outlier
    RB_A(Out_A,:) = [];
end
if length(Out_B)>.5                    % make sure there is at least 1 outlier
    RB_B(Out_B,:) = [];
RB_A(Out_A,:) = [];
RB_B(Out_B,:) = [];
```

After this step there will be fewer than N stimuli in each category, and the A and B categories will almost surely include different numbers of stimuli. This is why we set N to a larger value than needed. So if we want 300 stimuli and we had set N = 330, then after discarding outliers, we should be left with some number of stimuli in each category between 300 and 330. To conclude this step, select the first 300 of these samples in each category. For example, the Section B.1 code trims the category A stimuli to 300 samples via the following command

```
RB_A = RB_A(1:N_final,:).
```

We now have 300 samples from the two category distributions, all of which are no more than three standard deviations from the category mean.

### STEP 7. TRANSFORM THE SAMPLE SO THAT THE SAMPLE STATISTICS EXACTLY EQUAL THE POPULATION PARAMETERS

Of course, with any random sample, the sample means, variances, and covariance will not exactly equal the population values, no matter how large the sample size. If not, then the most accurate classifier for the sample will differ from the desired decision bound that was used to carefully select the population parameter values. To eliminate this problem, it is necessary to linearly transform the sample values so that the sample statistics exactly match the population values. Step 7 performs this transformation separately for each category.

Denote the vector of sample means for the selected category by $\underline{\bar{x}}$ and the sample variance-covariance matrix by S. The first problem is to construct the Cholesky matrix from the entries in S. If we call this matrix Q, then

$$Q = \begin{bmatrix} s_1 & 0 \\ \frac{\overline{\text{cov}}}{s_1} & \sqrt{s_2^2 - \frac{\overline{\text{cov}}^2}{s_1^2}} \end{bmatrix}, \tag{5.15}$$

where $\overline{\text{cov}}$ is the sample covariance, and $s_1^2$ and $s_2^2$ are the sample variances. Matlab includes a one-line command that constructs the transpose of this matrix. So Q can be constructed from the command: Q=(chol(S))′.

The transformation that converts $\underline{\bar{x}}$ to $\underline{\mu}$ and S to $\Sigma$ is

$$\underline{y} = PQ^{-1}(\underline{x} - \underline{\bar{x}}) + \underline{\mu}, \tag{5.16}$$

where P is the Cholesky matrix of the population variance-covariance matrix defined in Eq. 5.13. To use equation 5.16, simply substitute each random sample in for $\underline{x}$ and then perform the matrix operations to produce a new random sample $\underline{y}$. The sample mean of the $\underline{y}$'s created in this fashion will be exactly $\underline{\mu}$ and the sample variance-covariance matrix will be exactly $\Sigma$.

The section B.1 code that does this for the category A samples is as follows

```
S_A = cov(RB_A);
P_RB = (chol(Sig_RB))';
Q_A = (chol(S_A))';
for i=1:N_final
    RB_A_pop(i,:) = (P_RB*inv(Q_A)*(RB_A(i,:)'-mean(RB_A)'))' + mu_RB_A';
end
```

## STEP 8. CONSTRUCT THE II STIMULI BY ROTATING THE RB STIMULI BY 45°

Steps 5–7 could be used to create stimuli for both the RB and II categories. However, another approach, which has some design advantages, is to follow Steps 5–7 only for the RB categories and then rotate the resulting stimuli 45° counterclockwise in stimulus space to create the II categories. This is the approach taken in the Matlab code listed in section B.1. The advantage of this approach is that the stimuli in the two conditions will be identical in every way, except for their orientation in stimulus space.

The first step is to reset the mean of each RB category to (0,0), because the rotation we will perform is a rotation around the origin (i.e., the origin is the fixed point of the rotation). The second step is to then rotate each RB category by 45°, and the last step is to reset the means to the II means that we computed in step 2.

To zero-mean the category A RB stimuli, we simply subtract the category A population mean from each sample. Let $\underline{y}_{A_0}$ denote any one of these zero-meaned samples (i.e., so $\underline{y}_{A_0}$ is a $2 \times 1$ vector). Then the following operation, which is applied to every category A sample, rotates them by 45° counterclockwise:

$$\underline{w}_{A_0} = \begin{bmatrix} \cos(45°) & -\sin(45°) \\ \sin(45°) & \cos(45°) \end{bmatrix} \underline{y}_{A_0}. \tag{5.17}$$

For obvious reasons, this matrix is called the rotation matrix. Finally, to reset the mean, the II category A population mean is added to every one of these rotated samples. And of course, the analogous operations are performed on the RB category B stimuli.

The section B.1 code that performs this rotation to create category A in the II condition is as follows.

```
RB_A_zero=RB_A_pop-(ones(300,1)*mean(RB_A_pop));
R=[cos(pi/4) -sin(pi/4); sin(pi/4) cos(pi/4)];
II_A_zero = (R * RB_A_zero')';
II_A_pop=II_A_zero + (ones(300,1)*mu_II_A');
```

### STEP 9. GENERATE THE STIMULI

The final step is to convert each numerical sample into a physical stimulus. This requires using the transformation we identified in step 1 to convert the $100 \times 100$ abstract space to the $[\text{min}_1, \text{max}_1] \times [\text{min}_2, \text{max}_2]$ space in which the dimensions are in physical units, where $\text{min}_i$ and $\text{max}_i$ are the minimum and maximum values on dimension $i$ (for $i = 1$ or 2) that we identified in step 1.

## 5.7 PROOF OF THEOREM 5.1

When the base rates for the two categories are equal, accuracy is maximized by responding with the most likely alternative. Let $f_J(\underline{x})$ denote the likelihood that the specific stimulus or percept $\underline{x}$ is a random sample from the category $J$ distribution. Then accuracy is maximized by the following decision rule:

$$\text{Respond A if } \frac{f_A(\underline{X})}{f_B(\underline{X})} > 1; \text{otherwise respond B.} \tag{5.18}$$

Therefore, the decision bound, which is the set of points that separates stimuli or percepts that elicit A versus B responses, equals the set of $\underline{x}$ for which

$$\frac{f_A(\underline{x})}{f_B(\underline{x})} = 1. \tag{5.19}$$

Now, by assumption, $f_A(\underline{x})$ and $f_B(\underline{x})$ are both multivariate normal probability density functions. Therefore, equation 5.19 equals

$$1 = \frac{(2\pi)^{-r/2}|\Sigma_A^{-1/2}| \exp\left[-\frac{1}{2}(\underline{x} - \underline{\mu}_A)'\Sigma_A^{-1}(\underline{x} - \underline{\mu}_A)\right]}{(2\pi)^{-r/2}|\Sigma_B^{-1/2}| \exp\left[-\frac{1}{2}(\underline{x} - \underline{\mu}_B)'\Sigma_B^{-1}(\underline{x} - \underline{\mu}_B)\right]} \tag{5.20}$$

$$= \left(\frac{|\Sigma_B|}{|\Sigma_A|}\right)^{\frac{1}{2}} \exp\left[-\frac{1}{2}(\underline{x} - \underline{\mu}_A)'\Sigma_A^{-1}(\underline{x} - \underline{\mu}_A) + \frac{1}{2}(\underline{x} - \underline{\mu}_B)'\Sigma_B^{-1}(\underline{x} - \underline{\mu}_B)\right]. \tag{5.21}$$

Taking the natural log of both sides reduces this expression to equation 5.3. Setting $\Sigma_A = \Sigma_B = \Sigma$ reduces equation 5.3 to equation 5.2. ■

# 6

# THE SUMMARY STATISTICS APPROACH TO TESTING FOR PERCEPTUAL INTERACTIONS

GRT provides what are arguably the default current methods for studying perceptual interactions. There are two qualitatively different approaches. One uses summary statistics derived from the identification-confusion matrix to test for perceptual separability and perceptual independence. The other approach fits a variety of different GRT models directly to the confusion matrix and then makes inferences about perceptual separability, perceptual independence, and decisional separability based on statistical comparisons of goodness-of-fit. This chapter considers the summary-statistics approach, and chapter 7 considers the model-fitting approach.

The summary-statistics approach was developed first. The idea was to test perceptual assumptions by comparing various parametric and nonparametric summary statistics that are derived from the identification-confusion matrix—most typically from a $2 \times 2$ factorial identification experiment. The first such tests were proposed by Ashby and Townsend (1986). These were quickly augmented by a variety of statistics that estimated $d'$-like measures that were inspired by SDT. Overall, this latter approach became known as multidimensional signal detection analysis (Kadlec, 1995; Kadlec & Townsend, 1992a, 1992b; Silbert & Hawkins, 2016; Silbert & Thomas, 2013). Sometime later, a variety of different tests were proposed that use the observer's response times (RTs), rather than the accuracies that populate the confusion matrix (Ashby & Maddox, 1994; Townsend et al., 2012).

The principal advantage of summary-statistics tests is that they are computationally simple—so simple that, in most cases, they can be computed by hand. In contrast, the prominent alternative—fitting GRT models directly to the confusion matrix of each observer—requires estimating many parameters via an iterative search routine and each step typically requires numerical integration. So computationally, the summary-statistics approach is enormously easier.

The primary disadvantage of the summary-statistics approach is that virtually all current GRT-based tests of perceptual interactions that depend on summary statistics assume decisional separability. And unfortunately, although statistical tests of decisional separability are possible in the model-fitting approach, such tests are currently impossible using the summary-statistics approach—that is, there are no current summary-statistics tests of decisional separability. As a result, if a summary-statistics approach is used, decisional separability must be assumed without any way of assessing its possible validity. If participants use some decision strategy that violates decisional separability, then the summary statistics tests are invalid. For this reason, it is vital that if the goal is to use identification confusions to make perceptual inferences on the basis of summary statistics, then the experiment that collected those data should be designed in a way to make decisional separability as likely as possible. This issue was discussed in detail in the last chapter, so researchers interested in using a summary-statistics approach should carefully review section 5.5. For example, decisional separability is more likely if (1) the stimuli have the tic-tac-toe arrangement that characterizes designs in which the stimuli are constructed by factorially combining several levels of two different stimulus features or dimensions; (2) participants are shown the factorial nature of the stimuli before the experiment begins; and (3) a keypad is used for responding that maps onto this factorial structure in a natural way (e.g., a $2 \times 2$ keypad in the case of a $2 \times 2$ factorial design).

To illustrate the various tests discussed in this chapter, consider a $2 \times 2$ factorial identification experiment described by the GRT model illustrated in figure 6.1. Note that in this model, (1) decisional separability holds on both dimensions; (2) perceptual independence holds in all stimuli except $A_2B_2$; and (3) component B is perceptually separable from component A, but component A is integral with component B. The violations of perceptual separability for component A but not B mean that the perception of components $B_1$ and $B_2$ do not change when the level of component A changes from $A_1$ to $A_2$, but the perceived value of component $A_1$ is different when component B is at level $B_1$ versus $B_2$, and the same is true for the perceived value of component $A_2$. These properties are easily seen in the marginal distributions on each dimension. Note that on dimension $x_1$, the marginal distributions when A

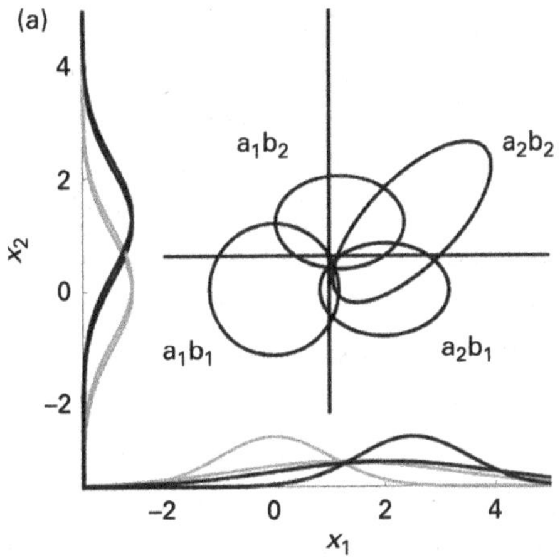


(b)

| | $a_1b_1$ | $a_1b_2$ | $a_2b_1$ | $a_2b_2$ |
|---|---|---|---|---|
| $A_1B_1$ | .61 | .23 | .12 | .04 |
| $A_1B_2$ | .12 | .33 | .15 | .40 |
| $A_2B_1$ | .17 | .07 | .55 | .21 |
| $A_2B_2$ | .06 | .01 | .22 | .71 |

(c)

| | $a_1b_1$ | $a_1b_2$ | $a_2b_1$ | $a_2b_2$ |
|---|---|---|---|---|
| $A_1B_1$ | .67 | .21 | .08 | .04 |
| $A_1B_2$ | .11 | .31 | .18 | .40 |
| $A_2B_1$ | .10 | .08 | .57 | .25 |
| $A_2B_2$ | .04 | .01 | .22 | .72 |

**6.1** (a) A GRT model of the 2 × 2 factorial identification experiment. The marginal distributions in gray signal that the relevant stimulus component is at level 1, whereas the marginal distributions in black signal that the relevant stimulus component is at level 2. (b) The confusion matrix predicted by the panel (a) model. Each entry is a conditional response probability. (c) A confusion matrix from a hypothetical participant described by the panel (a) GRT model from an experiment with 600 trials (150 for each stimulus). Each entry is a conditional response proportion.

is at level $A_i$ (for $i = 1$ and 2) are different when B is at level $B_1$ versus $B_2$. In contrast, on dimension $x_2$, the marginal distributions when B is at level $B_j$ (for $j = 1$ and 2) are identical for both levels of component A. Suppose further that the stimuli used in this experiment are the Gabor patches shown in figure 4.2. Therefore, because the perceptual dimensions of these stimuli are known (i.e., perceived bar orientation and bar width), the identifiability problems discussed in chapter 4 (i.e., in section 4.5) are not a concern in this experiment.

Figure 6.1b shows the confusion matrix predicted by this model (rounded to two decimal places), which was generated by integrating under each bivariate normal distribution in turn, over each of the four response regions. Each such integral equals a predicted response probability. The entry in row $A_iB_j$ and column $a_kb_m$ equals the conditional probability that response $a_kb_m$ is given on a trial when the stimulus is $A_iB_j$ (where $i, j, k$, and $m$ each equal 1 or 2). Therefore, note that each row sum equals 1.0, the entries on the main diagonal list the probabilities of each correct response, and the off-diagonal elements describe the various confusions.

Figure 6.1c shows a confusion matrix from a hypothetical participant whose perceptual and decisional processes are described by the figure 6.1a model in an experiment with 600 identification trials—150 of each stimulus type. This is about the maximum amount of data one could collect from one participant in a single one-hour experimental session. The entry in row $A_iB_j$ and column $a_kb_m$ is now the proportion of $A_iB_j$ trials on which the participant responded $a_kb_m$. These proportions were generated by drawing 150 random samples from each of the four perceptual distributions shown in panel (a) and then selecting a response by applying the decision bounds shown in that panel.

These confusion matrices have several features worth discussing. First, note that the sample and population confusion matrices display some substantial discrepancies. The largest of these is 0.07, which is likely large enough to skew a statistical test (as we will see). There are several possible antidotes. One is to collect much more data from each participant. This is an expensive solution though because it requires multiple experimental sessions, and therefore will almost always require paid participants. Another solution is to collect single-session data from many participants. The data from each participant is likely to be at least as variable as the hypothetical data shown in figure 6.1c, but standard statistical sampling models predict that this variability will not be systematic, and therefore the majority of tests should produce outcomes that are consistent with the perceptual properties displayed by the model that generated the data. The problem here, of course, is that there is no reliable way to determine whether the outcomes of some summary-statistic test differ across participants because of random variability in the data or because of perceptual and/or decisional processing differences across participants. One solution to this problem is to fit the GRT-wIND model described in the next chapter (in section 7.7) to the identification-confusion matrices—either in addition to or instead of relying on summary-statistic tests. GRT-wIND assumes that all observers share the same perceptual representation but that different observers use different decision bounds and divide their attention differently between the two perceptual dimensions. Because of the group-level perceptual representation, fitting GRT-wIND simultaneously to the confusion matrices of all observers leads to group-level conclusions about perceptual separability and perceptual independence. Another advantage of GRT-wIND is that, unlike almost all summary-statistics tests, it does not assume decisional separability.

Second, note that because perceptual separability is violated in figure 6.1 for component A, decisional separability is far from the optimal decision strategy. This is most easily seen in the responses to stimulus $A_1B_2$. Note that this hypothetical participant responded correctly on $A_1B_2$ trials only one-third of the time, and responded incorrectly that the stimulus was $A_2B_2$ on 40% of the $A_1B_2$ trials. This is easily seen in panel (a) of figure 6.1. Note that more of the $A_1B_2$ perceptual distribution spills into the $a_2b_2$ response region than is in the $a_1b_2$ region. The participant could correct this problem by increasing the $x_1$ intercept of the vertical decision bound, but note that this would decrease accuracy on $A_2B_1$ trials. The violations of perceptual separability that characterize the figure 6.1 perceptual representations imply that with these perceptual distributions there is an inevitable accuracy cost to any strategy that assumes decisional separability (i.e., relative to the optimal strategy). This also illustrates one of the dangers of all currently popular summary-statistics tests. They all assume decisional separability and even if all precautions are taken by ensuring an experimental design that makes decisional separability as likely as possible, there is no guarantee that participants will use such a strategy—especially if perceptual separability and/or perceptual independence are violated.

Third, note that accuracy is only around 57% correct in this experiment. Such reasonably low accuracy levels are a necessary feature of any experiment with the stated goal of studying perceptual or decisional processes. If accuracy is perfect, then we can conclude that observers were able to discriminate the stimuli perfectly, but we can't draw any inferences about how this discrimination occurred. There will generally always be many different perceptual representations and different decision strategies (an infinite number?) that could produce perfect performance. Errors constrain these possibilities. For example, the fact that the hypothetical observer in figure 6.1b responded correctly on 71% of stimulus $A_2B_2$ trials, but only on 33% of stimulus $A_1B_2$ trials implies there must be some profound differences in how these two stimuli are processed. So any experiment that uses SDT or GRT to analyze the data should arrange conditions so that there are many errors.

Although all but one of the summary-statistics tests described in this chapter assume decisional separability, in general, little or nothing is known about how sensitive the various tests are to violations of decisional separability. As a first step in addressing this issue, consider figure 6.2, which is identical to figure 6.1, except decisional separability is violated on dimension $x_1$. Specifically, the bivariate and marginal perceptual distributions are all identical in figures 6.2 and 6.1, and the horizontal decision bound is identical. Despite these similarities, note that the single change in decision strategy—from a vertical bound to a diagonal bound with positive slope—causes profound differences to the resulting confusion matrices. For example, note that expected $A_1B_2$ accuracy increases from 31% in figure 6.1 to about 57% in figure 6.2.

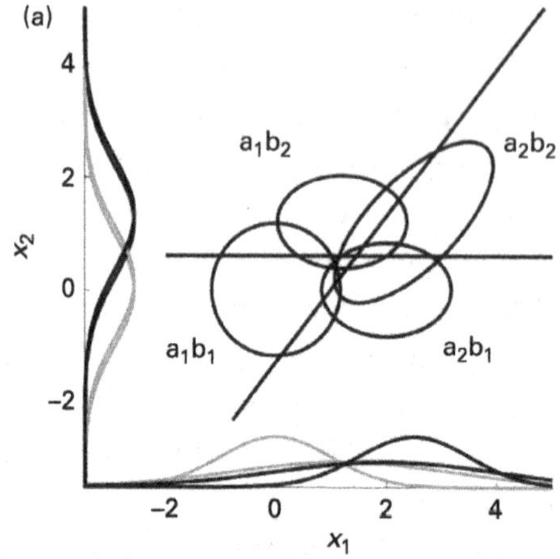


(b)

| | $a_1b_1$ | $a_1b_2$ | $a_2b_1$ | $a_2b_2$ |
|---|---|---|---|---|
| $A_1B_1$ | .52 | .26 | .21 | .01 |
| $A_1B_2$ | .12 | .55 | .16 | .17 |
| $A_2B_1$ | .14 | .13 | .59 | .14 |
| $A_2B_2$ | .05 | .16 | .23 | .56 |

(c)

| | $a_1b_1$ | $a_1b_2$ | $a_2b_1$ | $a_2b_2$ |
|---|---|---|---|---|
| $A_1B_1$ | .47 | .29 | .22 | .02 |
| $A_1B_2$ | .07 | .57 | .15 | .21 |
| $A_2B_1$ | .14 | .13 | .56 | .17 |
| $A_2B_2$ | .05 | .16 | .23 | .56 |

**6.2** A GRT model of the 2 × 2 factorial identification experiment. The model is identical to the model shown in figure 6.1a except that decisional separability is violated on dimension $x_1$. (b) The confusion matrix predicted by the panel (a) model. Each entry is a conditional response probability. (c) A confusion matrix from a hypothetical participant described by the panel (a) GRT model from an experiment with 600 trials (150 for each stimulus). Each entry is a conditional response proportion.

We will use the confusion matrices shown in figures 6.1 and 6.2 to illustrate the different summary statistics tests of perceptual independence and perceptual separability that have been proposed. Applications to the expected matrix shown in figure 6.1b will illustrate the ideal results that would be expected in the absence of any noise and when all assumptions of the summary-statistics tests are met, whereas application to the matrix shown in figure 6.1c will provide an example of a best-case, real-world scenario. It is best case because every response used to form the matrix came from the figure 6.1a GRT model. Finally, application to the matrix shown in 6.2b will illustrate one possible scenario in which all assumptions of the summary-statistics test are met, except that decisional separability is violated on one dimension.

## 6.1 NONPARAMETRIC TESTS

### 6.1.1 TESTS OF PERCEPTUAL SEPARABILITY

Perceptual separability was defined and discussed in section 4.3.2. Briefly though, stimulus component A is perceptually separable from component B if changing the level of B does not change the perception of component A. Thus, unlike perceptual independence, perceptual separability is a property of a group of stimuli. For example, perceptual separability requires that the perception of level 1 of component A is the same in stimuli $A_1B_1$ and $A_1B_2$.

The most popular summary-statistics test of perceptual separability uses a measure called *marginal response invariance* that was proposed by Ashby and Townsend (1986). The idea is that if the perception of component $A_i$ does not change when component B changes from level $B_1$ to level $B_2$, then on trials when component A is at level $A_i$, the probability that the observer reports the level of A to be $A_i$ should not depend on the level of B, and therefore should be the same on $A_iB_1$ and $A_iB_2$ trials. More formally,

**Definition of marginal response invariance** *Marginal response invariance holds at the $i^{th}$ level of stimulus component A if and only if:*

$$P(a_i|A_iB_1) = P(a_i|A_iB_2),$$

*or in other words, if and only if:*

$$P(a_ib_1|A_iB_1) + P(a_ib_2|A_iB_1) = P(a_ib_1|A_iB_2) + P(a_ib_2|A_iB_2). \tag{6.1}$$

Ashby and Townsend (1986) showed that if perceptual and decisional separability both hold, then marginal response invariance must also hold. Therefore, if the assumption of decisional separability is valid, then a failure of marginal response invariance at any level of a given dimension implies that perceptual separability fails on that dimension.

In the figure 6.1a model, and therefore in the figure 6.1b and 6.1c confusion matrices, component B is perceptually separable from component A, but component A is not perceptually separable from B. As a result, we expect marginal response invariance to hold for component B, but be violated for component A in the confusion matrix predicted by this GRT model. These predictions are tested in table 6.1. Each row describes a different test. If marginal response invariance holds then the value listed in column 2 must equal the value listed in column 3. As a result, any difference in these values represents a failure of marginal response invariance. The top half of the table tests marginal response invariance for stimulus component A and the bottom half tests this property for component B. Note that, as expected, marginal response invariance is strongly violated for component A. Changing the level of B substantially changes the predicted probability of correctly identifying the level of component A. For example, changing the level of B from $B_1$ to $B_2$ causes the probability that the level of A is correctly identified as $A_1$ to drop by almost half (i.e., from .84 to .45). In contrast, also as predicted, marginal response invariance holds for component B, because changing the level of A does not affect the probability of correctly identifying the level of B (the small effects shown in table 6.1 are due to rounding error).

Table 6.2 shows the same tests, except this time on the sample confusion matrix shown in figure 6.1c. Note that the violations of marginal response invariance are still large for component A, although for component $A_2$ the violation reduced in size from .17 (i.e., .93–.76) to .12 (i.e., .94–.82). For component B though, marginal response invariance holds almost exactly for component $B_2$ (the violation is only .02), but for component $B_1$ the difference in the response proportions when component A is at level $A_1$ versus $A_2$ is .08 (i.e., .75–.67), whereas, at the population level this difference is zero. A violation of marginal response invariance of .08 would make it difficult to conclude that perceptual separability holds for both levels of component B for this participant. Looking back at the confusion matrices shown in figure 6.1 makes it apparent that this .08 discrepancy is due to abnormal response proportions from two cells of the confusion matrix. Specifically, in the sample matrix, the proportion of $a_1b_1$ responses deviated from the response probabilities by .06 on both $A_1B_1$ and on $A_2B_1$ trials.

Table 6.3 shows the results of testing marginal response invariance on the confusion matrix shown in figure 6.2b. Recall that this matrix shows the predicted probabilities for a model that is identical to the model illustrated in figure 6.1a, except that decisional separability is violated on dimension $x_2$. The good news here is that marginal response invariance again holds for component B. This was expected since perceptual and decisional separability both hold on dimension $x_2$. As a result, according to the theorem of Ashby and Townsend (1986), marginal response invariance must also hold. Table 6.3 confirms that marginal response invariance

**Table 6.1** Marginal response invariance tests on the response probabilities reported in the figure 6.1b confusion matrix.

| $i$ | $P(a_i \mid A_iB_1)$ | $P(a_i \mid A_iB_2)$ |
|---|---|---|
| 1 | $.61+.23=.84$ | $.12+.33=.45$ |
| 2 | $.55+.21=.76$ | $.22+.71=.93$ |
| $j$ | $P(b_j \mid A_1B_j)$ | $P(b_j \mid A_2B_j)$ |
| 1 | $.61+.12=.73$ | $.17+.55=.72$ |
| 2 | $.33+.40=.73$ | $.01+.71=.72$ |

**Table 6.2** Marginal response invariance tests on the response proportions reported in the figure 6.1c sample confusion matrix.

| $i$ | $\hat{P}(a_i \mid A_iB_1)$ | $\hat{P}(a_i \mid A_iB_2)$ |
|---|---|---|
| 1 | $.67+.21=.88$ | $.11+.31=.42$ |
| 2 | $.57+.25=.82$ | $.22+.72=.94$ |
| $j$ | $\hat{P}(b_j \mid A_1B_j)$ | $\hat{P}(b_j \mid A_2B_j)$ |
| 1 | $.67+.08=.75$ | $.10+.57=.67$ |
| 2 | $.31+.40=.71$ | $.01+.72=.73$ |

**Table 6.3** Marginal response invariance tests on the response probabilities reported in the figure 6.2b confusion matrix.

| $i$ | $P(a_i \mid A_iB_1)$ | $P(a_i \mid A_iB_2)$ |
|---|---|---|
| 1 | $.52+.26=.78$ | $.12+.55=.67$ |
| 2 | $.59+.14=.73$ | $.23+.56=.79$ |
| $j$ | $P(b_j \mid A_1B_j)$ | $P(b_j \mid A_2B_j)$ |
| 1 | $.52+.21=.73$ | $.14+.59=.73$ |
| 2 | $.55+.17=.72$ | $.16+.56=.72$ |

provides a valid test of perceptual separability even if decisional separability is violated on the other dimension. The not-so-good news in table 6.3 is that the violations of marginal response invariance for component A are much smaller than in table 6.1. There, the deviations of marginal response invariance are .39 when A is at level $A_1$ (i.e., .84−.45) and .17 when A is at level $A_2$ (i.e., .93−.76). In contrast, in table 6.3 these violations drop to .11 and .06, respectively. This large difference, which was caused by a modest change in one decision bound, illustrates the dangers of using marginal response invariance as a test of perceptual separability when decisional separability is violated.

### 6.1.2 TESTS OF PERCEPTUAL INDEPENDENCE

Components A and B are perceived independently in stimulus $A_iB_j$ if the perceived value of component $A_i$ is statistically independent of the perceived value of component $B_j$ (see section 4.3.1). The most popular summary-statistics test of perceptual independence uses a statistic that was originally called sampling independence, but is now known as report independence. Report independence is a direct application of the well-known multiplication rule for computing the probability that two independent events co-occur. If events $E_1$ and $E_2$ are statistically independent, then the probability that both occur is

$$P(E_1 \; \& \; E_2) = P(E_1) \times P(E_2). \tag{6.2}$$

So for example, if two coins are tossed independently, the probability of observing two heads equals the product of the probabilities that each coin toss in isolation produces a head (so .5 × .5 = .25).

Because perceptual independence is a property of single stimuli, report independence applies the multiplication rule separately for every stimulus used in the experiment. Formally, report independence is defined as follows.

**Definition of report independence** *Report independence holds for stimulus $A_iB_j$ if and only if:*

$$\begin{aligned} P(a_ib_j|A_iB_j) &= P(a_i|A_iB_j) \times P(b_j|A_iB_j) \\ &= \left[P(a_ib_1|A_iB_j) + P(a_ib_2|A_iB_j)\right] \\ &\quad \times \left[P(a_1b_j|A_iB_j) + P(a_2b_j|A_iB_j)\right]. \end{aligned} \tag{6.3}$$

Ashby and Townsend (1986) showed that if decisional separability holds, then perceptual independence implies report independence, meaning that if decisional separability and perceptual independence both hold, then report independence must be satisfied. Furthermore, they also showed that if report independence holds for all possible values of the decision criteria, then perceptual independence must be satisfied. Therefore, if decisional separability holds, then a failure of report independence implies that perceptual independence is violated. On the other hand, decisional separability is critical here because Ashby and Townsend (1986) also showed that if decisional separability fails, then report independence and perceptual independence are logically unrelated.

Table 6.4 shows the results of testing report independence on the four stimuli associated with the confusion matrix shown in figure 6.1b. Column 2 shows the probability of a correct response for each stimulus, which corresponds to the left side of equation 6.3. Column 3 shows the value predicted if report independence holds, which is computed from the right side of equation 6.3. Report independence holds if the values in columns 2 and 3 are equal. Since the confusion matrix was generated from the figure 6.1a model, we know that perceptual independence holds for stimuli

**Table 6.4** Report independence tests on the response probabilities shown in the figure 6.1b population confusion matrix.

| Stimulus | $P(a_i b_j \mid A_i B_j)$ | $P(a_i \mid A_i B_j) \times P(b_j \mid A_i B_j)$ |
|---|---|---|
| $A_1B_1$ | .61 | $(.61 + .23)(.61 + .12) = .61$ |
| $A_1B_2$ | .33 | $(.12 + .33)(.33 + .4) = .33$ |
| $A_2B_1$ | .55 | $(.55 + .21)(.17 + .55) = .55$ |
| $A_2B_2$ | .71 | $(.22 + .71)(.01 + .71) = .67$ |

**Table 6.5** Report independence tests on the response proportions shown in the figure 6.1c sample confusion matrix.

| Stimulus | $\hat{P}(a_i b_j \mid A_i B_j)$ | $\hat{P}(a_i \mid A_i B_j) \times \hat{P}(b_j \mid A_i B_j)$ |
|---|---|---|
| $A_1B_1$ | .67 | $(.67 + .21)(.67 + .08) = .66$ |
| $A_1B_2$ | .31 | $(.12 + .31)(.31 + .40) = .31$ |
| $A_2B_1$ | .57 | $(.57 + .25)(.10 + .57) = .55$ |
| $A_2B_2$ | .72 | $(.22 + .72)(.01 + .72) = .69$ |

$A_1B_1$, $A_1B_2$, and $A_2B_1$, and is violated for stimulus $A_2B_2$. As expected, note that report independence is satisfied for these former three stimuli and violated for stimulus $A_2B_2$. Even so, the violation for stimulus $A_2B_2$ is only .04 (i.e., .71–.67), which would likely require an experiment with a large number of participants to detect.

With real data, of course, report independence will not hold exactly, even if perceptual independence and decisional separability both hold. Table 6.5, which tests report independence on the sample confusion matrix shown in figure 6.1c, provides an illustration of this best-case scenario. There is good news and bad news here. The good news is that the largest deviation of the equality tested by report independence occurs with stimulus $A_2B_2$, which we know is the single stimulus that violates perceptual independence. The bad news though, is that this deviation is only .03, and even worse, only .01 larger than the next greatest deviation.

There are two reasons, though, why the picture isn't quite as bleak as table 6.5 might make it appear. First, note that the violation of report independence for stimulus $A_2B_2$ in this hypothetical participant is smaller than expected, given the figure 6.1a GRT model. (i.e., compare tables 6.4 and 6.5). So with respect to testing for perceptual independence, sampling this particular participant was bad luck. With more participants, we would expect many $A_2B_2$ deviations of report independence to exceed .04.

The second reason why report independence might be a better test of perceptual independence than tables 6.4 and 6.5 suggest is that failures of perceptual independence might often lead to violations of report independence that are substantially larger than in these examples. For example, consider the GRT model shown in figure 6.3a. Note that this model is similar to the figure 6.1 model, in the sense that perceptual separability holds for component B in both models but not for component A, decisional separability holds on both dimensions, and perceptual independence is satisfied for three stimuli and violated for one.

The confusion matrix predicted by this model is shown in figure 6.3b and the report independence tests on these predicted probabilities are shown in table 6.6. Note that, as expected, report independence is satisfied for stimuli $A_1B_1$, $A_1B_2$, and $A_2B_2$, and is violated for stimulus $A_2B_1$. However, note that the violation for stimulus $A_2B_1$ is almost twice as large as the violation for stimulus $A_2B_2$ in table 6.4. This substantially larger violation should be considerably easier to verify statistically. It is also important to note here that I made no attempt to either minimize or maximize the magnitudes of these violations. The GRT models shown in figures 6.1 and 6.3 were simply the first two models I happened to construct to illustrate how report

(a)

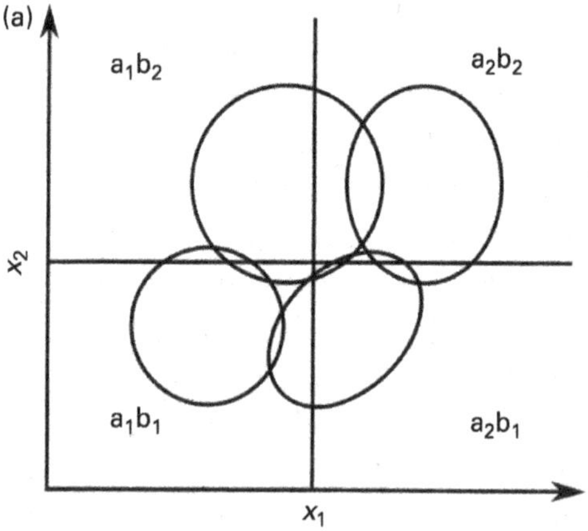


(b)

| | $a_1b_1$ | $a_1b_2$ | $a_2b_1$ | $a_2b_2$ |
|---|---|---|---|---|
| $A_1B_1$ | .62 | .20 | .14 | .04 |
| $A_1B_2$ | .18 | .50 | .09 | .23 |
| $A_2B_1$ | .30 | .00 | .46 | .24 |
| $A_2B_2$ | .05 | .14 | .21 | .60 |

**6.3** (a) Another GRT model of the 2 × 2 factorial identification experiment in which perceptual separability holds for component B, but not for component A. In this model, perceptual independence holds for stimuli $A_1B_1$, $A_1B_2$, and $A_2B_2$, and is violated for stimulus $A_2B_1$. (b) The confusion matrix predicted by the model shown in panel (a).

**Table 6.6** Report independence tests on the response probabilities predicted by the figure 6.3 GRT model.

| Stimulus | $P(a_i b_j \mid A_i B_j)$ | $P(a_i \mid A_i B_j) \times P(b_j \mid A_i B_j)$ |
|---|---|---|
| $A_1B_1$ | .62 | $(.62+.20)(.62+.14)=.62$ |
| $A_1B_2$ | .50 | $(.18+.50)(.50+.23)=.50$ |
| $A_2B_1$ | .46 | $(.46+.24)(.30+.46)=.53$ |
| $A_2B_2$ | .60 | $(.21+.60)(.14+.60)=.60$ |

**Table 6.7** Report independence tests on the response probabilities shown in the figure 6.2b population confusion matrix.

| Stimulus | $P(a_i b_j \mid A_i B_j)$ | $P(a_i \mid A_i B_j) \times P(b_j \mid A_i B_j)$ |
|---|---|---|
| $A_1B_1$ | .52 | $(.52+.26)(.52+.21)=.57$ |
| $A_1B_2$ | .55 | $(.12+.55)(.55+.17)=.48$ |
| $A_2B_1$ | .59 | $(.59+.14)(.14+.59)=.53$ |
| $A_2B_2$ | .56 | $(.23+.56)(.16+.56)=.57$ |

independence is tested. I have no idea whether the magnitudes of the violations shown in tables 6.4 and 6.6 are representative of violations that might be expected from other reasonable GRT models in which perceptual independence is violated in one or more stimuli. Answering this question would require a prodigious computational effort that systematically explores a high-dimensional parameter space (i.e., the space of parameters that define GRT models of the $2 \times 2$ factorial identification task).

The caution that these tests depend on decisional separability is worth repeating. As an illustration of what can occur when decisional separability is violated, consider table 6.7, which shows report independence tests for each stimulus from the confusion matrix predicted by the figure 6.2a GRT model. Recall that this model is identical to the model shown in figure 6.1 in every way, except the vertical decision bound in the figure 6.1a model is no longer vertical. As table 6.7 clearly shows, this single violation of decisional separability destroys the ability of report independence to identify stimuli for which perceptual independence is either satisfied or violated. Now the *smallest* violation of report independence is by the single stimulus that violates perceptual independence (i.e., stimulus $A_2B_2$), and the violations of report independence for each of the three stimuli that satisfy perceptual independence are all substantial. As a result, if decisional separability is violated, even if only on one dimension, then using report independence to test for perceptual independence is likely to lead to many invalid inferences.

## 6.2 TESTS INSPIRED BY SDT

Although marginal response invariance and report independence are the most widely used summary-statistics tests of GRT perceptual properties, other accuracy-based summary-statistics tests have also been proposed. These other tests use statistics that were inspired by SDT—for example, by measures that are similar to $d'$ (Kadlec, 1995; Kadlec & Townsend, 1992a, 1992b). As noted in previous chapters, in its most general form, it is not immediately clear how one should define the $d'$ between two GRT perceptual representations. Possible differences in perceptual variability and perceptual dependence across different perceptual representations suggest that multiple definitions could be equally plausible. If the goal though is to develop $d'$-related summary statistics tests of GRT perceptual properties, then our overriding concern should be to choose the definition that allows the strongest possible inferences about those properties, even if some other definition might better capture the qualitative properties associated with the SDT $d'$ measure.

Kadlec and Townsend (1992a, 1992b) proposed two qualitatively different $d'$ definitions—one that they described as macroanalytic and one they described as microanalytic. They used their macroanalytic $d'$s to test for perceptual separability and their microanalytic $d'$s to test for perceptual independence. As mentioned previously, these tests are collectively referred to as multidimensional signal detection analysis. We will consider their macroanalytic tests first.

### 6.2.1 MACROANALYTIC TESTS OF PERCEPTUAL SEPARABILITY

The Kadlec and Townsend (1992a, 1992b) macroanalytic definition of $d'$ is conceptually similar to the $d'$ measure used in SDT. Specifically, for any pair of stimuli, the macroanalytic $d'$ is a measure of the standardized distance between the two perceptual means. Even so, there are a variety of different ways to define such a distance. For example, consider figure 6.4, which shows a GRT model of the $2 \times 2$ factorial identification experiment. Now suppose we wish to define a $d'$ that measures the standardized distance between the perceptual representations of stimuli $A_1B_1$ and $A_2B_1$. Note that $\Delta_1$ denotes the distance between the means of the two distributions,

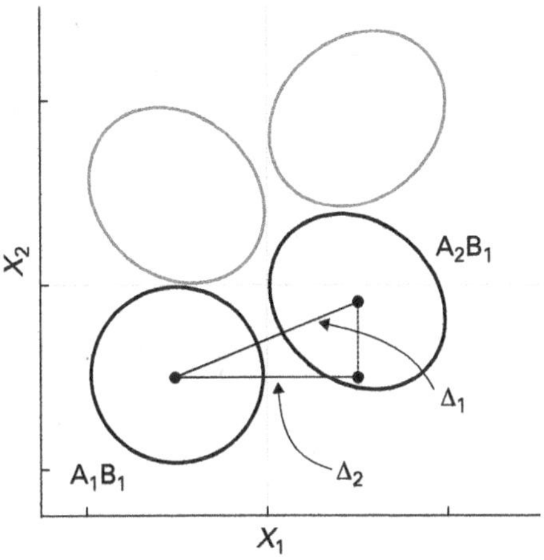


**6.4** A GRT model of the 2 × 2 factorial identification experiment and two different plausible distances $\Delta_1$ and $\Delta_2$ that could be used to define the $d'$ between the perceptual representations of stimuli $A_1B_1$ and $A_2B_1$.

whereas $\Delta_2$ denotes their distance apart on dimension $x_1$. Either distance could be used to define a useful $d'$. $\Delta_1$ could be used to define a $d'$ that measures the discriminability difference between the two stimuli, whereas $\Delta_2$ could be used to define a $d'$ that measures the discriminability difference between components $A_1$ and $A_2$ when component B is at level $B_1$.

If the goal is to define a $d'$ in a way that allows the strongest possible inferences to be made about perceptual separability, then the $\Delta_2$ distance seems more appropriate. For example, if component A is perceptually separable from component B, then we expect the discriminability difference between components $A_1$ and $A_2$ to be the same when B is at level $B_1$ as it is when B is at level $B_2$. Therefore, because Kadlec and Townsend (1992b) were interested in developing summary statistics tests of perceptual separability (and perceptual independence), they based their macroanalytic definition of $d'$ on the $\Delta_2$ distance. Specifically, they defined the $d'$ between the two levels of component A when component B is at level $B_j$ as

$$d'_{A|B_j} = \frac{\mu_{1,2j} - \mu_{1,1j}}{\sqrt{\left(\sigma^2_{1,2j} + \sigma^2_{1,1j}\right)/2}}, \tag{6.4}$$

where, for example, $\mu_{1,2j}$ and $\sigma^2_{1,2j}$ are the mean and variance on dimension $x_1$ of the perceptual representation for stimulus $A_2B_j$. Note that the denominator is just the square root of the mean of the two relevant variances.[1]

Equation 6.4 is a sensible definition. For example, note that in the case of the normal, equal-variance SDT model, equation 6.4 reduces to the standard $d'$ definition. Recall that in that model, the mean of the SN distribution equals $d'$, the mean of the N distribution is 0, and both variances equal 1 and so

$$\frac{d'-0}{\sqrt{(1+1)/2}}=d'. \tag{6.5}$$

In other words, $d'$ is the number of standard deviations between the means of the N and SN distributions.

The $d'$ defined by equation 6.4 can be estimated from the confusion matrix obtained from the $2 \times 2$ factorial identification experiment using standard SDT methods of estimation (i.e., using equations 2.11 and 2.13). As an example, consider the confusion matrix shown in figure 6.1b and the $d'$ for component A when B is at level $B_1$—that is, $d'_{A|B_1}$. In this case, we can arbitrarily treat level $A_2$ of component A as the signal and level $A_1$ as the noise. If so, then a hit occurs when stimulus $A_2B_1$ is presented and the observer gives a response that identifies the level of component A as $A_2$. Similarly, a false alarm occurs when stimulus $A_1B_1$ is presented and the observer responds that A is at level $A_2$. In other words, $d'_{A|B_1}$ and the corresponding $X_C$ can be estimated by first estimating the probability of a hit as

$$\begin{aligned}\hat{P}(\mathrm{H}) &= \hat{P}(a_2b_1|A_2B_1)+\hat{P}(a_2b_2|A_2B_1) \\ &= .55+.21=.76\end{aligned} \tag{6.6}$$

and the probability of a false alarm as

$$\begin{aligned}\hat{P}(\mathrm{FA}) &= \hat{P}(a_2b_1|A_1B_1)+\hat{P}(a_2b_2|A_1B_1) \\ &= .12+.04=.16,\end{aligned} \tag{6.7}$$

and then inserting these estimates into equations 2.11 and 2.13 to produce

$$\hat{X}_C=0.99 \text{ and } \hat{d}'_{A|B_1}=1.70. \tag{6.8}$$

Figure 6.5 shows the four component $\hat{d}'$s estimated in this way, along with the four population-level $d'$s computed from equation 6.4. Note that, except for $d'_{A|B_2}$, the estimated and true values of $d'$ are all nearly equal. The estimated value of $d'_{A|B_2}$ differs from the true value by 0.29, which seems substantial. There are two reasons for this discrepancy. First, the equation 6.4 definition is insensitive to the large positive correlation (and therefore the strong violation of perceptual independence) that characterizes the $A_2B_2$ perceptual distribution. This correlation affects the predicted response probabilities on all $A_2B_2$ trials, and therefore the predicted values of $P(\mathrm{H})$ and $P(\mathrm{FA})$. Second, the $\hat{d}'$s in figure 6.5 were estimated from the standard SDT

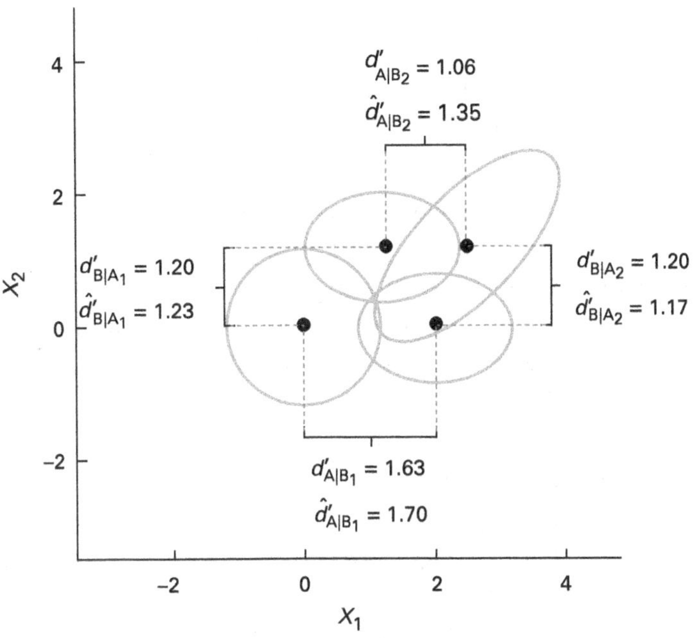


**6.5** Exact and estimated macroanalytic values of $d'$ for the model shown in figure 6.1a. The exact values were computed from equation 6.4 and the estimates were computed by estimating the probability of a hit and false alarm from the confusion matrix shown in figure 6.1b via equations 6.6 and 6.7 and then inserting these estimates into equations 2.11 and 2.13.

equations 2.11 and 2.13. These assume the normal, equal-variance SDT model. As a result, the $d'$ and $\hat{d}'$ values in figure 6.5 should not be expected to agree exactly when the two distributions used to compute the $d'$ value (i.e., from equation 6.4) have different variances on the relevant dimension. In the figure 6.5 example, it turns out that the relevant variances are equal in the case of $d'_{B|A_1}$ and $d'_{B|A_2}$, and as expected, in both cases $d'$ and $\hat{d}'$ differ by only 0.03. However, in the case of both $d'_{A|B_1}$ and $d'_{A|B_2}$ the relevant variances differ by a factor of 2-to-1, so it should not be too surprising that in both of these cases $d'$ and $\hat{d}'$ differ by more than 0.03.

Note that these estimates depend strongly on the assumption that decisional separability holds. To see this, note that if decisional separability holds on both dimensions, as it does for example in figure 6.1, then those vertical and horizontal decision bounds are compatible with an independent-decisions strategy (see section 5.1.1) that uses the rule:

"Respond $a_1b_1$ if the stimulus is small on both dimensions; respond $a_1b_2$ if the stimulus is small on dimension 1 and large on dimension 2; respond $a_2b_1$ if the stimulus is large on dimension 1 and small on dimension 2; and respond $a_2b_2$ if the stimulus is large on both dimensions."

In this case, equation 6.6 makes sense as an estimator of $P$(H) because on trials when stimulus $A_2B_1$ is presented, the observer responds either $a_2b_1$ or $a_2b_2$ only when an explicit decision was made that component A is at level $A_2$. In other words, whether the observer gives one of these two responses depends only on the perceived value of component A. The perceived value of component B is irrelevant. In contrast, if decisional separability is violated then a decision to respond either $a_2b_1$ or $a_2b_2$ will depend on the perceived value of component B, and therefore equation 6.6 will no longer equal the $P$(H) we should expect in a two-alternative identification task in which the two stimuli are $A_1$ and $A_2$.

Kadlec and Townsend (1992a, 1992b) did not formally define the diagonal $d'$s—that is, either the $d'$ that measures the discriminability difference between stimuli $A_1B_1$ and $A_2B_2$ or between stimuli $A_1B_2$ and $A_2B_1$. This is because there is no obvious way to estimate a diagonal $d'$ if the only available data are the entries in the confusion matrix from a $2 \times 2$ factorial identification experiment. For example, consider trying to estimate the $d'$ that measures the discriminability difference between stimuli $A_1B_1$ and $A_2B_2$. The first problem is to decide how to define this $d'$ at the population level. Equation 6.4 is inappropriate because it only considers discriminability differences on a single perceptual dimension. Clearly, the appropriate measure of $d'$ should be sensitive to the variability on both dimensions, and probably also to the correlation. As a solution to this problem, R. D. Thomas (1999) proposed a definition of the diagonal $d'$s that was inspired by the Mahalanobis distance between the two relevant means. Mahalanobis distance changes with the direction of the line connecting the two points in question and it depends on all parameters in the relevant variance-covariance matrix. Recall that we used Mahalanobis distance to discard outliers in the step-by-step algorithm described in section 5.6 for constructing matched RB and II categories. We will also see Mahalanobis distance again in chapter 9 (section 9.2), where we will consider its properties in some detail.

The second problem to solve if our goal is to estimate a diagonal $d'$ that measures the discriminability difference between stimuli $A_1B_1$ and $A_2B_2$ is to figure out how to estimate the relevant $P$(H) and $P$(FA) values that are needed for the equations 2.11 and 2.13 estimators of $X_C$ and $d'$. As Kadlec and Townsend (1992a, 1992b) noted, if all we have available is the confusion matrix from a $2 \times 2$ factorial identification experiment, then we're out of luck because there are no cells in this matrix that would be classified as either a hit or false alarm in a two-alternative identification task that uses stimuli $A_1B_1$ and $A_2B_2$. If the goal is to estimate the diagonal $d'$s, then they suggested running two extra two-alternative conditions—one in which observers are asked to discriminate between stimuli $A_1B_1$ and $A_2B_2$ and one in which they discriminate between stimuli $A_1B_2$ and $A_2B_1$. This is hardly ideal, however, because it requires collecting many more trials of data and if the diagonal $d'$s that are estimated from these new conditions are used to make inferences about the perceptual processing that occurred during the $2 \times 2$ factorial identification experiment, then it also requires an extra assumption that the perceptual representation of each stimulus remains the same when the experimental condition changes.

The primary benefit to estimating the component $d'$s shown in figure 6.5 is that they can be used to test for perceptual separability. Specifically, Kadlec and Townsend (1992a) proved the following result.

**Theorem 6.1** If stimulus component A is perceptually separable from component B, then the marginal $d'$s of component A must be equal across the two levels of component B; that is,

$$d'_{A|B_1} = d'_{A|B_2}. \tag{6.9}$$

Similarly, if component B is perceptually separable from component A, then

$$d'_{B|A_1} = d'_{B|A_2}. \tag{6.10}$$

*Proof* See Kadlec and Townsend (1992a). ∎

This theorem formalizes our earlier intuition. If component A is perceptually separable from component B then the discriminability difference between levels $A_1$ and $A_2$ of component A should not depend on the level of component B. So if the marginal $d'$s are unequal then perceptual separability must be violated.

As an initial application of this test, look back at figure 6.5. Recall that this figure shows the true marginal $d'$s computed from the population parameter values of the model shown in figure 6.1a along with their values that were estimated from the 6.1b confusion matrix, which is the exact confusion matrix predicted by the model. Recall also that in this model, component B is perceptually separable from component A, but A is not separable from B. At the population level, note that as the theorem therefore requires, $d'_{B|A_1} = d'_{B|A_2}$. Furthermore, as we would hope, the analogous equality is strongly violated for the integral dimensions because $d'_{A|B_1} = 1.63$ whereas $d'_{A|B_2} = 1.06$. Fortunately, the same pattern holds after the marginal $d'$s are estimated from the population confusion matrix, although in both cases, the picture is somewhat clouded. Specifically, the difference in marginal $d'$s for component B rises from 0 when computed directly from the model's parameters to 0.06 when the $d'$s are estimated from the population confusion matrix, and the difference for component A drops from 0.57 (i.e., 1.63–1.06) to 0.35 (i.e., 1.70–1.35).

These marginal $\hat{d}'$ differences of 0.06 for component B and 0.35 for component A were estimated from the population confusion matrix of figure 6.1b. Of course, the best any empirical application could do would be to collect a limited number of samples from the figure 6.1a model. In this case, if the same model described every observer, we would expect the marginal $\hat{d}'$ differences to fluctuate across observers around these two values (i.e., 0.06 and 0.35). A reasonably powered experiment therefore would likely correctly reject perceptual separability for component A for most observers. For component B however, the results would likely be more ambiguous. The most likely outcome would be that perceptual separability would be accepted for some observers and rejected for others. Increasing the power of the experiment, either by collecting more data from each observer or by increasing the number of observers, would paradoxically make matters worse because it would make rejections of perceptual separability for component B more likely (i.e., since the true estimated marginal $\hat{d}'$ difference for component B is positive). Of course, the marginal $\hat{d}'$ differences of 0.06 and 0.35 that we estimated from the population confusion matrix of figure 6.1b would be different in other models in which perceptual separability holds for one component but not the other. Even so, this initial application seems to suggest that the marginal $d'$ test of perceptual separability may be more successful at detecting violations of separability than at identifying instances in which perceptual separability is satisfied.

One attractive feature of this test is that technically it does not require an assumption of decisional separability. This is because the result says that $d'_{A|B_1} = d'_{A|B_2}$, where $d'_{A|B_1}$ and $d'_{A|B_2}$ are defined by equation 6.4—a definition that depends only on parameters of the sensory representations, and therefore makes no assumptions about decision processes. On the other hand, for any practical application, these $d'$s must be estimated from the available confusion matrix, and as we saw, reliable estimators will generally require that decisional separability is satisfied. One way to examine the sensitivity of the test to decisional separability is to apply it to the figure 6.2b confusion matrix. Recall that this is the population-level confusion matrix for the figure 6.2a model, which is identical to the figure 6.1a model in every way, except that decisional separability is violated on dimension $x_1$. For component B, the two marginal $\hat{d}'$s turn out to be exactly equal (i.e., to 1.20), so in this case, the test correctly concludes that component B is separable from component A. For component A, however, things are not so good because $\hat{d}'_{A|B_1} = 1.32$ and $\hat{d}'_{A|B_2} = 1.25$—a difference of only 0.07, despite the fact that component A is not perceptually separable from B. A powerful experiment might detect this difference, but the important conclusion is that the test appears to be sensitive to violations of decisional separability. The marginal $\hat{d}'$ difference for component A was 0.35 when decisional separability was satisfied, and it dropped to 0.07 when it was violated. So in this case, a violation of decisional separability made the test much less sensitive to violations of perceptual separability. In practice then, it is not clear whether the marginal $d'$ test of perceptual separability really does depend on decisional separability any less than the marginal response invariance test we considered in the previous section.

Theorem 6.1 states that a violation of equal marginal $d'$s implies some sort of perceptual integrality. Unfortunately, however, the inference fails in the opposite direction. There are many ways that equations 6.9 and 6.10 could hold even though perceptual separability is strongly violated. To see this, consider first the simple case in which all variance-covariance matrices equal the identity—so the contours of equal likelihood are all circles of equal diameter. In this case, we can ignore the denominator of equation 6.4. The marginal $d'$s will be equal if and only if the component A mean differences are the same when B is at level $B_1$ and $B_2$, and when the component B mean differences are the same when A is at level $A_1$ and $A_2$. But even this very strong condition could occur in many different ways. One possibility is that the four perceptual means fall on the corners of a perfect rectangle. If the sides of the rectangle are parallel to the coordinate axes, then perceptual separability holds on both dimensions. But if the rectangle is tilted in any direction by some angle that is not a multiple of 90° then perceptual separability is violated on both dimensions even though the marginal $d'$s are equal. Another way that the marginal $d'$s would all be equal is if the four means fall on the corners of a parallelogram in which the angles are not 90°. If any two sides of the parallelogram are parallel to one of the perceptual axes then perceptual separability holds on one dimension and is violated on the other. If neither pair of sides is parallel to an axis then perceptual separability is violated on both dimensions. Allowing the variances to differ across perceptual representations creates even more ways that the marginal $d'$s could be equal despite a failure of perceptual separability. In summary, if decisional separability holds, then unequal marginal $d'$s are strong evidence against perceptual separability, but if the marginal $d'$s are equal, then no strong inferences are possible.

### 6.2.2 MICROANALYTIC TESTS OF PERCEPTUAL INDEPENDENCE

The Kadlec and Townsend (1992a, 1992b) microanalytic $d'$s are illustrated in figure 6.6. Specifically, the figure illustrates how microanalytic $d'$s can be defined that measure the discriminability difference between levels $A_1$ and $A_2$ of component A, given the GRT model depicted in figure 6.1a. Panels (a) and (b) define these microanalytic $d'$s when component B is at level $B_2$ and panels (c) and (d) define the analogous $d'$s when B is at level $B_1$. As we just saw, there is only one macroanalytic $d'$ that measures the $A_1$ versus $A_2$ difference for each level of B. For example, when B is at level $B_2$, we denoted this macroanalytic $d'$ as $d'_{A|B_2}$. At the microanalytic level though, there are two of these for each level of B that differ according to how the observer responded to component B. This approach treats level $B_2$ of component B as signal and level $B_1$ as noise. As a result, when component B is at level $B_2$, as in figure 6.6a, then any $b_2$ response represents a hit on component B and any $b_1$ response represents a miss. Therefore, in figure 6.6a, a hit on B occurs to any percept that falls above the horizontal bound and a miss on B occurs to any percept that falls below this bound. Kadlec and Townsend defined separate microanalytic $d'$s for these two scenarios. These are illustrated in figures 6.6a and 6.6b. The idea is to examine the marginal distributions of the perceived value of component A—that is, of $x_1$—for percepts on the same side of the bound (either above or below) when A is at level $A_1$ versus $A_2$ (and in this case, when B is at level $B_2$).

These marginal distributions are shown in figure 6.6b. The two gray curves show the marginal distributions of the perceived values of component A when A is at levels $A_1$ and $A_2$, respectively *and* a hit was made about the level of component B. The left-most of these curves shows the pdf of the $x_1$ values when random samples are drawn from the light gray shaded region of the $A_1B_2$ perceptual distribution and the right-most gray curve shows the pdf when the random samples are drawn from the light gray shaded region of the $A_2B_2$ distribution. The microanalytic $d'$, denoted $d'_{A|\text{hit on B}}$, uses equation 6.4 to measure the separation between the means of these two marginal distributions (i.e., using the means and variances of the relevant figure 6.6b marginal distributions). The two black curves in figure 6.6b show the same thing as the gray curves, except for the case in which the observer responded with a miss on component B (i.e., trials on which the random samples are drawn from the dark gray regions of the $A_1B_2$ and $A_2B_2$ perceptual distributions that lie below the horizontal decision bound). These marginal distributions are then used to define the second microanalytic $d'$ measure of the discriminability difference between levels $A_1$ and $A_2$ of component A when a miss was made on the level of

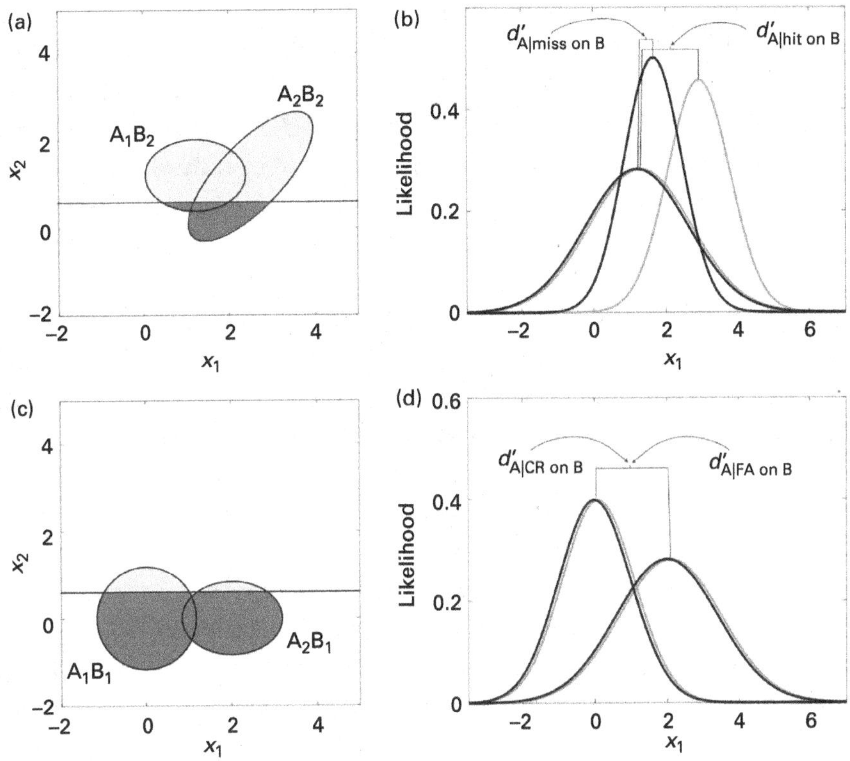


**6.6** An illustration of the Kadlec and Townsend (1992a, 1992b) definition of microanalytic *d*′s.

component B, denoted $d'_{\text{A|miss on B}}$. Note that $d'_{\text{A|hit on B}} > d'_{\text{A|miss on B}}$ and that this difference appears to be due to the positive perceptual dependence associated with stimulus $A_2B_2$.

Figures 6.6c and 6.6d show the same things as figures 6.6a and 6.6b, except for stimuli $A_1B_1$ and $A_2B_1$. Because B is at level $B_1$ for both of these stimuli, a response that B is at level $B_2$ represents a false alarm and a response that B is at level $B_1$ represents a correct rejection. As a result, the two microanalytic *d*′s depicted in figure 6.6d are denoted $d'_{\text{A|CR on B}}$ and $d'_{\text{A|FA on B}}$. Note that these two *d*′s are equal.

All of these microanalytic *d*′s can be estimated using standard SDT methods. For each of these four *d*′s, we need to estimate a *P*(H) and *P*(FA) and then estimate a *d*′ in the standard way via equations 2.11 and 2.13. At first glance, this can be confusing, because how can we estimate a *P*(FA) for $d'_{\text{A|hit on B}}$? The key is that the FA we need here is on the decision about the level of A, whereas the hit referred to in $d'_{A|\text{hit on B}}$ is a hit on the level of component B. So a less ambiguous notation might be to say that we need to estimate a *P*(H on A) and *P*(FA on A). In each case, this is a matter of selecting the appropriate cells from our confusion matrix. For each of the four microanalytic *d*'s, the appropriate estimators are listed in table 6.8. After these proportions are estimated, each of the *d*'s can then be estimated from equations 2.11 and 2.13. Table 6.8 also lists the numerical values that result when these estimators are applied to the confusion matrix shown in figure 6.1b.

Why is it useful to define $d'$ in this novel way? The key lies in the fact that in the figure 6.6 example, $d'_{A|\text{hit on B}} > d'_{A|\text{miss on B}}$ and that this difference is due entirely to the $A_2B_2$ perceptual distribution. For stimulus $A_1B_2$, the marginal distributions on dimension $x_1$ are identical above and below the horizontal decision bound (and therefore on trials when a hit and miss occurs on component B). In contrast, for stimulus $A_2B_2$, the mean of this marginal distribution is considerably larger above the bound than below the bound. In other words, the perceived component A values on $A_2B_2$ trials are larger when there is a hit on component B than when there is a miss. This difference occurs because perceptual independence holds for stimulus $A_1B_2$ but is violated for stimulus $A_2B_2$. If perceptual independence holds, then the major and minor axes of the contours of equal likelihood are parallel to the coordinate axes—so one will be horizontal and one will be vertical. As a result, if decisional separability holds then the vertical of these two axes will be orthogonal to the horizontal decision bound and so the marginal distributions on dimension $x_1$ above and below this bound will be equal. This intuition is formalized in the following result.

**Table 6.8** Columns 2 and 3 show the estimators of *P*(H on A) and *P*(FA on A) for each of the four microanalytic *d*'s on component A and columns 4 and 5 show the resulting estimates of $X_C$ and $d'$ that result when these estimators are applied to the confusion matrix shown in figure 6.1b.

| $d'$ | $\hat{P}$(H on A) | $\hat{P}$(FA on A) | $\hat{X}_C$ | $\hat{d}'$ |
|---|---|---|---|---|
| $d'_{A|\text{hit on B}}$ | $\hat{P}(a_2b_2|A_2B_2)$ | $\hat{P}(a_2b_2|A_1B_2)$ | 0.25 | 0.81 |
| $d'_{A|\text{miss on B}}$ | $\hat{P}(a_2b_1|A_2B_2)$ | $\hat{P}(a_2b_1|A_1B_2)$ | 1.04 | 0.26 |
| $d'_{A|\text{CR on B}}$ | $\hat{P}(a_2b_1|A_2B_1)$ | $\hat{P}(a_2b_1|A_1B_1)$ | 1.18 | 1.30 |
| $d'_{A|\text{FA on B}}$ | $\hat{P}(a_2b_2|A_2B_1)$ | $\hat{P}(a_2b_2|A_1B_1)$ | 1.75 | 0.95 |

**Theorem 6.2** Assume decisional separability holds on both dimensions.

(i) If perceptual independence holds for stimuli $A_1B_2$ and $A_2B_2$, then

$$d'_{\text{A|miss on B}} = d'_{\text{A|hit on B}}. \tag{6.11}$$

(ii) If perceptual independence holds for stimuli $A_1B_1$ and $A_2B_1$, then

$$d'_{\text{A|CR on B}} = d'_{\text{A|FA on B}}. \tag{6.12}$$

Analogous results hold for the microanalytic $d'$s that measure the discriminability difference between levels $B_1$ and $B_2$ of component B when component A is at level $A_i$ ($i = 1$ or 2).

*Proof* See Kadlec and Townsend (1992a). ■

Therefore, if decisional separability holds then a violation of either equations 6.11 or 6.12 implies at least one violation of perceptual independence. In the figure 6.1 GRT model, decisional separability holds on both dimensions, and perceptual independence holds for all stimuli except $A_2B_2$. Therefore, theorem 6.2 predicts that equation 6.11 should be violated by this model, whereas equation 6.12 will be satisfied. Note from figure 6.6 that when the microanalytic $d'$s are defined in terms of the population parameter values, then both predictions are correct—that is, $d'_{\text{A|hit on B}} > d'_{\text{A|miss on B}}$ and $d'_{\text{A|CR on B}} = d'_{\text{A|FA on B}}$. Unfortunately though, when these $d'$s are estimated from the population confusion matrix, the story complicates a bit. These estimated $d'$s are shown in the rightmost column of table 6.8. Note that in accordance with the theorem, $\hat{d}'_{\text{A|hit on B}}$ is substantially different from $\hat{d}'_{\text{A|miss on B}}$ (by a factor of more than 3-to-1), which strongly suggests a violation of perceptual independence in either stimulus $A_1B_2$ or $A_2B_2$ or in both. The problem though is that the difference between $\hat{d}'_{\text{A|CR on B}}$ and $\hat{d}'_{\text{A|FA on B}}$ is 0.35, whereas the theorem predicts equality. The problem, which is the same problem we encountered with the macroanalytic $d'$ test of perceptual separability, is that the estimation of all of these marginal $d'$s assumes the normal, equal-variance SDT model (because it uses equations 2.11 and 2.13). The figure 6.1 model violates the equal-variance assumption, and so the resulting $d'$ estimates are biased.

Note that theorem 6.2 assumes that decisional separability holds on both dimensions. In general, this is a requirement for virtually all summary statistics tests of perceptual separability and perceptual independence. Theorem 6.1 did not explicitly assume decisional separability, but even so, we saw that accurate estimates of $d'_{\text{A|B}_1}$ and $d'_{\text{A|B}_2}$ did require decisional separability to hold. However, if decisional separability does hold, then theorem 6.2 guarantees that if, for example, $d'_{\text{A|miss on B}} \neq d'_{\text{A|hit on B}}$, then we know that perceptual independence is violated for at least one of the stimuli $A_1B_2$ and $A_2B_2$. It could be violated in either one and not the other, or in both stimuli. Supplementing this analysis with tests of report independence could narrow these alternatives.

As one might expect, the reverse inference is not valid. It is logically possible that the microanalytic $d'$s are equal but perceptual independence is violated in one or both relevant stimuli. One way this could happen is because decisional separability fails. But even if decisional separability holds, the microanalytic $d'$s could be equal because the perceptual dependence is the same in both stimuli. For example, consider again figure 6.6. Note that $x_1$ and $x_2$ are positively correlated for stimulus $A_2B_2$ and statistically independent for stimulus $A_1B_2$. The positive correlation causes the $A_2B_2$ marginal $x_1$ distribution to have a greater mean for samples above the horizontal bound than for samples below this bound. In contrast, because perceptual independence is satisfied for stimulus $A_1B_2$, the analogous marginal distributions are equal. But suppose instead that the correlation between $x_1$ and $x_2$ on $A_1B_2$ trials was the same as on $A_2B_2$. Then the marginal distributions for stimulus $A_1B_2$ would be shifted by the same amount as the marginal distributions for stimulus $A_2B_2$. In this case the two microanalytic $d'$s would be equal—that is, $d'_{\text{A|hit on B}} = d'_{\text{A|miss on B}}$—despite the violations of perceptual independence.

In summary, the SDT-inspired summary-statistics tests that are based on some type of marginal $d'$ offer an elegant and familiar approach to testing perceptual separability and perceptual independence. Their greatest weakness is in the difficulty of estimating the requisite marginal $d'$s from the available confusion matrix. The very limited computational investigations reported here seem to suggest that if the confusion matrix is generated from a model that violates assumptions of the normal, equal-variance SDT model, then the resulting marginal $d'$ estimates will often be biased. As a result, the best option is probably to supplement these tests with tests of marginal response invariance and report independence.

## 6.3 RESPONSE-TIME TESTS

A number of different summary-statistics tests of perceptual separability and perceptual independence use the observer's response times (RTs), rather than their response accuracy. Statistical decision theory does not make enough assumptions to allow RT predictions to be made in any experimental task. As a result, extra assumptions are required to derive RT predictions from SDT or GRT. The primary advantage of this approach is that adding RTs to the empirical database greatly increases the degrees of freedom that are available for hypothesis testing, which should thereby increase our confidence in whatever inferences we make from our summary-statistics tests.

There have been a number of independent efforts to extend both SDT and GRT enough so that they make RT predictions. We briefly considered one of these in section 2.7. Chapter 10 reviews these approaches in detail. This chapter focuses on the summary-statistics tests of perceptual separability and perceptual independence that resulted from those extensions. See chapter 10 for a deeper discussion of the theoretical underpinnings and the more general RT predictions of each approach.

### 6.3.1 RT TESTS OF PERCEPTUAL SEPARABILITY

The first RT-based summary-statistics tests were created by adding one straightforward assumption to GRT—namely the same RT-distance hypothesis that we used in chapter 3 (section 3.1.4.1) to derive RT predictions from various recognition memory models. Recall that this hypothesis assumes that RT decreases monotonically with the distance from the percept to the nearest decision bound. The idea is that if decisions are made by comparing a percept to a decision bound or criterion, then the greater the distance between the two, the easier, and hence the faster the decision. As one might expect, there is considerable empirical support for this assumption (Ashby et al., 1994; Emmerich et al., 1972; Gescheider et al., 1969). Adding the RT-distance hypothesis to GRT allows the model to make accuracy and RT predictions in a wide variety of experimental tasks. This more general version of GRT is considered in detail in chapter 10. This section considers only the summary-statistics tests that this generalization makes possible.

Before GRT, the most popular method for assessing separability was via the *filtering task*, which uses the same stimuli as the $2 \times 2$ factorial identification task, but asks observers only to report the level of component A or the level of component B, rather than identify the stimulus uniquely (Garner, 1974). So in the filtering task, observers make categorization decisions, rather than identification responses. Ashby and Maddox (1994) proposed an RT version of marginal response invariance for this task that they called *marginal RT invariance*. Specifically, marginal RT invariance holds for component $A_i$ (for $i = 1$ or 2) if

$$P(RT_{i.} \leq t|A_iB_1) = P(RT_{i.} \leq t|A_iB_2), \text{ for all } t > 0, \tag{6.13}$$

where $RT_{i\cdot}$ is the RT on trials when the observer responds that stimulus component A is at level $i$. So whereas marginal response invariance requires that the probability that component A is reported to be at level $A_i$ is the same on $A_iB_1$ and $A_iB_2$ trials, marginal RT invariance requires instead that the RT probability distributions are equal on these trials. Ashby and Maddox (1994) showed that marginal RT invariance provides a powerful test of perceptual separability. Specifically, they proved the following result.

**Theorem 6.3** Consider a filtering task in which the observer is asked to categorize each stimulus according to the level of component A. If decisional separability holds and the RT-distance hypothesis is valid, then component A is perceptually separable from component B if and only if marginal RT invariance is satisfied for both correct and incorrect responses. Furthermore, this result holds even if a variable motor time is added to the time required for each categorization decision.

*Proof* See Ashby and Maddox (1994). ■

In other words, if decisional separability and the RT-distance hypothesis are both valid, then marginal RT invariance is a perfect predictor of perceptual separability. If marginal RT invariance holds then perceptual separability must also hold, and if marginal RT invariance is violated then so too is perceptual separability. Statistically, it is straightforward to test the null hypothesis that marginal RT invariance holds via the (nonparametric) two-sample Kolmogorov-Smirnov test.

Note that marginal RT invariance provides a much stronger test of perceptual separability than marginal response invariance. As we saw earlier, the combination of decisional separability and marginal response invariance is not enough to guarantee perceptual separbility. Marginal RT invariance provides a stronger test because the equation 6.13 equality must hold for all times $t$. If there is even one time $t$ for which the equality is violated then marginal RT invariance fails. In contrast, marginal response invariance (i.e., equation 6.1) only requires one single equality to hold. A similar strong result could be obtained with accuracy data if marginal response invariance was required to hold for all possible intercepts of the vertical decision bound (i.e., in the case where the observer's task is to report the level of component A).

Ashby and Maddox (1994) only investigated tasks with two response alternatives—specifically, the filtering and redundancy tasks popularized by Garner (1974). Townsend et al. (2012) applied a similar approach to the $2 \times 2$ factorial identification task. They defined an RT invariance condition similar to marginal RT invariance that they called *timed marginal response invariance*. This condition holds in the $2 \times 2$ identification task for level $i$ of component A if, for all $t > 0$

$$P(RT_{i1} \leq t|A_iB_1) + P(RT_{i2} \leq t|A_iB_1) \\ = P(RT_{i1} \leq t|A_iB_2) + P(RT_{i2} \leq t|A_iB_2). \quad (6.14)$$

Rather than assume the RT-distance hypothesis, Townsend et al. (2012) investigated predictions of a general class of models that assumes processing of the two stimulus components occurs in parallel. Specifically, they assumed separate channels for each level of each stimulus component. So in the $2 \times 2$ factorial identification experiment, their model would include two component A channels—one tuned to component $A_1$ and one tuned to component $A_2$—and two component B channels. Next, they assumed that on each trial the two component A channels race against each other, and so do the two component B channels. The level of component A is determined by whichever component A channel finishes first, and the level of component B is determined by the winner of the component B race. The overall identification response must wait for the slower of these two races to complete. So identification of the level of each component is parallel and self-terminating, whereas the overall identification response is parallel and exhaustive. Finally, Townsend et al. (2012) assumed that a variable motor time is added to this parallel, exhaustive identification time to determine observable RT. Within this general class of parallel, stochastic models, Townsend et al. (2012) showed that if perceptual and decisional separability hold then timed marginal response invariance must also hold.

The parallel models considered by Townsend et al. (2012) assume that activation in each channel is proportional to the evidence that the relevant stimulus component is at the level to which the channel is tuned (Townsend & Wenger, 2004; Wenger & Townsend, 2006). This approach is considered in detail in chapter 10. Note though that in GRT and SDT, if the likelihood ratio is monotonic, then evidence increases with distance from boundary (or criterion). As a result, assuming that activation in a channel is proportional to evidence is similar to assuming the RT-distance hypothesis, and so the parallel models considered by Townsend et al. (2012) are closely related to the RT-distance models investigated by Ashby and Maddox (1994).

Given this similarity, it is not surprising that marginal RT invariance and timed marginal response invariance are closely related. In fact, as part of their motivation for proposing marginal response invariance as a test of perceptual separability,

Townsend et al. (2012) highlighted the equivalence relation

$$P(RT_{i1} \leq t|A_iB_j) + P(RT_{i2} \leq t|A_iB_j) = P(RT_{i.} \leq t|A_iB_j). \tag{6.15}$$

Note that the left side of this equation is identical to the left side of equation 6.14 when $j = 1$ and the right side of this equation is identical to the left side of equation 6.13 when $j = 1$. Marginal RT invariance assumes the left side of equation 6.15 is the same, regardless of whether $j = 1$ or 2. Therefore, timed marginal response invariance is based on the assumption that marginal RT invariance is valid. Even so, it is important to note that the two tests were developed for different tasks. Marginal RT invariance was proposed as a test of RTs collected in a filtering task, whereas timed marginal response invariance was proposed as a test of RTs collected in a $2 \times 2$ factorial identification task. So timed marginal response invariance could be seen as an extension of marginal RT invariance to the $2 \times 2$ factorial identification experiment.

Note that this equivalence strengthens both tests because it means that neither test depends strongly on the RT assumptions that they make. In other words, marginal RT invariance and/or timed marginal response invariance are valid so long as either the RT-distance hypothesis is valid *or* the parallel, stochastic processing assumptions of the Townsend et al. (2012) model are satisfied.

### 6.3.2 RT TESTS OF PERCEPTUAL INDEPENDENCE

There are also RT tests of perceptual independence that assume either the RT-distance hypothesis or the parallel, stochastic processing model considered in the previous subsection. First, Ashby and Maddox (1991) proposed a test that assumed the decision time on each trial satisfies the RT-distance hypothesis. This test required data from an experiment that includes three conditions: (1) a filtering condition in which observers respond with the level of component A; (2) a filtering condition in which observers respond with the level of component B; and (3) an identification condition in which observers identify the stimulus uniquely. Also, the test is valid only if the motor time (i.e., the time for all nondecision processing) has no trial-by-trial variability. As a result, the Ashby and Maddox (1991) RT test of perceptual independence was far from optimal. Not only did it require augmenting the standard $2 \times 2$ factorial identification task with two extra filtering conditions, but it also required the unreasonable assumption that motor-time variability is negligible.

Townsend et al. (2012) solved both of these problems with a test they called *timed report independence*. This test only requires data from the standard $2 \times 2$

factorial identification experiment. Given such data, timed report independence holds for stimulus $A_iB_j$ if and only if

$$P(RT_{ij} \leq t|A_iB_j)P(RT_{..} \leq t|A_iB_j) = P(RT_{i.} \leq t|A_iB_j)P(RT_{.j} \leq t|A_iB_j). \tag{6.16}$$

The two terms on the right are defined by equation 6.15 and its component B analogue. The term $P(RT_{..} \leq t|A_iB_j)$ is the unconditional probability that the RT is less than or equal to $t$ on trials when stimulus $A_iB_j$ is presented. So to estimate this probability, one would collect all RTs on $A_iB_j$ trials (i.e., ignoring which response was made) and then for each time $t$, estimate the proportion of RTs that are less than or equal to this value of $t$. Given this definition, Townsend et al. (2012) proved the following result.

**Theorem 6.4** If (1) decisional separability holds on both stimulus dimensions; (2) perceptual independence is satisfied for stimulus $A_iB_j$; and (3) RT is determined by the parallel, stochastic processing assumptions described above; then timed report independence holds for stimulus $A_iB_j$.

*Proof* See Townsend et al. (2012). ∎

So assuming that decisional separability holds on both dimensions and that the parallel, stochastic processing model is valid then a violation of timed report independence for stimulus $A_iB_j$ implies that perceptual independence is violated for this stimulus.

## 6.4 OTHER TESTS

A few other summary-statistics tests have been developed, but since they have received little or no empirical application, they are described only briefly in this section.

R. D. Thomas (1996) proposed summary-statistics tests of perceptual separability and perceptual independence that used data from a variety of alternative same-different tasks. In each of these, two stimuli are presented on each trial, and the observer's task is to respond SAME or DIFFERENT depending on whether or not the stimuli are identical. In some conditions, a SAME response requires that the stimuli only match on one prespecified dimension, whereas in other conditions a SAME response requires a match on all dimensions. GRT models of this task are described in chapter 9 (section 9.3). R. D. Thomas (1996) proposed summary-statistics tests of perceptual separability and perceptual independence that were based on comparisons of performance in the different experimental conditions. All of these tests assumed decisional separability.

Perrin and Ashby (1991) proposed a summary-statistics test of perceptual independence that uses data from a similarity-rating experiment. This test is potentially quite important because it is the only existing summary-statistics test that does not assume decisional separability. The test uses the same stimuli as the $2 \times 2$ factorial identification experiment, except rather than presenting one stimulus on each trial and asking for an identification response, two stimuli are shown and the observer is asked to rate their dissimilarity (e.g., on a 1 to 7 rating scale). Chapter 9 describes applications of GRT to similarity-rating experiments in some detail (see section 9.1).

The $2 \times 2$ factorial similarity-rating experiment is described in figure 6.7. Note that with these four stimuli, there are six possible stimulus pairs, and therefore six possible trial types. The Perrin and Ashby (1991) summary-statistics test of perceptual independence uses a property of dissimilarity data proposed by Tversky and Gati (1982) that they called the dominance axiom. This axiom states that the dissimilarities associated with the two diagonal distances in figure 6.7 must be larger than the largest of the dissimilarities associated with the four one-dimensional distances (i.e., the two horizontal and two vertical distances). Perrin and Ashby (1991) proved that if perceptual independence holds and all perceptual distributions

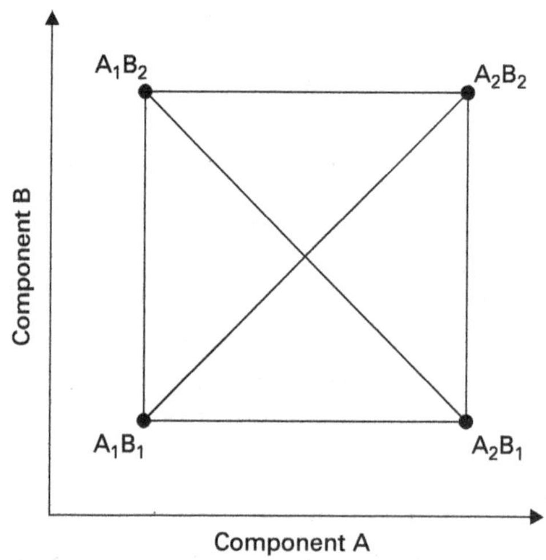


**6.7** The $2 \times 2$ factorial similarity-rating experiment.

have equal variance-covariance matrices, then the GRT-similarity model (described in section 9.1.1) predicts that dominance must be satisfied. So assuming that the conditions of this theorem are valid, a violation of dominance implies that perceptual independence is violated. Perrin and Ashby (1991) also reported that this test is robust with respect to violations of the equal variance-covariance matrix assumption. For example, the test remained valid so long as the ratio of the largest variance on a given dimension to the smallest variance on the same dimension did not exceed 5. This test assumes optimal responding, rather than decisional separability.

## 6.5 CONCLUSIONS

The various summary-statistics tests that were the primary focus of this chapter are listed in Table 6.9. Note that these tests are of three types. Marginal response invariance and report independence are nonparametric tests because they make no parametric assumptions about the perceptual distributions and therefore are valid for any family of distributions. The macro- and microanalytic $d'$ tests are parametric because they use estimators of $d'$ that assume underlying normal distributions with equal variance. Our limited computational investigations suggest these tests might be quite sensitive to violations of homogeneity of variance. To my knowledge though, no one has investigated how robust they are with respect to violations of normality. Finally, marginal RT invariance, timed marginal response invariance, and timed report independence use the observer's RT (rather than accuracy) to test for perceptual interactions. These tests require extra assumptions because GRT lacks the structure necessary for RT predictions. Even so, we saw that the tests are at least somewhat robust with respect to violations of these assumptions. This is because the tests derived from different RT assumptions are essentially equivalent.

**Table 6.9** Summary-statistics tests of perceptual separability (PS) and perceptual independence (PI).

| | Nonparametric | Parametric | RT |
|---|---|---|---|
| PS | MRI | Macroanalytic $d'$ test | Marginal RT invariance<br>timed MRI |
| PI | Report independence | Microanalytic $d'$ test | Timed report independence |

*Note*: MRI = marginal response invariance.

The strongest summary-statistics tests of perceptual separability or perceptual independence will use a converging operations approach that combines two or more of these tests. Because each type of test makes somewhat different assumptions, an analysis in which multiple tests agree about the perceptual separability of component A from B or about the perceptual independence of stimulus $A_2B_2$, for example, should raise our confidence in these conclusions. Even so, it is always important to keep in mind that all of the tests mentioned in table 6.9 assume decisional separability, and that the summary-statistics approach provides no way to test this assumption. As a result, if there is any doubt at all about whether decisional separability is satisfied, then the model-fitting approach described in the next chapter might be a better data-analysis choice.

# 7

# FITTING THE GAUSSIAN GRT MODEL TO IDENTIFICATION DATA

GRT can be used to learn about perceptual and decisional processing in two fundamentally different ways. One, which was the focus of the previous chapter, is to test assumptions about perceptual separability and perceptual independence using a variety of different summary statistics. A second approach, which is the focus of this and the next chapter, is to fit a variety of different GRT models directly to the confusion matrices that describe the observer's responses. Assumptions about perceptual and decisional processes can then be tested by comparing model fits of nested models in which the restricted model makes some specific assumption, such as perceptual separability, and the more general model does not (R. D. Thomas, 2001). The assumption is supported if the more general model does not fit significantly better than the restricted model. These two approaches can be used to complement each other. For example, Ashby and Perrin (1988) used the results of a variety of summary statistics tests to select candidate GRT models for model fitting.

GRT models are commonly fit to confusion matrices from both identification and categorization experiments. This chapter focuses on applications to identification data, and chapter 8 considers applications to categorization experiments. When fit to categorization data, GRT models are typically referred to as *decision-bound models*. As we will see in chapter 8, the most common goal here is to identify the type of decision strategy that participants used in the task. Decision-bound models typically make strong assumptions about the perceptual representation of each stimulus, and as a result, fitting decision-bound models to categorization data is a simple computational process. In particular, unlike model fits to identification data, no numerical integration is needed to fit decision-bound models to confusion matrices from a categorization experiment.

Compared to the summary-statistics approach, model fitting has its own advantages and disadvantages. The summary-statistics approach has at least two advantages over the model-fitting approach. First, and most significantly, computing summary statistics is less computationally intensive than model-fitting. As we will see, this is because model fitting requires an iterative search through a parameter space that is often high dimensional and each iteration frequently requires numerical integration. Even so, fast computers and the ready availability of a variety of free software packages make this concern much less serious than it was 20 years ago. Second, nonparametric summary-statistics tests are available, whereas model fitting is necessarily parametric. For example, all GRT model fitting to date has assumed that the underlying perceptual distributions are multivariate normal, whereas some summary statistic tests—most notably, tests of marginal response invariance and report independence—are nonparametric. On the other hand, I know of no empirical evidence that a nonparametric approach should be considered more valid than one that assumes normality.

In contrast, the model-fitting approach has many advantages over summary-statistics testing. First, although model fitting is necessarily parametric, it makes fewer assumptions about perceptual and decisional processes, and therefore should be less prone to false conclusions. For example, as we saw in chapter 6, virtually all summary-statistics tests of perceptual separability and perceptual independence require an assumption of decisional separability. If decisional separability is violated then these tests are all invalid. This is problematic since there are no known summary-statistic tests of decisional separability. In contrast, model fitting allows a variety of different decision strategies to be investigated and because of this, it allows for the assumption of decisional separability to be tested statistically.

Second, the model-fitting approach makes inferences about perceptual and decisional processes that are based on fitting all of the data simultaneously. In contrast, each summary-statistics test is based on only a subset of cells in the confusion matrix.

Third, and perhaps most important, a successful model-fitting analysis provides a much more detailed description of perceptual and decisional processing than a summary-statistics analysis. Specifically, a model fit will provide estimates of all perceptual representations, including means, variances, and covariances, as well as the decision bounds that best account for the observer's responses. In contrast, at best, a summary-statistics analysis provides only binary decisions about perceptual properties—for example, that perceptual separability was satisfied on one dimension but not on the other, and that perceptual independence was satisfied in three stimuli and violated in one.

Fitting GRT models to identification or categorization data is necessarily parametric since numerical predictions are possible only if a specific functional form is specified for the perceptual distributions. All previous applications of this approach have assumed that the perceptual distributions are multivariate normal. Furthermore, almost all applications have assumed that there are only two relevant sensory dimensions. In principle, the fitting algorithms (described below) are straightforward to extend to more than two dimensions, but such models could include many more free parameters and therefore significantly increased computation time. Thus, to date, applications that have fit GRT models to identification or categorization confusion matrices have typically assumed that the sensory distributions are bivariate normal pdfs, and the response regions are defined on a plane. Even so, chapter 9 describes an application in which a GRT model was fit to data from a task where the stimuli varied on six perceptual dimensions (see section 9.2).

All model-fitting algorithms follow similar steps. In step 1, initial guesses are made about the numerical values of all free parameters in the model. Step 2 inserts these guesses into the equations that define the model, which generates a predicted confusion matrix that specifies the numerical values of all response probabilities that should occur if the model is correct and the parameters have the chosen values. In step 3, the predicted and observed confusion matrices are compared and a numerical goodness-of-fit score is computed. Different methods use different measures of goodness-of-fit. For example, the method of least squares uses the sum of squared errors to determine how well the model accounts for the data, where the error is defined as the difference between the predicted and observed scores. However, in statistics, the method of maximum likelihood is seen as the superior method. This approach is described in the next section. In step 4, the parameter estimates are changed in some way and the whole process is repeated. Step 5 compares the new goodness-of-fit score (from step 4) to the old goodness-of-fit score. If the new score is better, it replaces the old score, and the new parameter estimates replace the old parameter estimates. Step 6 repeats steps 4 and 5 over and over until a decision is made that no better goodness-of-fit score is possible. At that point, the process terminates and the final goodness-of-fit score and parameter estimates are reported.

Steps 1 and 2 are unique to the specific model being fitted to the data, and, as a result, these steps must be programmed by hand. All other steps however, are essentially identical for all models, and therefore can use preprogrammed software. These are called optimization routines since their goal is to find parameter values that minimize differences between predicted and observed confusion matrices. Matlab has a whole toolbox filled with such routines (called the optimization toolbox). One popular choice is called "fmincon." Appendix B includes Matlab code that uses fmincon to fit several different decision-bound models to categorization data. See chapter 8 for details. In Excel, the optimization routine is called "solver." The different optimization routines differ primarily in how they change the parameter values from one iteration to the next. For example, some algorithms approximate a gradient descent search through parameter space by moving in the direction (i.e., during step 4) that causes goodness-of-fit to improve most quickly. These algorithms are highly efficient, in the sense that they can usually find the global minimum in the fewest number of steps, but they require partial derivatives of each parameter value with respect to the goodness-of-fit measure. In more complex models, these partial derivatives are usually analytically intractable, so other less clever algorithms are usually used.

This and the next chapter describe the model-fitting process in detail, with a focus on step 2. We begin by describing applications of the method of maximum likelihood to the problem of fitting GRT models either to identification or categorization data.

## 7.1 MODEL FITTING VIA THE METHOD OF MAXIMUM LIKELIHOOD

When a GRT model is fit to categorization or identification data, the best-fitting values of all free parameters must be found. The standard approach to model fitting uses the method of maximum likelihood in which numerical values of all parameters are found that maximize the likelihood of the data given the model. The idea is that we have faith in our experiment and therefore believe that the data we collected are representative or typical of the phenomenon we are studying. Therefore, when estimating the parameters of some model, it makes sense to choose estimates that cause the model to predict that our results are typical (and therefore likely), rather than atypical, given the experiment we ran and assuming that the parameters we are estimating are from the correct model of the phenomenon under study. For example, if we collect data with a sample mean of 50, then these data are most likely if the population mean is also 50. The samples could be from a population with mean 60, but if they were then our results would be unusual (i.e., unlikely) because we would expect most samples from this population to have a sample mean close to 60. For this reason, the sample mean is the maximum likelihood estimator of the population mean.

Let $S_1, S_2, \ldots, S_{N_S}$ denote the $N_S$ stimuli in the block of data to be modeled and let $R_1, R_2, \ldots, R_{N_R}$ denote the $N_R$ alternative responses. In identification experiments $N_R = N_S$, whereas in categorization experiments $N_R < N_S$. Let $r_{ij}$ denote the frequency with which the participant responded $R_j$ on trials when stimulus $S_i$ was presented. Note that the $r_{ij}$ are random variables. For any particular stimulus, the $r_{ij}$ have a multinomial distribution. When we fit a model to data, our null hypothesis is that the model is correct. If so, then the true probability $P(R_j|S_i)$ that response $R_j$ is given on trials when stimulus $S_i$ was presented can be computed from the equations that describe the model we are fitting to the data. In particular, once we set numerical values for each free parameter in the model, then the model's equations can be used to compute a predicted probability for every cell of our confusion matrix. Given these predictions, then the predicted probability of observing the response frequencies $r_{i1}, r_{i2}, \ldots, r_{iN_R}$ equals

$$P[r_{i1}, r_{i2}, \ldots, r_{iN_R}|S_i] = \frac{n_i!}{r_{i1}!r_{i2}!\ldots r_{iN_R}!} P(R_1|S_i)^{r_{i1}} P(R_2|S_i)^{r_{i2}} \ldots P(R_{N_R}|S_i)^{r_{N_R}} \tag{7.1}$$

where $n_i$ is the total number of times that stimulus $S_i$ was presented during the course of the experiment. The probability or joint likelihood of observing the entire dataset is the product of the probabilities of observing the various responses to each stimulus; that is,

$$\begin{aligned} L &= \prod_{i=1}^{N_S} P[r_{i1}, r_{i2}, \ldots, r_{iN_R}|S_i] \\ &= \prod_{i=1}^{N_S} \frac{n_i!}{\prod_{j=1}^{N_R} r_{ij}!} \prod_{j=1}^{N_R} P(R_j|S_i)^{r_{ij}}. \end{aligned} \tag{7.2}$$

Note that if we change the numerical value of any model parameters then at least some of the predicted probabilities [i.e., the $P(R_j|S_i)$] will change, and therefore, so will $L$. The method of maximum likelihood says that the best guesses for all the unknown parameter values are those specific values that make $L$ as large as possible—that is, those values that make our observed data as likely as possible.

Note that the first term in line two of Eq. 7.2 (i.e., the first product) does not depend on the values of any model parameters. Rather, it only depends on the data. Thus, the parameter values that maximize the second term (i.e., the second product) in this line also maximize the whole expression. For this reason, the first term can be ignored during the parameter estimation process. Another common practice is to take logs of both sides of equation 7.2. This is done because some of the $P(\mathrm{R}_j|\mathrm{S}_i)$ could be very small numbers. Since log is a monotonic (i.e., order-preserving) transformation, the parameter values that maximize $L$ will also maximize $\ln L$. So, the standard approach is to find values of the free parameters that maximize

$$\ln L^* = \sum_{i=1}^{N_S} \sum_{j=1}^{N_R} r_{ij} \ln P(\mathrm{R}_j|\mathrm{S}_i). \tag{7.3}$$

This equation will generate negative values (the maximum possible value is zero), but the maximum likelihood estimators are still those values of the parameters that make ln $L^*$ as large as possible (i.e., as close to zero as possible). Most optimization routines, including fmincon, find parameter estimates that minimize, rather than maximize the user-supplied objective function. In these cases, $-\ln L^*$ is used to find the maximum likelihood estimates, rather than $\ln L^*$. This works just as well though because the parameter values that maximize the likelihood of our data are also the parameter values that maximize $\ln L^*$ and that minimize $-\ln L^*$.

Although finding the values of the free parameters that maximize equation 7.3 produces maximum likelihood estimates of all model parameters, this equation is not a good choice for deciding which model provides the best account of the data. This is because adding more parameters to a model can never cause a decrease in $\ln L^*$. Therefore, to decide which model provides the most parsimonious account of the data, it is vital to choose a goodness-of-fit measure that penalizes models for extra free parameters (e.g., Myung et al., 2016). The current consensus in the literature is to choose the model that has the smallest Bayesian information criterion (BIC) score:

$$BIC = r \ln N - 2 \ln L^*, \tag{7.4}$$

where $N$ is the sample size, $r$ is the number of free parameters, and $\ln L^*$ is as in equation 7.3 (Schwarz, 1978). Note that for each given model, $r$ and $N$ are fixed, so the parameter estimates that maximize $\ln L^*$ in equation 7.3 or that minimize $-\ln L^*$ will also minimize BIC in equation 7.4. So equations 7.3 and 7.4 will lead to exactly the same parameter estimates, but the BIC values can also be used to compare different models. Note that the BIC statistic penalizes a model for bad fit and for extra free parameters. Specifically, decreasing $\ln L^*$ or increasing $r$ both increase BIC. Therefore, to find the best model among a set of competitors, one simply computes a BIC value for each model and then chooses the model with the smallest BIC.

As an example, consider two GRT models of a $2 \times 2$ factorial identification experiment. Suppose one model assumes that decisional separability holds on both dimensions, whereas the other assumes that decisional separability holds on dimension 1 but not on dimension 2. Specifically, both models assume that the decision bound is a vertical line on dimension 1, whereas they differ about dimension 2. The first model assumes that the decision bound on dimension 2 is always a horizontal line, whereas the second model assumes it is a line with arbitrary slope and intercept (see section 5.1 for details). Note that the two models are identical, except that the first model needs only two parameters to describe its decision bounds (i.e., a criterion or intercept on each dimension), whereas the second model needs three parameters (a criterion on dimension 1 and a slope and intercept on dimension 2). Suppose that the parameter-estimation algorithm reports a final BIC value of 205 for the best-fitting version of the simpler model, and 208 for the best best-fitting version of the more complex model. Then the conclusion would be that the simpler model provides the best account of these data, and therefore that decisional separability holds on both dimensions. Note though that the more complex model can never fit worse than the simpler model in an absolute sense, because the more complex model could always set the slope of its decision bound on dimension 2 to zero. In this case, the BIC statistic is suggesting that despite this fact, the best account of the data is provided by the simpler model that assumes decisional separability because the absolute fits of the simple and complex models are almost identical (i.e., the second term in equation 7.4) but the simple model has fewer free parameters and therefore incurs a smaller penalty (i.e., the first term in equation 7.4). Thus, BIC implements a parsimony criterion. The simpler model predicts that the decision bound on dimension 2 must be horizontal. The more complex model assumes only that the decision bound on dimension 2 is linear. Therefore, if the data show evidence that the decision bound on dimension 2 is horizontal, then BIC rewards the model that predicted that this is the only possible outcome.

The BIC values identify which model provides the best account of the participant's responses, but this fact alone does not indicate whether the fit was good or bad. It is possible that all models provided poor fits and the best-fitting model just happened to provide the *least* poor fit. Unfortunately, the numerical value of the raw BIC score does not help with this problem because BIC scores increase with sample size, regardless of the quality of fit.

Any GRT model will provide a poor fit to randomly generated data. With random data, a model that assumes random responding (i.e., random guessing) should provide the best fit. So one way to assess how well a GRT model fits the data is to compare its fit to the fit of a model that assumes random guessing (described in the next paragraph). Bayesian statistics includes a method to make such comparisons via the so-called Bayes factor. If the prior probability that a GRT model, denoted by $M_{GRT}$, is correct is equal to the prior probability that the guessing model $M_G$ is correct, then under certain technical conditions (e.g., Raftery, 1995), it can be shown that

$$P(M_{GRT}|\text{Data}) \doteq \frac{1}{1+\exp\left[-\frac{1}{2}(BIC_G - BIC_{GRT})\right]}, \tag{7.5}$$

where $P(M_{GRT}|\text{Data})$ is the probability that the GRT model is correct, assuming that either the GRT model or the guessing model are correct, and $\doteq$ means "is approximately equal to." Thus, for example, if the GRT model is favored over the guessing model by a BIC difference of 2, then the probability that the GRT model is correct is approximately .73. In other words, even though the GRT model fits better than the guessing model, the fit is not very good because there is a better than 1-in-4 chance that the data were just generated by random coin tossing. In contrast, if the BIC difference is 10, then the probability that the GRT model is correct is approximately 0.99, which means that we can be very confident that this participant was consistently using a single decision strategy that is at least reasonably well described by our GRT model. In this case, the GRT model provides an excellent fit to the data.

In an identification experiment with $N_S$ stimuli (and therefore $N_S$ response alternatives) and random guessing, $P(R_j|S_i) = 1/N_S$ for all $i$ and $j$. Therefore, note that, in this case, equation 7.3 reduces to

$$\ln L^* = -\sum_{i=1}^{N_S}\sum_{j=1}^{N_R} r_{ij} \ln N_S = -N_T \ln N_S, \tag{7.6}$$

where $N_T$ is the total number of trials used to estimate the confusion matrix. This guessing model has no free parameters, and, as a result, its BIC is

$$BIC = 2N_T \ln N_S. \tag{7.7}$$

When BIC is chosen as the goodness-of-fit measure, the maximum likelihood estimators of the parameters are those numerical values that minimize equation 7.4. These values are found numerically using any one of the many available optimization algorithms mentioned earlier. In the case of any GRT model, the user would: (1) write code that computes all of the relevant predicted response probabilities for any given set of numerical parameter values, (2) select initial guesses for all parameters, and (3) specify BIC as the goodness-of-fit measure. The algorithms would then use the user-provided code to generate an initial fit value from those initial guesses (e.g., an initial numerical value for BIC as defined in equation 7.4), and then proceed to change the parameter guesses in some systematic way until no better fit can be found. At this point, the routine stops and reports the parameter estimates that led to the smallest possible value of BIC as well as the numerical value of this best fit. Fitting any GRT model, either to identification or categorization data, is therefore a matter of deriving equations or at least an algorithm that makes it possible to generate numerical predictions for the various $P(R_j|S_i)$ specified in equation 7.3. This problem is solved for fits to identification data in section 7.3, and for fits to categorization data in section 8.2.

We saw in chapter 4 that an identification confusion matrix has $N_S \times (N_S - 1)$ degrees of freedom, where $N_S$ is the number of stimuli. This value places an upper limit on the number of free parameters in any model of the data that can be tested. As a result, the experimenter's choice of $N_S$ is a trade-off. A large $N_S$ guarantees many degrees of freedom, which eliminates the need to make simplifying assumptions when building the models and provides extra degrees of freedom to test the validity of competing models. But a large $N_S$ also means that the data require much more time to collect. Not only are there more stimuli to present, but there are also more stimulus-response associations for the participant to learn. The goal of most identification experiments is to study how the stimuli are perceived. As a result, any error that participants make because they don't remember which response is associated with some stimulus is just a nuisance. Furthermore, GRT is not a model of learning, and therefore should only be applied to stationary data. For these reasons, more time must often be spent training participants when $N_S$ is large, and the data from this training period usually must be discarded.

A bivariate normal distribution has a maximum of five free parameters—a mean and variance on both dimensions, and a covariance. Therefore, the smallest value of $N_S$ for which the most general possible GRT model can be estimated is $N_S = 6$. In this case, the $6 \times 6$ confusion matrix has 30 degrees of freedom, and the six perceptual distributions needed to model the perceptual effects of the six stimuli have 30 parameters. However, the origin and unit of measurement on each perceptual dimension are arbitrary. Therefore, without loss of generality, the means on both dimensions can be set to zero in any one perceptual distribution (to set the origin) and the variances in that distribution can be set to one (to set the unit of measurement on each dimension). This reduces the number of free parameters to 26 (i.e., to $5N_S - 4$), which leaves a maximum of four parameters to model the decision bounds and assess the validity of the model. The number four rules out some decision models (e.g., the general quadratic classifier), but not all. For example, the minimum distance and optimal classifiers have no free decision parameters, and the model that assumes decisional separability has only one free parameter for each decision bound (i.e., the intercept).

As the order of the confusion matrix increases above six, the degrees of freedom increases faster than the number of free parameters in the full GRT model. As a result, the larger the matrix, the more extra degrees of freedom there are to estimate decision-bound parameters and to test the validity of the model. For example, Ashby et al. (2001) fit the full GRT model to a variety of different $9 \times 9$ confusion matrices, which each have 72 degrees of freedom, and with nine stimuli the full model has only 41 free perceptual parameters.

On the other hand, note that the $2 \times 2$ factorial design, which is arguably the most popular identification experiment, includes only four stimuli. Therefore, the full model includes 16 free perceptual parameters (i.e., $5 \times 4 - 4$) and each confusion matrix includes only 12 degrees of freedom (i.e, $4 \times 3$). As a result, the full GRT model is not estimable in these experiments. So when GRT models are fit to single confusion matrices from $2 \times 2$ factorial designs, some assumptions must be made to reduce the number of free parameters.

## 7.2 MODELING THE DECISION PROCESS

A variety of different assumptions about decision processes are possible. For example, any decision model that is applied to categorization data can also be used to build an identification model. It makes sense that the same decision strategies would mediate identification and categorization. If there is any noise in the stimulus or percept, then in both cases, the mapping from percepts to responses is many-to-one. In other words, at the level of the percept, both are categorization tasks. Not surprisingly, there have been many proposals that identification and categorization are mediated by the same decision processes (e.g., Nosofsky, 1986; Shepard & Chang, 1963; Shepard et al., 1961), and there is also evidence that the same two explicit-rule and procedural-learning systems thought to dominate during category learning are also used during stimulus identification (Ashby et al., 2001). As a result, the decision bounds used to model identification and categorization data are typically similar.

Figure 7.1 illustrates the six decision models that are most widely used in GRT modeling of identification data. The application here is to a task with only four

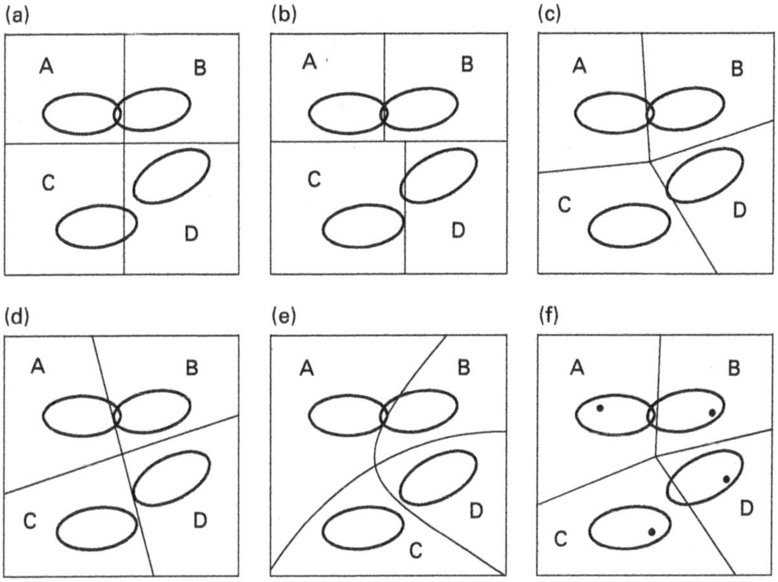


**7.1** Different types of decision bounds used in GRT modeling. (a) Decisional separability is satisfied on both dimensions. (b) Decisional separability is satisfied on dimension 2, but not on dimension 1. (c) Decision bounds of the minimum distance classifier. (d) Decision bounds of the general linear classifier. (e) Decision bounds of the general quadratic classifier. (f) Decision bounds of the striatal pattern classifier.

stimuli—an obvious example is a 2 × 2 factorial identification experiment. Even so, all of these models can easily be scaled up for application to tasks with any number of stimuli. The models were all considered in detail in chapter 5, so here only a brief review will be given.

Perhaps the simplest model, illustrated in figure 7.1a, assumes decisional separability on both dimensions. This model has two free decision parameters—namely, a criterion value on each dimension. Not surprisingly, this was the first GRT model fit to an identification confusion matrix (Ashby & Perrin, 1988). A slightly more complex model assumes that decisional separability holds locally, but not globally. The figure 7.1b example assumes decisional separability holds globally on dimension 2, but only locally on dimension 1. This model has three free parameters—a single criterion on dimension 2 and two separate criteria on dimension 1. The models illustrated in figures 7.1a and 7.1b are both compatible with explicit rule use.

The other four models illustrated in figure 7.1 are compatible with a strategy that might evolve from procedural learning. Panel (c) shows bounds predicted by the minimum-distance classifier, which always gives the response associated with the nearest perceptual mean (or mode or prototype). This model has no free decision parameters. Figure 7.1d shows bounds predicted by the GLC. This model has four free parameters—two that describe each linear bound. Figure 7.1e shows bounds predicted by the general quadratic classifier, which is identical to the GLC, except the two linear bounds of the GLC are replaced by quadratic curves. This model is rarely used because it has 10 free decision parameters—five parameters are needed to describe each quadratic curve. Finally, figure 7.1f shows bounds of the striatal pattern classifier (Ashby & Waldron, 1999; Ashby et al., 2001). See chapter 5 for details (section 5.4.2), but from a computational perspective, this model is identical to the minimum-distance classifier of figure 7.1c, except the location of the perceptual prototypes are free parameters. The idea is that the decision process does not have direct access to the perceptual distributions and so must estimate the locations of the perceptual means from trial-by-trial samples. This model has eight free parameters—two parameters are needed to estimate the coordinates of each perceptual prototype.

## 7.3 COMPUTING PREDICTED PROBABILITIES

When fitting any GRT model to identification data, parameter estimation is accomplished via the method of maximum likelihood that was described in section 7.1. As a result, the goal is to maximize the equation 7.3 $\ln L^*$ statistic, or equivalently to minimize the equation 7.4 BIC statistic. In all Gaussian GRT identification models, each $P(\mathrm{R}_j|\mathrm{S}_i)$ in equation 7.3 is computed by integrating a multivariate normal pdf over some response region. This is step 2 of the six-step model-fitting process described earlier. Different models make different assumptions about the pdf and about the shape of the region. For each separate model being fit, the researcher's step 2 task is to write code that computes each of these predicted probabilities, and to set the goodness-of-fit measure used by the off-the-shelf optimization routine to BIC. The optimization routine will then find the numerical values of all model parameters that minimize BIC (which will find the maximum likelihood parameter estimates by maximizing $\ln L^*$).

The difficulty of computing the integrals required for this minimization procedure depends on whether or not perceptual independence is assumed and on the nature of the decision bounds assumed by the model. The combination of perceptual independence and decisional separability greatly simplifies the model-fitting process because under these conditions each double integral reduces to the product of two single integrals. Furthermore, each single integral reduces to some area under the Z distribution. As a result, all of these single integrals usually can be evaluated with a single line of code. For example, in Matlab the command "normcdf(x)" returns the cumulative distribution function of the Z distribution evaluated at the value x (i.e., the area under the Z pdf from $-\infty$ to $x$).

Violations of perceptual independence complicate the process, but models in which decisional separability is satisfied on both dimensions are still quicker to fit than models in which decisional separability is violated on one or both dimensions. First, Owen (1956) showed how to reduce integrals under the general bivariate normal pdf over the rectangular regions that characterize the response regions of models that assume decisional separability from double integrals to single integrals. Second, for this general class of models, Wickens (1992) derived the first and second partial derivatives of the likelihood function with respect to each model parameter. Given these partial derivatives, the Newton-Raphson method can be used to quickly find the maximum-likelihood estimates of all parameters. This process still requires an iterative search procedure and numerical integration, but compared to the general GRT model that does not assume decisional separability, the numerical integrations should be quicker and the search for the maximum likelihood estimates should require fewer iterations.

In models that do not assume decisional separability, the integrals are under the bivariate normal pdf over irregularly shaped regions of the plane, and therefore are double integrals that, in general, cannot be reduced to single integrals. As a result, numerical integration under the bivariate normal pdf is required. Ennis and Ashby (2003) proposed an efficient algorithm for evaluating these integrals that can be used to estimate the parameters of virtually any GRT model via standard minimization software. The remainder of this section describes this method. Readers not interested in these technical details can skip to the next section.

The Ennis and Ashby (2003) method is illustrated in figure 7.2. The left side of the figure shows a contour of equal likelihood from a bivariate normal distribution that describes the distribution of percepts that occur on trials when some stimulus $S_i$ is presented, and the solid lines denote two different linear decision bounds. The two bounds divide the perceptual space into four response regions. Suppose the region on the right that includes the shaded part of the perceptual distribution corresponds to some response $R_j$. Therefore, the predicted probability of responding $R_j$ on $S_i$ trials equals the volume of the bivariate normal distribution that lies in this shaded

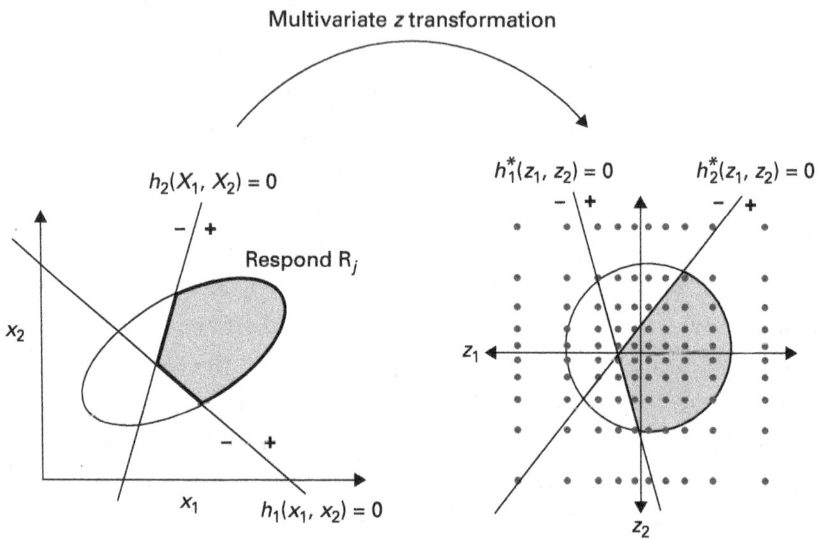


**7.2** A schematic illustration of how numerical integration is performed in the multivariate normal GRT model via Cholesky factorization.

response region. As a result, the shaded region in the left panel of figure 7.2 illustrates the volume that must be integrated numerically.

The algorithm identifies response regions by associating a discriminant function with each decision bound in the same way as in decision-bound models. In the figure 7.2 example, there are two linear bounds and two linear discriminant functions $h_1(x_1, x_2)$ and $h_2(x_1, x_2)$ (see section 5.1.4 for a review). Suppose we select the signs of the coefficients of these two discriminant functions so that any point in the shaded region of the left panel satisfies $h_1(x_1, x_2) > 0$ and $h_2(x_1, x_2) > 0$. Therefore, if we denote the multivariate normal pdf for stimulus $S_i$ as $\text{mvn}(\underline{\mu}_i, \Sigma_i)$, then

$$P(R_j|S_i) = \iint_{\{(x_1,x_2)\,|\,h_1(x_1,x_2)>0,\, h_2(x_1,x_2)>0\}} \text{mvn}(\underline{\mu}_i, \Sigma_i)\, dx_1\, dx_2. \tag{7.8}$$

Ennis and Ashby (2003) showed how to approximate integrals of this type quickly. The basic idea is to transform the problem using a multivariate form of the well-known Z transformation. This allows the integral to be evaluated in Z-space, rather than in perceptual space. This greatly speeds computation because the bivariate Z pdf can be preloaded and therefore, no pdf computations are required.

How can the problem be transformed to Z space? Briefly, the key is to note that converting the integral to Z space can be done with a linear transformation, and linear transformations are order preserving, in the sense that no points in the original $(x_1, x_2)$ space cross a decision bound during the transformation (more on this below). As a result, the volumes of the multivariate normal pdf for stimulus $S_i$ that fall in each response region are preserved. So the shape of each response region changes during the transformation, but the predicted response probabilities do not.

To formalize this process, note that any random vector $\underline{X}$ that has a multivariate normal distribution with mean vector $\underline{\mu}$ and variance-covariance matrix $\Sigma$ can always be rewritten as

$$\underline{X} = P\,\underline{Z} + \underline{\mu}, \tag{7.9}$$

where $\underline{Z}$ is a random vector with a multivariate Z distribution (i.e., a multivariate normal distribution with mean vector $\underline{0}$ and variance-covariance matrix equal to the identity matrix I), and P is a square matrix such that $PP' = \Sigma$, where $'$ denotes matrix transpose and $\Sigma$ is the variance-covariance matrix of $\underline{x}$. Given such a matrix P, then

$$\underline{Z} = P^{-1}(\underline{X} - \underline{\mu}). \tag{7.10}$$

Therefore, subtracting the mean vector and then premultiplying the result by $P^{-1}$ transforms any bivariate normal random vector $\underline{X}$ to a bivariate Z. As a result, equation 7.10 is the multivariate form of the well-known Z transformation (e.g., if $\underline{X}$ is $1 \times 1$, then $P^{-1} = 1/\sigma$).

Several choices for P are possible. Perhaps the best known is $P = QD^{1/2}$, where Q is the matrix in which the columns equal the eigenvectors of $\Sigma$ and D is the diagonal matrix in which the entry in row $i$ and column $i$ equals the eigenvalue of $\Sigma$ that corresponds to the eigenvector in column $i$ of Q.[1] Ennis and Ashby (2003) proposed instead to use Cholesky factorization, which requires less computation. If $\underline{x}$ is bivariate normal then in this approach P is the lower triangular matrix

$$P = \begin{bmatrix} \sigma_1 & 0 \\ \frac{\text{cov}_{12}}{\sigma_1} & \sqrt{\sigma_2^2 - \frac{\text{cov}_{12}^2}{\sigma_1^2}} \end{bmatrix}. \tag{7.11}$$

Note that the multivariate Z transformation of equation 7.10 is linear, and, as a result, it transforms linear bounds in $\underline{x}$ space to linear bounds in $\underline{z}$ space. In the figure 7.2 example, it transforms the line $h_1(x_1, x_2) = 0$ to the line $h_1^*(z_1, z_2) = 0$ and the line $h_2(x_1, x_2) = 0$ to the line $h_2^*(z_1, z_2) = 0$. As mentioned previously, linear transformations are order preserving. As a result, they also preserve parallelism, in the sense that parallel lines remain parallel after any linear transformation. Why is this important? Well, imagine a slew of lines all parallel to the vertical-tending $h_2$ decision bound shown in the left panel of figure 7.2. Note that all points on all such lines to the right of the bound have a positive $h_2$ discriminant value, and all points on all lines to the left of the bound have a negative $h_2$ discriminant value. Because the multivariate Z transformation preserves parallelism, the imaginary lines that were parallel in $\underline{x}$ space remain parallel in $\underline{z}$ space. As a result, the points on these lines that had a positive $h_2$ discriminant value will now have a positive $h_2^*$ value, and the points that had a negative $h_2$ discriminant value will now have a negative $h_2^*$ value. Therefore the multivariate Z trasnformation maps all points in the shaded region of the figure 7.2 $\underline{x}$ space to the shaded region of $\underline{z}$ space, and as a result, the integrals over the two shaded regions in figure 7.2 are identical.

These facts suggest the following algorithm for evaluating the integral of equation 7.8.

*Step 1*: Find the matrix P and the vector $\underline{\mu}$ of the equation 7.10 multivariate Z transformation that converts the multivariate normal pdf for stimulus $S_i$ to a multivariate Z distribution.

*Step 2*: Apply equation 7.9 to convert $h_1(x_1, x_2)$ to $h_1^*(z_1, z_2)$ and $h_2(x_1, x_2)$ to $h_2^*(z_1, z_2)$.

*Step 3*: Step through a preloaded grid of points in Z space. For each point, insert the pair $(z_1, z_2)$ into $h_1^*(z_1, z_2)$ and $h_2^*(z_1, z_2)$, and note the signs that result. For any such point for which both signs are positive, increment the integral by the preloaded volume associated with that Z point.

The discriminant functions $h_1(x_1, x_2)$ and $h_2(x_1, x_2)$ are transformed to $h_1^*(z_1, z_2)$ and $h_2^*(z_1, z_2)$ as follows. The first step is to write the discriminant function in matrix form. Note that in equation 5.4 of chapter 5 we wrote the discriminant function of a linear bound as

$$h(x_1, x_2) = a_1 x_1 + a_2 x_2 + b, \tag{7.12}$$

where $a_1$, $a_2$, and $b$ are parameters that determine the bound's slope and intercept. In matrix form, this equation becomes

$$h(x_1, x_2) = \begin{bmatrix} a_1 & a_2 \end{bmatrix} \begin{bmatrix} x_1 \\ x_2 \end{bmatrix} + b \tag{7.13}$$

or

$$h(\underline{x}) = \underline{a}'\underline{x} + b. \tag{7.14}$$

Given this, note that the decision bound

$h(\underline{x}) = \underline{a}'\underline{x} + b = 0,$

is equivalent to

$$h(P\underline{z} + \underline{\mu}) = \underline{a}'(P\underline{z} + \underline{\mu}) + b = 0. \quad (7.15)$$

Therefore, after the multivariate z transformation, the linear bound $h(\underline{x}) = 0$ in $\underline{x}$ space, becomes the linear bound in $\underline{z}$ space

$$h^*(\underline{z}) = (\underline{a}'P)\underline{z} + (\underline{a}'\underline{\mu} + b) = 0. \quad (7.16)$$

Thus, in this way we can transform the equation 7.8 integral to

$$\begin{aligned} P(R_j|S_i) &= \iint_{\{(x_1,x_2)\,|\,h_1(x_1,x_2)>0,\,h_2(x_1,x_2)>0\}} \text{mvn}(\underline{\mu}_i, \Sigma_i)\, dx_1\, dx_2 \\ &= \iint_{\{(z_1,z_2)\,|\,h_1^*(z_1,z_2)>0,\,h_2^*(z_1,z_2)>0\}} \text{mvn}(\underline{0}, I)\, dz_1\, dz_2. \end{aligned} \quad (7.17)$$

The key to evaluating this second integral quickly is to preload Z values that are each the center of an interval that all have equal area under the Z distribution (i.e., a normal distribution with mean 0 and variance 1). In figure 7.2, each gray point in the right panel has a $z_1$ coordinate that is the center of an interval with area 0.10 under the Z distribution (since there are 10 points). Taking the Cartesian product of these 10 points produces a table of 100 ordered pairs $(z_1, z_2)$ that are each the center of a rectangle with volume 0.01 (i.e., $0.10 \times 0.10$) under the bivariate z distribution.[2] Given such a table, the equation 7.17 integral is evaluated by stepping through all $(z_1, z_2)$ points in the table. Each point is substituted into the analogues of equation 7.16 and the signs of $h_1^*(z_1, z_2)$ and $h_2^*(z_1, z_2)$ are determined. If both signs are positive then the equation 7.17 integral is incremented by 0.01. If either sign is negative then the value of the integral is unchanged. So the value of the integral is approximately equal to the number of $(z_1, z_2)$ points that are in the response region that is being integrated over, divided by the total number of $(z_1, z_2)$ points in the table. Figure 7.2 shows a $10 \times 10$ grid of $(z_1, z_2)$ points, but better results can be expected from a grid with higher resolution. For example, Ashby et al. (2001) used a $100 \times 100$ grid, which should produce approximations to the integral that are accurate to within 0.0001.

Note that GRT assumes that all entries in each row of a confusion matrix are computed by integrating under the same perceptual distribution. Different columns are associated with different response regions. The algorithm therefore works row-by-row. The idea is to transform the perceptual distribution associated with the current row to a multivariate z-distribution by computing the appropriate Cholesky matrix P using equation 7.11. The next step is to apply this affine transformation to the decision bounds using equation 7.16. This step transforms the decision bounds from $(x_1, x_2)$ space to $(z_1, z_2)$ space. Next, the algorithm steps through all 10,000 of the z grid points (e.g., if a $100 \times 100$ grid is used). Each point is substituted into all of the z-transformed discriminant functions and the signs of the resulting discriminant values are recorded. The pattern of signs that are produced identifies the decision region of the grid point. If the current grid point is identified as belonging to response region $J$, then the final step is to increment the integral associated with response $J$ by 1 over the number of grid points (e.g., 1/10,000). After stepping through all 10,000 points, estimation of the predicted probabilities for all columns in the current row is complete. The algorithm now moves to the next row in the confusion matrix, and these steps are all repeated until the entire predicted confusion matrix is computed.

## 7.4 TESTING ASSUMPTIONS ABOUT PERCEPTUAL AND DECISIONAL PROCESSING

Assumptions about perceptual or decisional processing are tested using this approach by fitting two versions of a GRT model to the data using the procedures outlined in the previous two sections. In the first, restricted version of the model, a number of parameters are set to values that reflect the assumption being tested. For example, fixing all correlations to zero would test perceptual independence. In the second, unrestricted version of the model, the same parameters are free to vary. Once the restricted and unrestricted versions of the model have been fit, they can be compared through a likelihood ratio test:

$$\Lambda = 2 \ln L^*_U - 2 \ln L^*_R, \tag{7.18}$$

where $\ln L^*_U$ and $\ln L^*_R$ are computed from equation 7.3 for the restricted and unrestricted models, respectively. The restricted model can never fit better than the unrestricted model, so note that this statistic can never be negative. In fact, under the null hypothesis that the restricted model is correct, $\Lambda$ has a chi-squared distribution with degrees of freedom equal to the difference in the number of parameters between the restricted and unrestricted models. The mean of a chi-squared distribution equals its degrees of freedom, so the more extra parameters that the unrestricted model has the better its fit should be, relative to the restricted model, even if the restricted model is correct. This is because the unrestricted model will use its extra parameters to overfit the data, so the more extra parameters it has, the more overfitting we should expect.

If several non-nested models are fit to the data, the standard goal is to select the best candidate from this set. The likelihood ratio test cannot be used to select among non-nested models. Instead, we can compute the BIC statistic of equation 7.4 and then choose the model with the smallest BIC score.

## 7.5 THE SIMILARITY-CHOICE MODEL

For many years, the most successful model of identification was the similarity-choice model (SCM; Luce, 1963; Shepard, 1957), which assumes that

$$P(\mathrm{R}_j|\mathrm{S}_i) = \frac{\eta_{ij}\beta_j}{\sum_{k=1}^{N_\mathrm{R}} \eta_{ik}\beta_k}, \tag{7.19}$$

where $\eta_{ij}$ is the similarity between stimuli $\mathrm{S}_i$ and $\mathrm{S}_j$ and $\beta_j$ is the bias toward response $\mathrm{R}_j$ (without loss of generality, one can set $\eta_{ii} = 1$ for all values of $i$ and $\sum \beta_j = 1$). To ensure that the model is testable, similarity is assumed to be symmetric (i.e., so that $\eta_{ij} = \eta_{ji}$ for all values of $i$ and $j$).

We saw the SCM in chapter 2 (i.e., section 2.5.1) where we considered it as an alternative to the normal, equal-variance SDT model. In that section, the SCM was applied to the YES-NO detection task, in which there are only two stimuli, N and SN, and two responses, YES and NO. As a result, applications of the SCM to the YES-NO detection task require only one similarity parameter (i.e., $\eta$) and one response bias parameter (i.e., $\gamma$). Specifically, in the SCM YES-NO detection model

$\eta_{\mathrm{N,N}} = \eta_{\mathrm{SN,SN}} = 1$, $\eta_{\mathrm{N,SN}} = \eta_{\mathrm{SN,N}} = \eta$, $\beta_{\mathrm{YES}} = \gamma$, and $\beta_{\mathrm{NO}} = 1 - \gamma$.

When used as a model of the confusions that occur in identification experiments, the SCM has been remarkably successful—so successful that for many years, it was literally the standard against which competing models were compared. For example, in 1992, J. K. Smith summarized its performance by concluding that it "has never had a serious competitor as a model of identification data. Even when it has provided a poor model of such data, other models have done even less well" (p. 199, J. K. Smith, 1992). When I was in graduate school and Jim Townsend and I were fitting feature-analytic models of letter identification to various confusion matrices, we used the fit of the SCM as our benchmark to test the models. We had no hope that any of the models we were testing would fit better than the SCM (this was before we developed GRT). Instead, we would evaluate the success of the different models by recording how close they came to the SCM goodness-of-fit score.

Despite the universal success of the SCM at fitting identification-confusion data, the model was never considered completely satisfactory—primarily because a good fit provides little insight into the psychological processes of the observer producing the data. The model merely says that the probability of confusing stimulus $S_j$ for $S_i$ is proportional to the product of the similarity between the two stimuli and the bias toward response $R_j$ (the denominator in equation 7.19 is just a normalizing constant). Also, note that the model makes no predictions about how a decision is reached. It simply predicts the proportion of $R_j$ responses to expect over the course of a large number of $S_i$ trials.

GRT provided the first models that ended the dominance of the SCM, at least for identification data collected from experiments with stimuli that differed on only a couple of stimulus dimensions. In virtually every such comparison, the GRT model provided a substantially better fit than the SCM, in many cases with fewer free parameters (Ashby et al., 2001). An example of this dominance is illustrated in figure 7.3, which compares goodness-of-fit statistics from best-fitting versions of the SCM and GRT.

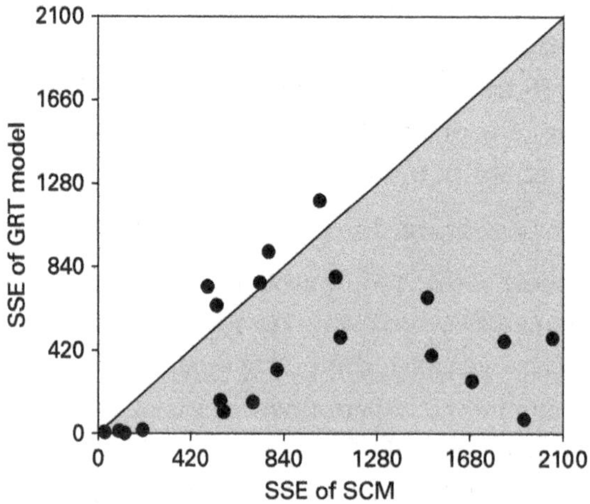


**7.3** SSEs from best-fitting versions of the SCM and GRT to 21 different 9 × 9 confusion matrices collected in a variety of different experiments. Each dot summarizes the performance of a different individual observer (adapted from Ashby et al., 2001).

Each point in figure 7.3 depicts the goodness-of-fit value of the two models to a $9 \times 9$ confusion matrix that summarizes the identification performance of a single observer over several thousand trials. Results from a variety of different experiments are included. In every case, the nine stimuli had an obvious two-dimensional structure, but the different experiments used a variety of different stimulus dimensions, including: (a) circles that varied in size and orientation of a radial line, (b) squares that varied in size and brightness, (c) red squares that varied in saturation and brightness, (d) parallelograms that varied in length and width, and (e) colored rectangles that varied in shape and saturation. The GRT models assumed (piece-wise) linear or quadratic decision bounds.

The goodness-of-fit measure in these applications was the sum of squared deviations between the observed and predicted response frequencies (SSE). The abscissa is the SSE from the best-fitting SCM, and the ordinate is the SSE from the best-fitting GRT model. Points on the main diagonal represent equal fits by the two models, points above the diagonal represent a superior fit by the SCM, and points below the diagonal represent a better fit by the GRT model. The results strongly favor the GRT model over the SCM. The SSE goodness-of-fit statistic does not penalize a model for extra free parameters, and in all these cases, the SCM and the GRT model had a different number of parameters. Nevertheless, in two-thirds of the applications, the GRT model had 72 fewer free parameters than the SCM, yet it still almost always fit better. These results suggest that, at least when the stimuli have a two-dimensional structure, GRT should be considered a better model of the confusions that observers make in identification experiments than the SCM. Not only does it usually provide a better overall goodness-of-fit but it also provides insights into the observer's perceptual and decision processes that are impossible with the SCM.

Despite the superiority of GRT in these applications, it is important to note that the SCM is still valuable, especially in the case of identification experiments in which the stimuli vary on many unknown stimulus dimensions.

## 7.6 AN EMPIRICAL APPLICATION

As an illustrative empirical application, consider the modeling results reported by Ashby et al. (2001). They fit many GRT models to identification data from two experiments that used stimuli like those shown in figure 7.4a. Note that the nine stimuli, which have a $3 \times 3$ factorial structure, vary on two dimensions—the orientation of the radial line segment and the diameter of the half circle. The experiments were

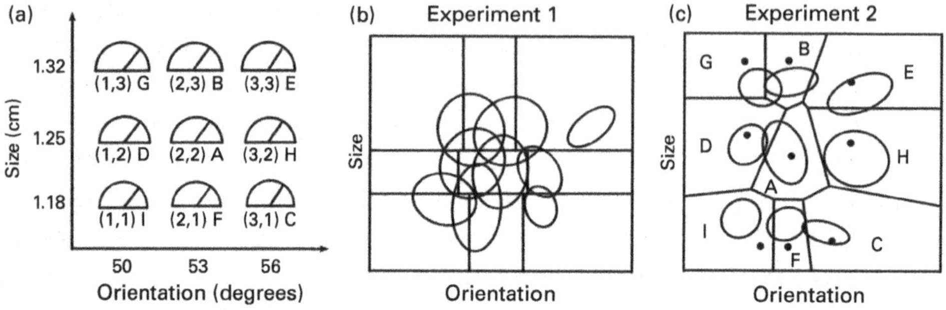


**7.4** Results from the identification experiment reported by Ashby et al. (2001). (a) The nine stimuli used in both experiments. The ordered pair beneath each stimulus denotes the required response in experiment 1, and the letter is the required response in experiment 2. (b) Best-fitting GRT model for observer 2 of experiment 1. (c) Best-fitting GRT model for observer 2 of experiment 2.

essentially identical, with the primary differences being in the instructions to the participants and the layout of the response devices. In experiment 1, which was run by Ashby and Lee (1991), two participants were told about the 3 × 3 factorial structure of the stimuli and they were instructed to identify each stimulus by responding first with its level of orientation (i.e., 1, 2, or 3) and then its level of size (1, 2, or 3). Each participant completed one practice session and four experimental sessions of 300 trials each. In experiment 2, each stimulus was randomly assigned (without replacement) one of the first nine letters of the alphabet, participants were not told about the factorial structure of the stimuli, and they were asked to identify the presented stimulus with its assigned letter. This experiment was run by Ashby et al. (2001) and also included two participants.[3] The task of learning the nine stimulus-response associations was considerably more difficult in experiment 2, and to compensate, the experiment 2 stimuli were slightly more discriminable and participants completed five practice sessions and three experimental sessions.

The results are shown in table 7.1. The focus of these studies was on how the different response instructions might affect the decision strategies used by participants. As a result, all models allowed for violations of perceptual independence and perceptual separability. The models differed only in the type of decision bounds they assumed. All decision models illustrated in figure 7.1 were tested. GRT(DS) assumed decisional separability on both dimensions (as in figure 7.1a). GRT($\text{DS}_\text{S}$) assumed decisional separability on the size dimension and piecewise decisional separability on the orientation dimension—that is, this model assumed decisional separability

**Table 7.1** The sum of squared errors (SSE) between observed and predicted response frequencies reported by Ashby et al. (2001).

| | | Experiment 1 | | Experiment 2 | |
|---|---|---|---|---|---|
| **Model** | **Parameters** | **Observer 1** | **Observer 2** | **Observer 1** | **Observer 2** |
| SCM | 44 | ~~224~~ | ~~1948~~ | ~~113~~ | ~~142~~ |
| GRT(DS) | 43 | **59** | 189 | 84 | ~~94~~ |
| GRT($DS_S$) | 47 | ~~58~~ | **93** | 58 | ~~92~~ |
| GRT(MDC) | 42 | ~~684~~ | 1430 | 90 | 69 |
| GRT(GLC) | 47 | 45 | ~~131~~ | ~~84~~ | ~~94~~ |
| GRT(GQC) | 59 | 29 | 66 | ~~84~~ | ~~93~~ |
| GRT(Optimal) | 39 | 331 | 1445 | 170 | 137 |
| GRT(SPC) | 57 | 34 | 89 | 23 | 11 |

Models with fit scores that have been struck out can be rejected without statistical testing because they are worse than an alternative model that has the same number or fewer free parameters. Scores of experiment 1 models that had the lowest AIC score are in bold. SCM = similarity-choice model; DS = decisional separability; $DS_S$ = decisional separability only on the size dimension; MDC = minimum distance classifier; GLC = general linear classifier; GQC = general quadratic classifier; SPC = striatal pattern classifier.

on the orientation dimension, but allowed the decision criteria to be different for each level of judged size (as in figure 7.1b). GRT(MDC) assumed that participants used a minimum distance or nearest prototype rule in which they responded with whichever stimulus had the nearest perceptual mean (as in figure 7.1c). GRT(GLC) and GRT(GQC) both assumed the participant divided the perceptual plane into a tic-tac-toe pattern via four decision bounds—two that had a vertical-tending orientation and two that had a horizontal-tending orientation (as in figures 7.1d and 7.1e). GRT(GLC) assumed these bounds were linear, whereas GRT(GQC) assumed they were quadratic. GRT(Optimal) assumed participants used the optimal strategy in which they responded with whichever stimulus was most likely to have produced the perceived values (not shown in figure 7.1). Note that neither GRT(Optimal) nor GRT(MDC) have any free decision parameters. GRT(SPC) was similar to GRT(MDC) in the sense that it also assumed that participants responded with the stimulus that had the nearest prototype. The difference however, was that GRT(SPC) allowed the coordinates of the prototype that defined each response region to be free parameters.

Thus, the locations of the prototypes were not constrained to lie at the perceived means (as in figure 7.1f). Finally, the SCM is the similarity-choice model of equation 7.19. As described in the previous section, the SCM is known for its universally excellent fits to virtually any identification confusion matrix. As a result, it provides a useful benchmark via which to assess the quality of the GRT model fits. Any GRT model that outperforms the SCM is providing an outstanding account of the data.

All models listed in table 7.1 were fit to the confusion matrix of each individual observer using the method of least squares, and in the case of experiment 1 the analysis was also re-done using maximum likelihood estimation. Because the stimuli used in experiment 2 were more discriminable than in experiment 1, overall accuracy was eventually higher (i.e., after five practice sessions of 300 trials each), and therefore the experiment 2 confusion matrices had many observed response frequencies near zero. Ashby et al. (2001) used a numerical integration algorithm to fit the GRT models that was more primitive than the one described in section 7.3, and for technical reasons, the many zeros caused this algorithm to fail.[4] As a result, in experiment 2 they only used the method of least squares for model fitting.

The goodness-of-fit statistic that results from least squares parameter estimation is the sum of squared errors (SSE) between observed and predicted response frequencies. The SSEs for the best-fitting versions of each model are shown in table 7.1. Model comparison is more problematic with least squares than with maximum likelihood parameter estimation, which is one reason that maximum likelihood is the preferred method of parameter estimation. Even so, some comparisons are possible. One of the most obvious is that a model can be rejected if some other model provides a lower SSE with the same number or with fewer free parameters. In table 7.1 all such rejected models are indicated with a strike through their SSE. In experiment 1, we can also appeal to the maximum likelihood fits for model comparison. The SSEs in bold indicate the models that had the lowest AIC goodness-of-fit scores.[5]

First note that the GRT models provided a substantially better fit than the SCM. In all cases, the SCM can be rejected because some GRT model provided a better overall fit with fewer free parameters. Second, note that in experiment 1, in which observers were instructed about the factorial nature of the stimuli and the response instructions encouraged decisional separability, the best-fitting models (according to AIC) assumed some form of decisional separability. Specifically, for observer 1, the best-fitting model assumed decisional separability on both dimensions. For observer 2, the best fit assumed decisional separability on size and local decisional separability on orientation. This latter model is described in panel (b) of figure 7.4. It is also important to note that in both cases, these fits were excellent. For example, in both cases, the best-fitting GRT models accounted for 99.6% of the variance in the response frequencies.

For experiment 2, the story was slightly different for the two observers. Recall that in this experiment, observers were given no instruction or hint (e.g., via the construction of the response device) about the factorial structure of the stimuli. For observer 2, all but a few models can be rejected because some other model with fewer parameters fit better. Of the three that are left, a statistical analysis reported by Ashby et al. (2001) indicated that the best fit was by the striatal pattern classifier [i.e., by GRT(SPC)]. This model is illustrated in figure 7.4c. This fit was almost perfect, accounting for 99.9% of the variance in observer 2's response frequencies.

For observer 1, a statistical analysis could not decide whether the best fit was by GRT(SPC) or GRT($DS_S$). GRT(SPC) fit better, but it has more parameters, and since the models are not nested a statistical comparison of SSEs is not possible. However, Ashby et al. (2001) reported the very interesting observation that midway through the experimental sessions, Observer 1 spontaneously (i.e., without being queried) reported to the experimenter that he had discovered the $3 \times 3$ factorial structure of the stimuli and that he had begun making his responses by deciding whether the sizes and orientations were small, medium, or large—exactly the strategy assumed by decisional separability. This report, along with the almost equally good fits of GRT(SPC) and GRT($DS_S$) suggests that Observer 1 may have switched strategies midway through the experimental sessions—beginning with a procedural strategy that was best described by GRT(SPC) and ending with an explicit rule strategy best described by GRT($DS_S$).

There are a number of important overall conclusions to draw from this application of GRT. First, for all observers, GRT models accounted for the identification confusions substantially better than the SCM. This reinforces the conclusions of figure 7.3 that, at least when the stimuli have a well-defined two-dimensional structure, GRT is a better model of identification than the SCM. Not only does it provide better absolute fits, but it also provides critical insights into the perceptual and decisional processes that mediated the observers' performance.

Second, these results support the hypothesis that humans have access to alternative decision strategies that are of qualitatively different types. One type appears to use explicit rules that logically combine independent decisions about the level of each stimulus component. These strategies are consistent with decision bounds constructed from piecewise vertical and horizontal line segments, and therefore with decisional separability or local decisional separability. A second type, which appears to result from procedural learning, is more similarity based, and leads to responding that is well fit by the GRT striatal pattern classifier.

Third, the results suggest that the choice of which decision strategy to use does not depend, at least not exclusively, on the stimuli and their configuration in stimulus space. Experiments 1 and 2 used essentially the same stimuli, yet seemed to recruit qualitatively different strategies. A fundamental assumption of SDT is that the response criterion $X_C$ is under the observer's control. The present results suggest that in identification tasks with multidimensional stimuli, observers have even more flexibility in how they make decisions.

In everyday life, one would not expect to encounter nine perceptually similar, novel stimuli that fall into a perfect $3 \times 3$ factorial arrangement. In the absence of such an arrangement, there is no reason to expect decisional separability to be an effective identification strategy. Experiments 1 and 2 support this conclusion. When explicitly instructed about the $3 \times 3$ stimulus arrangement, both observers used a strategy based on decisional separability, but when given no instructions about the stimuli, observers appeared to use a similarity-based strategy instead. Ashby et al. (2001) also reported the results of a third experiment that provided more support for this hypothesis. In this third experiment, nine stimuli were constructed by pseudorandomly sampling nine stimulus coordinates from the same two-dimensional (orientation, size) stimulus space used in experiments 1 and 2. In other words, the stimuli in this experiment did not have a $3 \times 3$ factorial arrangement, and instead had no systematic relationship to each other. As expected, the striatal pattern classifier again provided excellent fits to the data—accounting for 99.9% of the variance in each observer's identification confusions.

Finally, these results provide a caution to researchers who wish to use any of the summary statistics described in chapter 6 to test for perceptual interactions. All of those tests assume decisional separability, and as a result, the results reported in table 7.1 suggest that all currently popular summary-statistics tests would be invalid for three of the four observers in these two experiments. The finding that violations of decisional separability were more profound in experiment 2 reinforces the recommendation that summary-statistics tests should only be used in experiments where care is taken during experimental design to encourage observers to use explicit rules. In particular: (1) the stimuli should have a factorial structure, (2) observers should be informed explicitly of this structure, and (3) the response device should mimic the factorial structure of the stimuli.

## 7.7 GRT-wIND

As we have seen, GRT provides a powerful theoretical framework within which to analyze identification data. It not only provides extremely accurate accounts of the confusions that observers make in identification tasks but also allows strong tests of perceptual interactions and the chance to identify the observer's decision strategy, at least in experiments that use stimuli varying on only a couple of stimulus dimensions. Even so, the classical GRT modeling approach has a number of significant limitations.

First, classical GRT does not offer any easy way to make group-level inferences. In the classical approach, GRT models are fit separately to the confusion matrix of each individual observer. For each observer, researchers can ask whether two dimensions interact, but what conclusion can be drawn if the data of say, 13 observers suggest perceptual separability and the data of seven participants do not? Although we expect that individual observers could feasibly adopt different decision strategies, decades of research in visual perception suggests that individual differences in the perceptual processing of simple visual stimuli should be small.

Second, as we have seen, in the popular $2 \times 2$ factorial identification experiment, the full GRT model has more free parameters than there are degrees of freedom in the data. As a result, it is impossible to fit the full model to these data and so some restrictive assumptions must be imposed. And unfortunately, there is usually no way to test whether these extra assumptions are valid. Furthermore, even when they are reasonably valid, the small number of degrees of freedom left over to test the model increases the risk of over-fitting. A related problem that was discussed in section 4.5 of chapter 4 is that decisional separability is not identifiable in the traditional GRT analyses of results from $2 \times 2$ factorial designs.

All of these problems disappear if the data are analyzed instead with the group-level version of GRT called GRT-wIND, which is an acronym for GRT with INdivual Differences (Soto et al., 2015). GRT-wIND is fit simultaneously to the individual confusion matrices of all observers, and it allows group-level inferences to be made about the underlying perceptual representations—for example, about whether they satisfy perceptual independence and/or perceptual separability. In the case of the $2 \times 2$ factorial design, the model is trivial to fit because there is a freely available R package that fits this model using only a few commands (Soto et al., 2017).

GRT-wIND assumes that all observers share the same group perceptual representation. In other words, there is one perceptual distribution associated with each stimulus that describes the percepts experienced by all observers on trials when that stimulus is presented. No restrictions need to be made to constrain the number of free parameters needed to describe these distributions, even in 2 × 2 factorial designs. Thus, GRT-wIND assumes that basic perceptual properties, such as perceptual separability and perceptual independence, or their violations, are shared by all observers.

GRT-wIND allows for three different types of individual difference. Specifically, it assumes that different observers produce different confusion matrices for three reasons: (1) they allocate different amounts of overall attention to the task; (2) they divide their attention between the two dimensions differently; and (3) they use different decision bounds. The assumption that there are individual differences in attention is popular in multidimensional scaling (MDS; e.g., Carroll & Chang, 1970; Nosofsky, 1986), but, of course, MDS does not explicitly model decision processes, so it (implicitly) assumes that all observers use their percepts to select responses in exactly the same way. Even so, decades of SDT research has revealed significant individual differences in how people make even simple decisions (e.g., Aminoff et al., 2012; Layher et al., 2020). GRT-wIND accommodates this research by allowing every observer to use completely different decision bounds.

Fitting GRT-wIND to a set of individual observer confusion matrices returns estimates of (1) the group perceptual representation (i.e., the full GRT perceptual model), (2) the total amount of attention allocated by each observer to the task, (3) the proportion of attention allocated by each observer to the two perceptual dimensions, and (4) unique decision bounds for each observer.

GRT-wIND was inspired by the INDSCAL model of multidimensional scaling (Carroll & Chang, 1970). INDSCAL assumes all observers share the same group perceptual representation and that individual differences in similarity ratings occur because different observers allocate different amounts of attention to each perceptual dimension. For example, one observer might allocate most of their attention to stimulus dimension 1 and only a small amount to dimension 2, whereas another observer might allocate equal amounts of attention to the two dimensions. Allocating more attention to a dimension should increase perceptual differences. In MDS, this is modeled by stretching the distances along the more heavily attended dimensions. GRT-wIND includes the assumption that different observers divide their attention differently among the stimulus dimensions, and the assumption that different observers bring different amounts of overall attention to the task—for example, some observers try harder than others. However, unlike INDSCAL, GRT-wIND also allows for the possibility that different observers might use different decision strategies.

Attentional differences are modeled in GRT by allowing the perceptual variances to change with attention—specifically, the more attention allocated to a task, the smaller the perceptual variances, and therefore, the higher the accuracy. Although there is good evidence that attention increases perceptual sensitivity (e.g., Reynolds et al., 2000), GRT does not assume that this effect is necessarily purely perceptual. This is because memory (e.g., criterial) and perceptual noise are generally nonidentifiable in GRT (and SDT). Therefore, there is generally no way to tell whether a reduced variance is due to a reduction in memory noise or perceptual noise.

The formal identification of attentional changes with differences in perceptual variance came from equivalence relations that were established between GRT and various models that assume an MDS representation of percepts. In MDS, each percept is represented by a single point, rather than a probability distribution, and confusions are assumed to be inversely related to the distance between point representations (e.g., Kruskal & Wish, 1978). To account for errors, MDS-based models use some function that converts distances to response probabilities. By far, the most common choice is the SCM of equation 7.19, in which the $\eta_{ij}$ similarity terms are computed by measuring the relevant interpoint distances in the MDS representation.

More specifically, consider an MDS-based model in which the perceptual representation has two dimensions. In many such models, the similarity between two stimuli $i$ and $k$ is assumed to be inversely related to the weighted Euclidean distance $\delta_{ik}$ between the stimuli $i$ and $k$ point representations:

$$\delta_{ik} = c\sqrt{w(x_{i1} - x_{k1})^2 + (1-w)(x_{i2} - x_{k2})^2}, \tag{7.20}$$

where $c$ is the amount of overall or global attention allocated to the task, $w$ is the proportion of attention allocated to dimension 1, and $x_{ij}$ is the coordinate value of stimulus $i$ on the $j$th perceptual dimension. The INDSCAL model of MDS allows $c$ and $w$ both to vary across participants.[6] GRT-wIND also allows for both types of individual difference, as well as allowing for each observer to use different decision bounds.

A popular approach (e.g., Nosofsky, 1986; Shepard, 1957), called the MDS-choice model, combines the equation 7.20 distance measure with the SCM of equation 7.19 by assuming that

$$\eta_{ij} = \exp(-\delta_{ij}) \tag{7.21}$$

or

$$\eta_{ij} = \exp(-\delta_{ij}^2), \tag{7.22}$$

which are known as the exponential and Gaussian similarity functions, respectively.

Ashby and Maddox (1993) showed that MDS-choice models are formally equivalent to a special class of GRT models that assume optimal responding (i.e., see proposition 3 of Ashby & Maddox, 1993). In the case in which all GRT covariances equal zero (so perceptual independence holds for all stimuli), all stimuli have variance on dimension 1 equal to $\sigma_1^2$, and all stimuli have variance on dimension 2 equal to $\sigma_2^2$, then Ashby and Maddox (1993) showed that the GRT and MDS-choice models are mathematically equivalent when

$$c^2 = \frac{\sigma_1^2 + \sigma_2^2}{2\sigma_1^2\sigma_2^2} \text{ and } w = \frac{\sigma_2^2}{\sigma_1^2 + \sigma_2^2}. \tag{7.23}$$

Note that when $\sigma_1^2 = \sigma_2^2 = \sigma^2$, the first equation reduces to $c = 1/\sigma$, so the more global attention an observer allocates to the task the smaller the perceptual variances. And the second equation shows that the greater the proportion of attention allocated to dimension 1, the smaller the dimension 1 perceptual variance, relative to the variance on dimension 2.

The attentional assumptions are implemented in GRT-wIND via the way in which the perceptual variances are parameterized for each stimulus and observer. Specifically, GRT-wIND sets the stimulus $A_iB_j$ perceptual variances for observer $k$ to:

$$\frac{\sigma^2_{A_iB_j,1}}{\kappa_k\lambda_k} \tag{7.24}$$

and

$$\frac{\sigma^2_{A_iB_j,2}}{\kappa_k(1-\lambda_k)}, \tag{7.25}$$

where $\sigma^2_{A_iB_j,1}$ and $\sigma^2_{A_iB_j,2}$ are the stimulus $A_iB_j$ variances in the group representation on dimensions 1 and 2, respectively, $\kappa_k$ is the total amount of global attention that observer $k$ allocates to the task (i.e., $\kappa_k > 0$), and $\lambda_k$, which is constrained to the interval [0,1], is the proportion of attention that observer $k$ allocates to dimension 1. Note that $\kappa_k$ here plays an identical role as $c$ in equation 7.20 and $\lambda_k$ plays the same role as $w$.

The model is illustrated in Figure 7.5. The top panel shows a hypothetical group perceptual representation from a $2 \times 2$ factorial identification experiment. The four ellipses are contours of equal likelihood from the four perceptual distributions that best represent the group's perceptual experience with each of the four stimuli. Note that perceptual separability holds on dimension 2 but not on dimension 1, and that perceptual independence holds for stimuli $A_1B_2$ and $A_2B_1$, but not for stimuli $A_1B_1$ and $A_2B_2$. Because this is the group-level perceptual representation, these are therefore group-level inferences, and in this way GRT-wIND avoids awkward

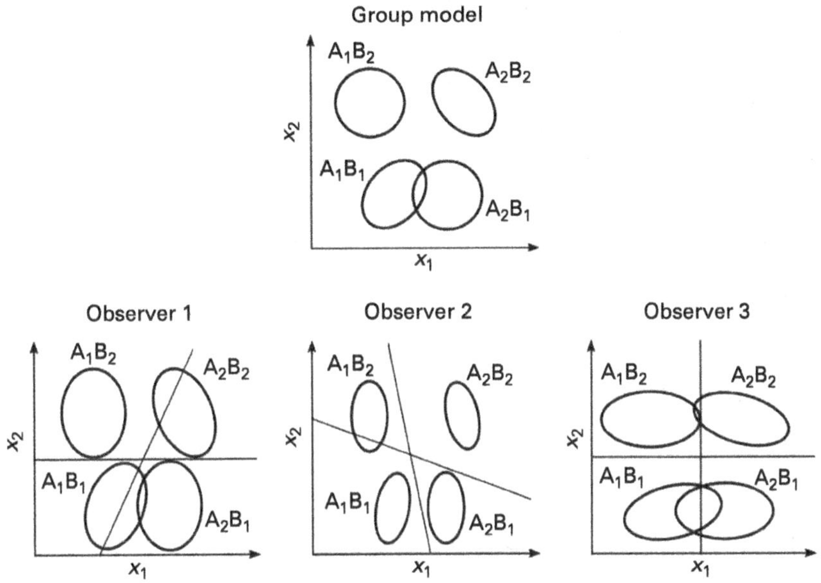


**7.5** The GRT-wIND model. The top panel shows the group perceptual representation that is common to all observers. The bottom half shows GRT-wIND models for three different observers. This figure was adapted from Soto et al. (2015).

conclusions, such as that perceptual independence holds for some observers, but not for all.

The bottom half of figure 7.5 shows the GRT models that best describe the identification performance of three different hypothetical observers. First, note that all three use substantially different decision strategies. For observer 3, decisional separability is satisfied on both dimensions, for observer 1, decisional separability holds on dimension 2 but is violated on dimension 1, and for observer 2 decisional separability is violated on both dimensions. So GRT-wIND predicts that one reason these observers have different confusion matrices is because they are all using qualitatively different decision strategies.

Second, note that although all three observers share the same group perceptual representation, they nevertheless express large individual differences in attention. First, note that observer 2 is allocating more overall attention to the task than observers 1 and 3. This can be seen in the smaller contours of equal likelihood in the observer 2 model than in the models for observers 1 or 3. Smaller contours mean smaller variances, which mean higher overall accuracy and therefore greater overall attention.

Second, note that there are large differences among these observers in how they divide their attention between the two perceptual dimensions. Observers 1 and 2 are allocating more attention to dimension 1 than dimension 2, whereas observer 3 is doing the opposite. To see this, note that in the group model, the dimension 1 and 2 variances are equal for stimuli $A_1B_2$ and $A_2B_1$ (evident from the circular contours of equal likelihood). For observers 1 and 2, however, these same stimuli have smaller variance on dimension 1, reflecting increased attention to this dimension, whereas for observer 3, the smaller variance is on dimension 2.

GRT-wIND also solves the problem that with a standard GRT analysis, decisional separability is not identifiable in $2 \times 2$ factorial designs. As noted by Silbert and Thomas (2013), for any GRT model of the $2 \times 2$ factorial task in which the four response regions are defined by two intersecting lines that violate decisional separability, there is always another GRT model that makes identical predictions for all entries in the confusion matrix, but in which decisional separability is satisfied. Furthermore, the two models are related by a linear transformation. The idea is similar to the trick we used in figure 7.2, except rather than find the linear transformation that converts the perceptual distribution to a multivariate Z distribution, as in figure 7.2, the goal is to find the linear transformation that makes the vertical-trending linear bound in the left panel of figure 7.2 perfectly vertical and the horizontal-trending linear bound perfectly horizontal. As in figure 7.2, this linear transformation preserves all response probabilities (because parallel lines remain parallel).

Now consider an application of GRT-wIND to a $2 \times 2$ factorial identification experiment with multiple observers. Suppose that after we fit the model, we find that none of the observers used bounds that satisfy decisional separability. As with any standard GRT model, we can always find a linear transformation that rotates the observer 1 bounds to vertical and horizontal, so that after this transformation, decisional separability is satisfied for this observer. However, because of the GRT-wIND assumption that all observers share the same group perceptual representation, if we linearly transform the best-fitting model for one observer, we must apply the exact same linear transformation to the best-fitting models for all other observers. Otherwise, they would no longer share the same group representation (e.g., note from figure 7.2 that such transformations change perceptual variances and covariances). Figure 7.6 illustrates the problem. The left column shows individual observer models for two hypothetical observers after we fit GRT-wIND to data from some $2 \times 2$ factorial experiment. Note that decisional separability is violated for both observers. The right column shows the equivalent models after a linear transformation that rotates

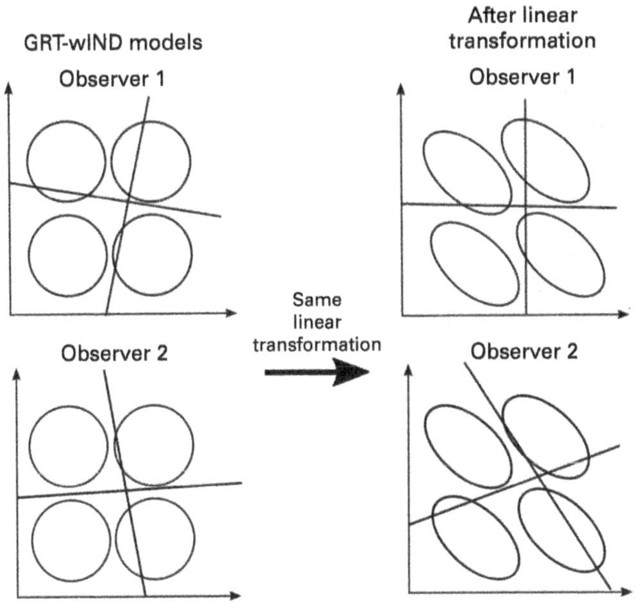


**7.6** The left column shows individual observer models from a hypothetical GRT-wIND analysis. Note that decisional separability is violated for both observers. The right column shows the equivalent models after a linear transformation that rotates the observer 1 bounds so that they satisfy decisional separability.

the observer 1 bounds so that they satisfy decisional separability. Note that this works for observer 1, but not for observer 2. In fact, for observer 2 the violations of decisional separability are even more extreme after the transformation. The only way to end up with models in which decisional separability is satisfied for all observers is to apply a different linear transformation to each individual observer model. But this is not allowed in GRT-wIND because it would violate the core assumption of the model that all observers share the same group perceptual representation. As a result, decisional separability is observable and can be tested in GRT-wIND, even in $2 \times 2$ factorial designs.

Soto et al. (2015) fit GRT-wIND to the confusion matrices of 24 different observers in a face identification experiment that used a $2 \times 2$ factorial design in which the four stimulus faces were created by crossing two facial identities with two emotional expressions. The stimuli in this experiment were similar to those shown in figure 4.2b. The 24 matrices included a total of 288 degrees of freedom (i.e., $24 \times 12$). The GRT-wIND model included an average of 6.67 free parameters for each individual confusion matrix, which is less than typical applications of traditional GRT models to $2 \times 2$ factorial designs. GRT-wIND accounted for 99.52% of the variance in the 24 confusion matrices, so the fit was almost perfect. In fact, a follow-up statistical analysis found that the model accounted for all systematic variability in the data.

An obvious and important question is to ask how GRT-wIND compares to a traditional GRT analysis, in which a separate GRT model is fit to the confusion matrix of each individual observer. If the data had an unlimited number of degrees of freedom, then GRT-wIND could never fit better because, at worst, the individual GRT models could provide the identical fits simply by setting the free parameters for each individual GRT model to the values of the best-fitting parameters from GRT-wIND. But, as we have seen, each confusion matrix in a $2 \times 2$ factorial design has only 12 degrees of freedom, and the full individual-observer GRT model has 16 free perceptual parameters. Therefore, to compare GRT-wIND to traditional GRT, Soto et al. (2015) fit many different GRT models separately to each individual observer confusion matrix. The different models made different assumptions about perceptual independence and perceptual separability, and for each observer, the best-fitting of these models was selected and then the performance of all 24 of these best-fitting models was compared to GRT-wIND. Note that if the GRT-wIND assumption that all observers share the same group perceptual representation is invalid, then the individual GRT models should outperform GRT-wIND.

The results of comparing GRT-wIND to a traditional GRT analysis are summarized in figure 7.7. This scatterplot shows the percent of variance accounted for by each model in all 24 individual observer confusion matrices. Points below the diagonal (in the shaded region) denote confusion matrices better fit by GRT-wIND, and points above the diagonal denote matrices better fit by an individual GRT model. Note that only a few points are above the diagonal, indicating that GRT-wIND provided a better overall fit to the data than the individual GRT models. This goodness-of-fit measure does not penalize for the number of free parameters, but collectively, the individual GRT models had more free parameters than GRT-wIND, making the excellent fits of GRT-wIND even more impressive.

The group perceptual representation from the best-fitting GRT-wIND model is shown in figure 7.8. None of the perceptual parameters were constrained in this model, so these contours represent the best fit to all 24 confusion matrices. The most striking result illustrated in figure 7.8 is that perceptual separability appears to be strongly violated on the identity dimension and to hold approximately for

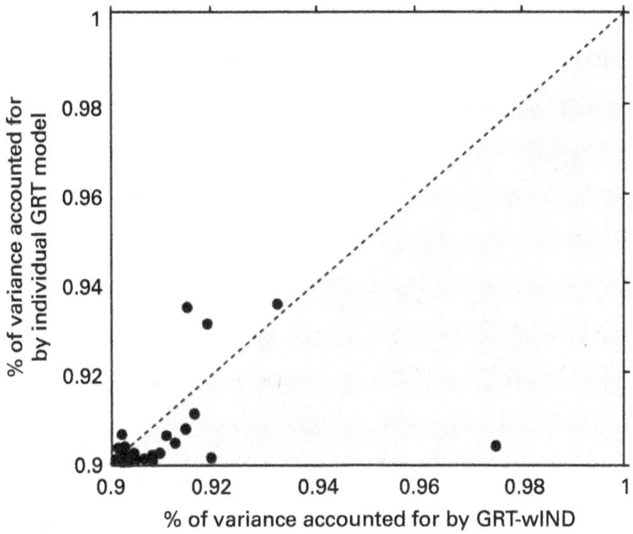


**7.7** Percent of variance accounted for in each of the 24 individual observer confusion matrices by GRT-wIND (on the abscissa) and by the best-fitting individual GRT model (on the ordinate; adapted from Soto et al., 2015).

emotional expression. In other words, GRT-wIND suggests that changing emotional expression from neutral to sad changes the apparent identity of the face, but changing facial identity has little or no effect on perceived emotional expression. Neutral faces look neutral, and sad faces look sad, regardless of identity. This conclusion would have been impossible from a traditional GRT analysis because among the best-fitting individual observer GRT models were some in which perceptual separability was violated on both dimensions, was violated on identity but not emotional expression, was violated on emotional expression but not identity, and was satisfied on both dimensions.

The individual difference parameters that were estimated uniquely for each observer showed considerable variability, suggesting large individual differences. First, the global attention parameter $\kappa$ varied from the most attentive participant to the least attentive by a factor of 5, suggesting that some observers put considerably more effort into the task than others. Second, the selective attention parameter $\lambda$ varied from about 0.2 to more than 0.7. So some observers attended much more heavily to identity and some attended much more heavily to emotional expression. Finally, there were also large individual differences in the decision bounds that observers used—in both their slopes and intercepts. Even so, most observers used bounds that

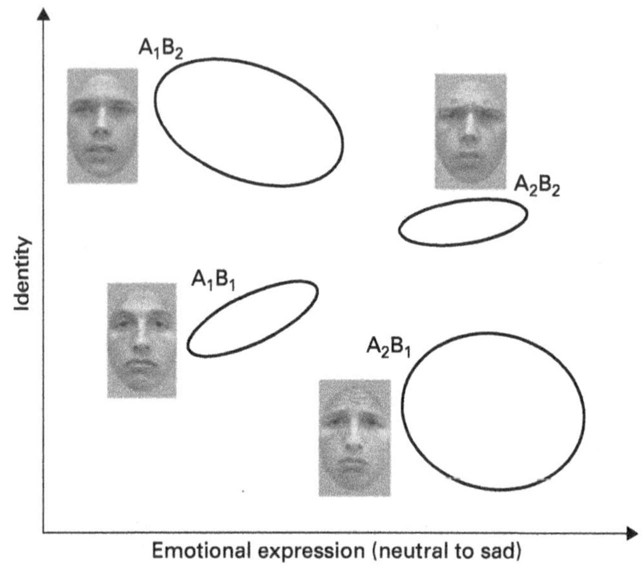


**7.8** Group representation of the best-fitting version of GRT-wIND to the data from the Soto et al. (2015) identification experiment (adapted from Soto et al., 2015).

violated decisional separability and, as a result, summary-statistics tests would be invalid for most observers in this experiment.

Note that INDSCAL, which assumes that all individual differences are attentional, fails to account for the observed individual differences in decision strategy. To their credit, the MDS-choice models do have a limited ability to account for decisional differences across observers. Even so, this class of models is problematic for several reasons. First, as shown in figure 7.3 and table 7.1, GRT provides a much better account of identification data when the stimuli vary on two dimensions. And second, the SCM assumes that all individual differences in decision strategy can be attributed to individual differences in response bias (see equation 7.19). This is formally equivalent to assuming that all observers use decision bounds with the same slopes but with intercepts that vary across observers. Therefore, MDS-choice models are presumably unable to account for individual differences that occur because different observers use decision bounds that have different slopes. Most importantly, though, INDSCAL and the MDS-choice models assume point representations for all stimuli and therefore allow for only very limited inferences about perceptual independence and perceptual separability.

In summary, GRT-wIND is an attractive model for analyzing data from $2 \times 2$ factorial identification experiments. First, it should provide excellent fits to the data. In fact, based on what we know today, one might expect GRT-wIND to provide a better account of the whole data set than any other existing identification model. Second, GRT-wIND allows group inferences about perceptual representations that are impossible with a traditional GRT analysis. Third, GRT-wIND allows for formal tests of decisional separability for each individual participant. Finally, analyzing $2 \times 2$ factorial data with GRT-wIND is a simple matter because of the freely available R package grtools (Soto et al., 2017).

# 8

# FITTING THE GAUSSIAN GRT MODEL TO CATEGORIZATION DATA

Chapter 7 described methods for fitting GRT models to confusion matrices from identification experiments. This chapter describes the computationally simpler problem of fitting GRT models to categorization data. As mentioned in chapter 7, when GRT models are applied to data from categorization experiments, they are typically called decision-bound models because the most common goal of this type of model fitting is to identify the type of decision strategy that participants used in the task. A strategy analysis of this type is important because in most categorization experiments, a variety of different strategies can all lead to similar response accuracies. Furthermore, many previous studies have documented that individual differences in strategy selection are common. As a result, stronger inferences are often possible if the standard accuracy-based data analyses (e.g., t-tests, ANOVA) are done after participants have been partitioned according to the type of strategy they used. Fortunately, because most categorization experiments only include two categories, fitting decision-bound models to categorization data is a simple computational process. In particular, no numerical integration is needed. To facilitate this process, Matlab code is given in sections B.2 and B.3 of appendix B that fits two of the most widely used decision-bound models to categorization data using the optimization routine fmincon. Section 8.7 below describes an application of both model-fitting programs to some sample data.

## 8.1 THE METHOD OF MAXIMUM LIKELIHOOD WITH CATEGORIZATION DATA

The fitting of decision-bound models to categorization data is conceptually identical to the procedures described in chapter 7 that fit GRT models to identification data. Both rely on the method of maximum likelihood for parameter estimation and model selection. And both typically rely on identical optimization routines to find numerical values of the model's free parameters that maximize $\ln L^*$ of equation 7.3 or equivalently, that minimize *BIC* of equation 7.4. Readers who skipped chapter 7 because their primary interest is in categorization modeling are encouraged to review section 7.1 for a brief tutorial on the method of maximum likelihood.

In some categorization experiments, for example those that use the randomization technique described in sections 5.2 and 5.6 of chapter 5, it is typical to present each stimulus only one time in a session. So if a block includes 100 trials, then 100 different stimuli are presented. In this case, the number of stimuli $N_S = 100$, and the number of times each stimulus was presented is $n_i = 1$. If there are only two categories then the number of responses is $N_R = 2$, and the frequencies of A and B responses on trials when stimulus $S_i$ was presented equal $r_{iA} + r_{iB} = 1$, which means that one of $r_{iA}$ and $r_{iB}$ equals 1 and the other equals 0. Under these conditions, equation 7.3 reduces to

$$\ln L^* = \sum_{i=1}^{N_S} \ln P(R_i|S_i), \tag{8.1}$$

where $R_i$ is the response (i.e., either A or B) made on the trial when stimulus $S_i$ was presented.

Fitting decision-bound models to categorization data is therefore a process of finding parameter values that cause the model to generate numerical predictions for the various $P(R_i|S_i)$ specified either by equation 7.3 or equation 8.1 that result in the largest value of $\ln L^*$ that is possible—or equivalently, that result in the smallest possible value of *BIC* in equation 7.4. Fortunately, the problem of deriving these predicted probabilities is computationally simpler for decision-bound models than for GRT identification models. There are two reasons for this reduced difficulty. First, most categorization experiments use only two categories, and therefore participants have only two response options. And second, decision-bound models make some extreme simplifying assumptions about the perceptual representations. The next section derives these predicted probabilities.

## 8.2 COMPUTING PREDICTED PROBABILITIES

As described in chapter 4, the choice between whether to run an identification or categorization experiment is typically a choice about whether the primary interest is in perceptual or decisional processes. An identification experiment is a good choice if the primary interest is in perception, and, as a result, the lion's share of parameters are allocated to modeling the perceptual representations. For example, as we saw in chapter 7, it is not uncommon to allow the means of each perceptual distribution to be free parameters and to allow the perceptual distributions associated with the different stimuli to all have different variances and covariances. Of course, perceptual confusions between two stimuli are just as likely in categorization experiments as in identification experiments. However, as noted in chapter 4, most categorization errors are not caused by perceptual confusions, but rather by the application of a suboptimal decision strategy. Furthermore, in categorization tasks with two categories, which are the most common types of categorization task, the resulting data have only one degree of freedom for each stimulus, which greatly limits the number of parameters that can be estimated. For these reasons, GRT models of categorization data most commonly assume a highly simplified perceptual representation relative to the most general versions of GRT. In particular, most such GRT models assume that the mean of each perceptual distribution is known, that all perceptual distributions have equal variances on every perceptual dimension, and that all covariances equal zero. These assumptions leave only one free perceptual parameter—namely the common perceptual variance, denoted by $\sigma_p^2$. Decision-bound models are defined as the subset of GRT models that make these simplifying assumptions.

How could an experimenter know the perceptual mean of every stimulus? In some cases, these means could come from some previous multidimensional scaling experiment (e.g., Nosofsky, 1986) or from psychophysical modeling of the stimuli. For example, in the case of sine-wave gratings (such as Gabor patches) that vary in spatial frequency and orientation, a psychophysical model that describes the transformation from stimulus space to perceptual space was provided by Treutwein et al. (1989). Another possibility, especially for dimensions that are perceptually separable, is to use Stevens' exponent. For example, the Stevens exponent for brightness is 0.33, so the mean brightness of each stimulus could be computed from $kI^{0.33}$, where $I$ is the physical intensity of the stimulus and $k$ is an arbitrary constant that can be set for convenience. Furthermore, the Stevens exponent for several popular stimulus dimensions is 1.0, including, for example, length and orientation. As a result, in these cases, the perceptual means can simply be set to the coordinates of the stimuli in stimulus space.

Note that decision-bound models assume perceptual independence for all stimuli, and if the stimuli have a rectangular, tic-tac-toe arrangement in stimulus space, then they will also assume perceptual separability on both stimulus dimensions. If so, then why apply GRT models to categorization data? What can be learned from this exercise?

The main goal of decision-bound modeling is to identify the strategy that participants used in the task. This is typically impossible by examining accuracy alone, because in most experiments, many qualitatively different strategies will lead to the same accuracy. The importance of a strategy analysis is magnified by the now overwhelming evidence that humans have multiple learning systems, which learn different things in qualitatively different ways (see chapters 5 and 11 for details). A strategy analysis is usually required to identify which learning system is used by each individual participant.

The best-studied learning systems are explicit-reasoning and procedural-learning systems. Evidence suggests that the explicit-reasoning system uses executive attention and working memory to learn explicit rules, whereas the procedural system uses dopamine-mediated reinforcement learning to establish associations between specific stimuli and their appropriate response (e.g., Ashby & Valentin, 2017; Ashby et al., 1998). The most popular method for studying properties of these systems is to compare performance in the RB and II category-learning tasks described in chapter 5 under different levels of some independent variable that is thought to affect one system more than the other. A strategy analysis in such an experiment is vital because participants can and often do use simple explicit rules in II categorization tasks, even though such rules are suboptimal. A failure to identify rule-using participants in II conditions and separating their data from the data of II participants who used a procedural strategy will dilute any possible difference between the two conditions.

A statistical approach to strategy analysis is illustrated in figure 8.1. Panel (a) shows two II categories, where each stimulus is coded according to its category membership. During an experiment, the participant assigns each of these stimuli to a category by depressing a response key (e.g., either the key associated with a category A response or the key associated with a B response). So an alternative representation is to code each stimulus according to the response the participant made on the trial when that stimulus was presented. An example for a hypothetical participant is shown in figure 8.1b.

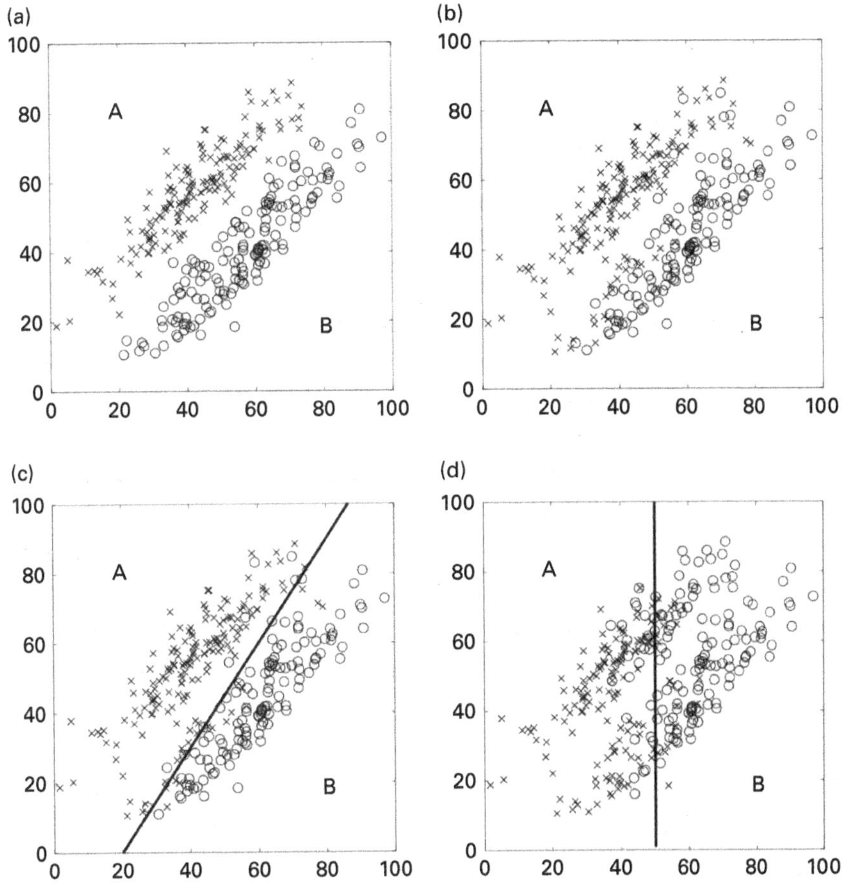


**8.1** (a): Stimuli in a hypothetical II categorization experiment coded by category membership. (b): Data from a hypothetical participant in the panel (a) experiment. Stimuli are now coded by the participant's response. (c): Same as in (b), except also showing the decision bound that provides the best statistical account of the participant's responses. (d): Responses from a different hypothetical participant in the panel (a) task along with the best-fitting decision bound.

Note that this participant performed well, but nevertheless appeared to be using a slightly suboptimal response strategy. A statistical method for identifying this strategy is provided by decision-bound modeling (Ashby, 1992; Ashby & Maddox, 1993).

In decision-bound modeling, the experimenter fits a number of statistical models to the responses of individual participants in an attempt to determine the type of decision strategy that each participant used. Two different types of decision-bound models are typically fit to the responses of each individual participant:

models that assume an explicit rule-learning strategy and models that assume a procedural strategy. It is also common to fit other models that assume the participant guesses at random on every trial. The rule- and procedural-learning models make no detailed process assumptions, in the sense that a number of different process accounts are compatible with each of the models (e.g., Ashby, 1992). For example, if a procedural-strategy model fits significantly better than a rule-learning model, then we can be confident that participants did not use a simple explicit rule, but we could not specify which specific non-rule-based strategy was used (e.g., a weighted combination of the two dimensions versus more holistic memory-based processing).

For example, consider figure 8.1c, which shows the decision bound of the best-fitting decision-bound model to the responses of the hypothetical participant illustrated in figure 8.1b. Note that the best-fitting bound requires integrating information from the two dimensions in a way that is impossible to describe verbally. Thus, the decision-bound analysis would conclude that this participant is using some type of procedural strategy. In contrast, note that the best-fitting bound for the different hypothetical participant shown in figure 8.1d is a vertical line, which corresponds to the explicit rule "Respond A if the stimulus has a small value on dimension 1 and B if it has a large value." Therefore, this participant would be classified as using an explicit rule, despite the fact that this was an II task.

Predictions for each alternative decision-bound model are derived via the model's discriminant function. Suppose the stimulus is two dimensional and denote the numerical value of the stimulus on these two dimensions by $(x_1, x_2)$. Then for any decision bound, we can always define a discriminant function $h(x_1, x_2)$ with the property that $h(x_1, x_2) > 0$ for any stimulus $(x_1, x_2)$ falling on one side of the bound, $h(x_1, x_2) = 0$ for any stimulus falling exactly on the bound, and $h(x_1, x_2) < 0$ for any stimulus falling on the other side of the bound (e.g., see figure 5.3). For example, for the vertical bound in figure 8.1d, the corresponding discriminant function is

$$h(x_1, x_2) = 50 - x_1. \tag{8.2}$$

Note that this function is positive for any stimulus in the A response region, negative for any stimulus falling in the B region, and 0 for any point on the bound. Similarly, the diagonal-trending bound shown in figure 8.1c corresponds to the discriminant function

$$h(x_1, x_2) = x_2 - 1.5x_1 + 30, \tag{8.3}$$

which is also positive in the A region and negative in the B region (e.g., note that $h(0,0)=30$ and $h(100,0)=-120$).

If we set up the discriminant function in this way—that is, with positive values of $h$ signaling an A response—then note that the predicted probability of responding A on a trial when stimulus $(x_1, x_2)$ is presented equals the probability that the discriminant function $h(x_1, x_2)$ is positive. As we saw in section 2.2.3, perceptual and criterial noise are not separately identifiable in SDT or in decision-bound models with linear bounds (Ashby & Maddox, 1993). Because of this, it makes no difference whether we assume that the noise is perceptual or decisional (or some combination of the two). In other words, if the bound is linear, then

$$\begin{aligned} P[\mathrm{A}|(x_1, x_2)] &= P[h(x_1+\epsilon_1, x_2+\epsilon_2) > \epsilon_C] \\ &= P[h(x_1, x_2) > \epsilon], \end{aligned} \tag{8.4}$$

where $\epsilon_1$ and $\epsilon_2$ model perceptual noise on stimulus dimensions 1 and 2 and are normally distributed with mean 0 and variances $\sigma_1^2$ and $\sigma_2^2$, respectively. Criterial noise is modeled by $\epsilon_C$, which is normally distributed with mean 0 and variance $\sigma_C^2$. The second equality establishes that this model makes identical predictions to a model in which there is one noise term, on the right side of the inequality, that is normally distributed with mean 0 and some variance $\sigma_p^2$. The equality holds because $h$ is a linear function and linear combinations of normal random variables are normally distributed.

Note that the only random variable in the last line of equation 8.4 is $\epsilon$ and that this random variable has a normal distribution. As a result, to evaluate this probability, we can take Z-transforms of both sides of the inequality. This converts the probability to

$$P[\mathrm{A}|(x_1, x_2)] = P\left[Z \leq \frac{h(x_1, x_2)}{\sigma_p}\right], \tag{8.5}$$

where $Z$ has a standard normal distribution (with mean 0 and variance 1). In two-category experiments, $P[\mathrm{B}|(x_1, x_2)] = 1 - P[\mathrm{A}|(x_1, x_2)]$.

All decision-bound models that assume a linear decision bound are described by equation 8.5. Two different classes of decision-bound models can be constructed depending on what assumptions are made about the decision process. These classes, along with the guessing models, are described in the following sections.

## 8.3 EXPLICIT-RULE MODELS

Explicit rule models assume the participant uses an explicit rule that is easy to describe verbally (Ashby et al., 1998). When the stimulus dimensions are perceptually separable and in incommensurable units, then rule models are equivalent to the independent-decisions models that were reviewed in section 5.1.1. In both cases, the decision bounds are restricted to piecewise vertical and horizontal line segments that are each perpendicular to some stimulus dimension—that is, to bounds that satisfy decisional separability, at least locally. For example, with the stimuli shown in figure 5.5, the only possible explicit rules are (1) give one response if the bars are thick and the contrasting response if the bars are thin; (2) give one response if the orientation is steep and the contrasting response if the orientation is shallow; and (3) some Boolean algebra combination of rules 1 and 2—for example, a logical conjunction, disjunction, or exclusive-or rule.

Suppose bar width is dimension 1 and bar orientation is dimension 2. Then the discriminant function that describes a one-dimensional rule on bar width in which the A response region is on the left and the B region is on the right (i.e., a type 1 explicit rule) is:

$$h(x_1, x_2) = c_1 - x_1, \tag{8.6}$$

where $c_1$ is the numerical value of the criterion that separates thin bars from thick bars. When fitting this model, equation 8.6 is substituted into equation 8.5 and a search algorithm is implemented (consisting of the six steps described in the introduction to chapter 7) that finds values of the two free parameters, $\sigma_p$ and $c_1$, that allow the model to give the best possible account of the participant's responses. Similarly, consider a one-dimensional rule on bar orientation in which the A response region is above the bound and the B region is below (i.e., a type 2 explicit rule). The discriminant function that describes this rule is:

$$h(x_1, x_2) = x_2 - c_2. \tag{8.7}$$

Matlab code that fits these two models to data from a randomization experiment is given in section B.2 of appendix B (see section 8.7 for details).

Models that assume a rule that is some logical combination of these two one-dimensional rules are only slightly more difficult to fit. For example, consider the conjunction rule: "Respond A if the bars are narrow and steep; otherwise respond B." This is equivalent to the following rule: "Respond A if $x_1 < c_1$ and $x_2 > c_2$; otherwise respond B." Therefore,

$$\begin{aligned}
P[\mathrm{A}|(x_1,x_2)] &= P\left(x_1-c_1<\epsilon_1 \text{ and } x_2-c_2>\epsilon_2\right)\\
&= P\left(x_1-c_1<\epsilon_1, x_2-c_2>\epsilon_2\right)\\
&= P(x_1-c_1<\epsilon_1)P(x_2-c_2>\epsilon_2)\\
&= \left[1-P\left(Z\le\frac{x_1-c_1}{\sigma_p}\right)\right]P\left(Z\le\frac{x_2-c_2}{\sigma_p}\right).
\end{aligned} \tag{8.8}$$

The joint probability described in the first line equals the product of the two marginal probabilities because we assume that the noise terms $\epsilon_1$ and $\epsilon_2$ are statistically independent.

Similarly, consider the disjunction rule: "Respond A if the bars are either narrow or wide; otherwise respond B," which is equivalent to: "Respond A if $x_1 < c_1$ or $x_1 > c_2$; otherwise respond B." Predictions for this model are as follows:

$$\begin{aligned}
P[\mathrm{A}|(x_1,x_2)] &= P\left(x_1-c_1<\epsilon_1 \text{ or } x_1-c_2>\epsilon_2\right)\\
&= P(x_1-c_1<\epsilon_1)+P(x_1-c_2>\epsilon_2)\\
&= [1-P(\epsilon_1\le x_1-c_1)]+P(\epsilon_2\le x_1-c_2)\\
&= \left[1-P\left(Z\le\frac{x_1-c_1}{\sigma_p}\right)\right]+P\left(Z\le\frac{x_1-c_2}{\sigma_p}\right).
\end{aligned} \tag{8.9}$$

If the dimensions are perceptually integral or in commensurable units, then it could be considerably more difficult to identify the set of all explicit rules. For example, consider rectangles that vary across trials in height and width. Since these dimensions are measured in the same units (and therefore are commensurable) other explicit rules can also be formed. For example, the rule "give one response if the rectangle is taller than it is wide, and give the contrasting response if it is wider than it is tall" corresponds to a linear bound with slope +1. If the dimensions are integral—such as the saturation and brightness of a color patch—then it is not clear what if any explicit rules can be formed. For these reasons, if a goal is to discriminate between explicit and procedural categorization strategies, then it makes sense to use stimuli constructed from perceptually separable dimensions measured in incommensurable units.

## 8.4 PROCEDURAL-LEARNING MODELS

Explicit-reasoning models assume participants make separate decisions about each relevant stimulus dimension, and then these decisions are combined if more than one dimension is relevant. In contrast, procedural-learning models assume perceptual information from two or more relevant dimensions is integrated *before* a decision is made. This integration could be linear or nonlinear. The most common application assumes linear integration, and the resulting model is known as the general linear classifier (GLC). The GLC assumes that participants divide the stimulus space using a linear decision bound. One side of the bound is associated with an "A" response, and the other side is associated with a "B" response. These decision bounds require linear integration of both stimulus dimensions, thereby producing a procedural decision strategy.

Before considering the computational details of the GLC, it is important to recall that section 5.4.2 reviewed evidence that people do not learn decision bounds in II tasks—of any type (Ashby & Waldron, 1999; Casale et al., 2012). Instead, the evidence suggests that people simply learn stimulus-response associations. For example, if they learned a decision bound, then they should be able to use that bound to respond accurately to novel stimuli. In fact, this is exactly what happens in RB tasks, but in II tasks, responding drops to chance when new (perceptually distinct) stimuli are introduced, even if those new stimuli could be categorized perfectly with the bound that separated the training stimuli (Casale et al., 2012). For this reason, the GLC is not a good model of the psychological processes participants use in II tasks, even if an explicit-rule model does accurately describe the psychological processes participants use in RB tasks. Even so, countless applications show that the GLC provides a good statistical model of responding in II tasks. Therefore, its use here is more like how one would use discriminant analysis—not as a psychological model, but as a statistical tool. Specifically, our only expectation is that of the three model classes, the GLC will provide the best account of the responses of a participant using a procedural strategy, even if the GLC does not accurately describe the psychological processes used by that participant.

The GLC decision rule is equivalent to: "Respond A if $a_1x_1 + a_2x_2 + b > 0$; otherwise respond B," and therefore the GLC decision bound is the set of points $(x_1, x_2)$ that satisfy $a_1x_1 + a_2x_2 + b = 0$. This is just the equation of a line. Of course, a line only has two free parameters, not three. For example, the most common characterization of a line is in its slope-intercept form $y = ax + b$. Unfortunately, though, slope-intercept form is insufficient for our purposes because it is impossible to express a vertical line in slope-intercept form (i.e., because the slope is infinite). Section B.3 of appendix B gives Matlab code that fits the GLC to data from a randomization experiment. That code parameterizes the decision bound as $a_1x_1 + a_2x_2 + b = 0$. The parameter vector in that code (i.e., param) therefore includes four parameters $a_1, a_2, b$, and $\sigma_p$, and fmincon will find values of all four that minimize BIC. Even so, it is important to note that only three of these four parameters are free. To see this, note that at least one of $a_1$ and $a_2$ must be nonzero, and that the decision rule "Respond A if $a_1x_1 + a_2x_2 + b > 0$" is unchanged if we divide both sides by a positive constant. Therefore, without loss of generality, we could always divide both sides of $a_1x_1 + a_2x_2 + b > 0$ by $|a_1|$ if $a_1$ is nonzero, and otherwise divide both sides by $|a_2|$. Either way, after this division, the line is characterized by only two parameters.

In summary, the GLC predicts that

$$\begin{aligned} P[\mathrm{A}|(x_1, x_2)] &= P(a_1x_1 + a_2x_2 + b > \epsilon) \\ &= P\left[Z \le \frac{a_1x_1 + a_2x_2 + b}{\sigma_p}\right]. \end{aligned} \tag{8.10}$$

Only three of these four parameters are free, however, so when computing the BIC of the GLC, $r$ should be set to $r = 3$.

## 8.5 GUESSING MODELS

Guessing models assume that the participant guesses randomly on every trial. All versions assume the probability of responding "A" (and therefore also the probability of responding "B") is the same for every stimulus. As a result, perceptual noise cannot change these predicted probabilities and so there is no need to account for perceptual noise in the guessing models. Because of this, guessing models do not include a noise variance parameter. They are not GRT models because they do not include perceptual distributions or decision bounds. Even so, they are useful for identifying participants who fail to learn for one reason or another, or who simply make no effort to follow the experimenter's instructions (e.g., by pressing the same response key on every trial).

Two types of guessing models are common. One version assumes that each response is selected with equal probability, or in other words that $P[\mathrm{A}|(x_1, x_2)] = 0.5$ for all stimuli. This model has no free parameters, and as a result, its goodness-of-fit can be calculated without the need for an optimization routine like fmincon. For example, note from equations 8.1 and 7.4 that in a randomization experiment, the model predicts

$$BIC = 0 \ln N_S - 2 \sum_{i=1}^{N_S} \ln \frac{1}{2} = 2N_S \ln 2. \tag{8.11}$$

The zero multiplier on $\ln N_S$ is because this model has no free parameters and as a result $r = 0$ in equation 7.4.

A second model, with one free parameter, assumes that the participant guesses response "A" with probability $p$ and guesses "B" with probability $1 - p$, where $p$ is a free parameter. This model is useful, for example, in identifying participants who ignore the stimulus and just press the same response key on every trial (in which case, the estimated value of $p$ would be either $\hat{p} = 0$ or $\hat{p} = 1$).

## 8.6 ITERATIVE DECISION-BOUND MODELING

The models described above all assume that the participant uses the same explicit-rule, procedural, or guessing strategy on every trial. In experiments where learning is expected, this assumption will be violated, so one common practice is to break the data into blocks of at least 50 trials each and then fit the models separately to each block of data. Another common approach is to fit the models only to the last block of data because we expect the participant's decision strategy to be most stable at the end of the session (in this case a block size of 100 or more trials is common). An iterative version of decision-bound modeling (called iDBM) accounts for learning by allowing strategy switches by individual participants during the course of the experimental session (Hélie et al., 2017). iDBM iteratively fits a series of decision-bound models to all trial-by-trial responses of individual participants in an attempt to identify: (1) all response strategies used by a participant, (2) changes in response strategy and, (3) the trial number at which each change occurs.

## 8.7 AN EMPIRICAL APPLICATION

As an empirical application, consider the data illustrated in figure 8.2. Panel (a) shows 300 stimuli from a hypothetical II experiment. Each circle denotes an exemplar of category A, whereas each asterisk denotes an exemplar from category B. The solid line is the optimal decision bound. The categories do not overlap, so perfect performance is possible in this task.

Figure 8.2b shows the responses of a hypothetical participant from this experiment. For example, these might be from the last 300 trials of an experimental session that included a total of 600 trials. Note that the stimuli are the same as in panel (a), except now they are coded according to the responses that were made. Now each circle denotes an A response and each asterisk denotes a B response. To generate these

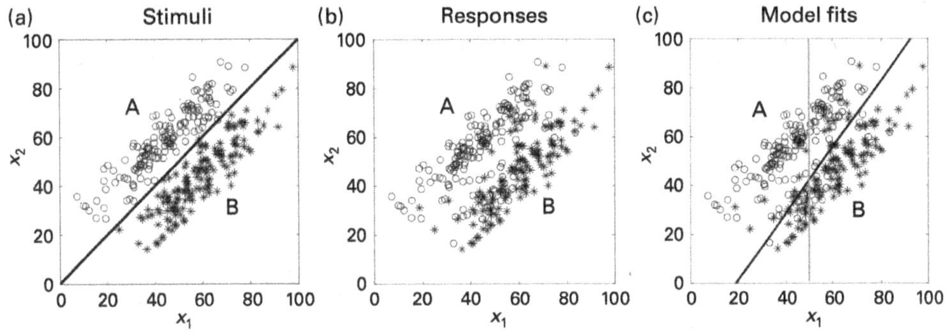


**8.2** (a) Stimuli from a hypothetical II experiment. Circles denote exemplars from category A, whereas asterisks denote exemplars from category B. The diagonal line denotes the optimal category bound. (b) The same stimuli, except now coded for the responses of a hypothetical participant. Circles denote stimuli that received an A response, whereas asterisks denote stimuli that received an B response. (c) Same as in (b), except with the best-fitting decision bounds of the GLC (black line) and the best one-dimensional rule model (gray line).

data, I imagined a participant who used the decision bound

$$x_2 = 1.5x_1 - 30, \tag{8.12}$$

which is steeper than the optimal bound (i.e., the optimal bound has a slope of 1). I generated responses from this hypothetical participant for three different perceptual noise values: $\sigma_p = 15$, $\sigma_p = 30$, and $\sigma_p = 45$. Given these parameter values, the accuracy of this hypothetical participant was 84% when $\sigma_p = 15$, 74% correct when $\sigma_p = 30$, and 66% correct when $\sigma_p = 45$. Figures 8.2b and 8.2c only show results when $\sigma_p = 15$.

I used the Matlab code listed in sections B.2 and B.3 to fit the GLC and both one-dimensional rule models to these responses (i.e., one rule model that assumes a vertical bound and one that assumes a horizontal bound). First, I loaded the three datasets into text files called "Data15.txt," "Data30.txt," and "Data45.txt," where the numbers indicate the values of $\sigma_p$ used to generate each set. These files are called by both model-fitting programs. Each file contained a row for every trial, and each row included four columns. Column 1 listed the category membership of the stimulus presented on that trial (i.e., denoted as a 1 or 2), columns 2 and 3 listed the numerical values of the presented stimulus on the two stimulus dimensions, and column 4 listed the participant's response to that stimulus (again as a 1 or 2).

In both routines, x0 contains initial guesses for all model parameters. If the A response region is meant to be above and/or to the left of the decision bound, then it is important to choose initial parameter values that guarantee a positive discriminant value in the A region. For example, note that the section B.2 code that fits the GLC begins with the vector x0 = [−1; 1; 0; 10]. The first three entries give the initial guesses for the parameters $a_1$, $a_2$, and $b$, whereas the last entry is the initial guess for $\sigma_p$. Therefore, note that the discriminant function begins as

$$h(x_1, x_2) = -x_1 + x_2, \tag{8.13}$$

which is positive above and to the left of the bound (e.g., note that the discriminant value for the point (40,70) equals +30). The vectors lb and ub contain lower and upper bounds on all parameters, respectively. The commands

[param,fval] = fmincon(@OneD,x0,[],[],[],[],lb,ub)

and

[param,fval] = fmincon(@GLC,x0,[],[],[],[],lb,ub)

call the fmincon optimization routine. The terms @OneD and @GLC refer to subroutines or functions that appear immediately below, which compute the BIC of each model. In both subroutines, the line that begins "Z = ..." computes the value of $h(x_1, x_2)/\sigma_p$ that appears in the right side of equation 8.5, and the line that begins "PC = ..." computes the probability specified in that equation. The output of these programs is specified by [param,fval], where param is a vector containing the final values of all parameter estimates and fval denotes the smallest value of BIC found by fmincon.

The results of this model fitting are shown in table 8.1, along with the BIC of the simple guessing model (computed using equation 8.11). The black line shown in figure 8.2c is the bound of the best-fitting GLC, and the gray line is the bound of the best-fitting one-dimensional rule model.

There are a variety of points worth noting. First, the best fits for all values of $\sigma_p$, by far, are by the GLC. This is the best possible outcome because the data were generated by a version of the GLC. Therefore, in all cases, fmincon succeeded in identifying the correct model. Furthermore, it accurately estimated the category bound used by this hypothetical participant. Recall that the true bound was $x_2 = 1.5x_1 - 30$. When fit to the $\sigma_p = 15$ data, the bound of the best-fitting GLC was $x_2 = 1.44x_1 - 27.97$, which is very close. In fact, even when fit to the noisiest data, the GLC accurately estimated the bound that generated the data (i.e., $x_2 = 1.62x_1 - 35.0$).

Second, note that the fits of all decision-bound models get worse as the value of $\sigma_p$ increases. This is expected since as $\sigma_p$ increases, the responses become more and more random, and as a result, when $\sigma_p$ is large, any model that assumes A and B responses are partitioned by a decision bound will mispredict many responses.

Third, note that the better fitting of the two one-dimensional rule models was always the one that assumed a vertical bound. The data were generated from a model that used a linear bound with a slope of 1.5, which is closer to vertical than horizontal. So the fact that the vertical-bound model fit better than the horizontal-bound model represents a success of decision-bound modeling.

In summary, this decision-bound modeling analysis correctly identified the strategy that was used to generate these data, and it also correctly discovered that this participant responded suboptimally with a decision bound that was steeper than the optimal bound. Furthermore, the clear superiority of the GLC, relative to either one-dimensional rule model, strongly suggests that this participant used some kind of procedural or information-integration strategy and is also strong evidence against any kind of explicit rule use.

## 8.8 CRITICISMS OF DECISION-BOUND MODELING

Overall, the simulations described in table 8.1 show that decision-bound modeling can be highly successful at identifying the strategy that was used to generate the responses, even when perceptual noise levels are high. Furthermore, a huge literature provides strong indirect support for this type of strategy analysis. For example, scores of studies have used decision-bound modeling to partition participants according to whether they used a rule-based or procedural strategy. If these partitions were invalid, then the two groups would be statistically homogeneous. However, in contrast to this prediction, these studies all reported a significant performance difference between

**Table 8.1** Results of fitting various decision-bound models to the data shown in figure 8.2.

| | **BIC** | | |
|---|---|---|---|
| **Model** | $\sigma_p = 15$ | $\sigma_p = 30$ | $\sigma_p = 45$ |
| 1D (horizontal bound) | 406.5 | 409.3 | 425.3 |
| 1D (vertical bound) | 330.0 | 398.6 | 407.5 |
| GLC | 211.5 | 339.3 | 400.9 |
| Guessing | 415.9 | 415.9 | 415.9 |

the two groups, and furthermore, in every case the difference was in the direction predicted by COVIS. These results would be impossible if decision-bound modeling was ineffective at strategy analysis.

Despite these successes, Edmunds et al. (2018) questioned the validity of decision-bound modeling, and even argued that "the potentially low validity of this strategy analysis casts doubt on the COVIS research that relies on it as a manipulation check" (p. 853). This is a surprising claim given that their own simulation results seem to support the exact opposite conclusion. The first simulations described by Edmunds et al. (2018) were of an II experiment, much like the one illustrated in figure 8.2, except with more widely separated categories. Given these categories, they simulated responses from a variety of different decision-bound models. For each model, they used a range of values for $\sigma_p$, and lumped all these results together, rather than report separate results for each value of $\sigma_p$ (as in table 8.1). The largest values of $\sigma_p$ must have been very large because the guessing model provided a better fit than the GLC that generated the responses on 6% of all simulations. This contrasts sharply with table 8.1, which shows that even when $\sigma_p = 45$, which causes accuracy to drop to only 66% correct, the GLC still produced a much better fit than the guessing model. Edmunds et al. (2018) interpreted the result that the guessing model fit better than the GLC that generated the data on 6% of all simulations as a failure of decision-bound modeling. However, with noise levels as high as Edmunds et al. (2018) used, any type of strategy analysis is of questionable utility. If accuracy is near chance, does it make sense to conclude that the participant is using any consistent strategy—at least, any strategy other than guessing? And regardless of how the responses were actually generated, wouldn't we want a strategy analysis to signal to us in this case that participants may have been guessing? Despite this unfavorable environment, the decision-bound models performed well. When the responses were simulated from a one-dimensional rule model, a one-dimensional rule model fit best 99% of the time. And when the responses were simulated from the GLC, the GLC fit best 83% of the time.

In a second set of simulations, Edmunds et al. (2018) began with RB categories that could be separated perfectly by a vertical bound. They then simulated responses from a one-dimensional rule model that used a vertical bound, and from the GLC. Importantly, though, they chose a version of the GLC with a bound that perfectly separated the categories, and so was essentially vertical. In other words, they simulated responses from a version of the GLC that used a strategy that was qualitatively identical to a one-dimensional rule. Not surprisingly, their results showed that in all but a few cases, a one-dimensional rule model provided the best fit. The few exceptions were all cases in which the best fit was by the guessing model. So even when the responses were generated by the GLC, the best fit was by a one-dimensional rule model. The authors interpreted this as another failure of decision-bound modeling, but in my opinion, these results are exactly what we would want a strategy analysis to tell us. The one-dimensional rule model and the GLC were implementing strategies that were qualitatively identical and the decision-bound modeling discovered this equivalence with near perfect accuracy. Edmunds et al. (2018) used BIC as their goodness-of-fit measure, which as we have seen, implements a parsimony criterion. If the one-dimensional rule model and the GLC both fit equally well, as in the simulations of Edmunds et al. (2018), then BIC favors the one-dimensional rule model, because this model predicts a priori that the decision bound must be vertical, whereas the GLC simply predicts that it must be linear. Given that Edmunds et al. (2018) chose to construct a simulation where the one-dimensional rule model and the GLC predict identical (or nearly identical) decision bounds, their choice to implement the BIC parsimony criterion during model selection made their results inevitable. The one-dimensional rule model did always provide the most parsimonious account of their simulated data, and this is exactly what the decision-bound modeling signaled.

In summary, the performance of decision-bound modeling in the Edmunds et al. (2018) simulations was about as good as one could expect from any statistical methodology. Despite extremely high levels of noise, it almost always correctly identified the strategy that was used to generate the data. It made some inevitable errors, but so do all statistical methods. For example, suppose we run t-tests on simulated data collected from two populations with different means. Suppose further that these simulations are run with a variety of different population variances. Obviously, as the population variance increases, the mean difference will become more difficult to detect. So any such set of simulations will include many nonsignificant t-tests, despite the difference in population means. No one would infer from such results that the validity of t-tests is so low that doubt should be cast on all research that established statistical significance via a t-test. Decision-bound modeling should be evaluated using similar criteria.

# 9

# GRT ACCOUNTS OF SIMILARITY, PREFERENCE, AND SAME-DIFFERENT JUDGMENTS

By far, the most common applications of GRT are to identification and categorization tasks. Even so, the theory has also been applied to many other tasks. Included in this list are tasks where participants judge the similarity of pairs of stimuli, where they rate their preference for various stimuli, where they rate the sensory magnitude of the stimulus on multiple stimulus dimensions, and where they judge whether a pair of stimuli are the same or different. This chapter describes many of these applications, with a particular focus on applications to similarity, preference, and same-different judgments.

## 9.1 SIMILARITY

Similarity is an important construct in almost every scientific field. For example, in mathematics, similarity plays a key role in the fields of congruence and homothety. In physics, computing similarity is common in optics, astrophysics, and quantum computing. Topological methods are used to measure similarity between semantic concepts. Graph theory is widely used for assessing cladistic similarities in taxonomy. Fuzzy set theory has developed its own measures of similarity, which are commonly used in fields such as management, medicine, and meteorology. Molecular biologists measure the sequence similarity of pairs of proteins.

Not surprisingly, similarity also plays a key role in many areas of psychology. One of the oldest and most influential theoretical assumptions in psychology is that perceived similarity is inversely related to psychological distance. This is the core assumption of multidimensional scaling (MDS; e.g., F. W. Young & Hamer, 2013). The idea is that the mental representation of an object, a concept, an ideal, a position on an issue, or any other mental entity can be modeled as a point in a psychological space that includes a separate dimension for each psychological feature. A fundamental assumption of MDS is that percepts that are close together in this space will be perceived as similar and percepts that are far apart will be perceived as dissimilar.

The popularity of MDS and other similarity-based theories led to many experiments in which participants were asked to directly or indirectly judge the similarity of pairs of stimuli. A variety of experimental techniques are used in these studies, but the most common is probably to simply ask participants to produce a number, between say 1 and 7, that matches their feelings about how similar the objects appear (e.g., with 1 meaning very dissimilar and 7 meaning very similar).

The earliest form of MDS, now called metric MDS, assumes that the distances between points in the MDS space are linearly related to the dissimilarities collected in these psychological experiments (Torgerson, 1952). This approach was later generalized to nonmetric MDS by replacing the assumption of a linear relationship between distance and dissimilarity with the weaker assumption of an ordinal relationship (Kruskal, 1964; Shepard, 1962). Many different computer packages are available for performing MDS. For example, Matlab includes commands to perform both metric and nonmetric MDS within its statistical and machine-learning toolbox. Nonmetric MDS is performed via the command

```
Y = mdscale(X,r),
```

where X is a matrix of dissimilarities and r is the number of dimensions of the resulting MDS similarity space. The value of r must be specified by the user. The standard approach is to try a number of different values of r, and then select the value that leads to the best account of the observed dissimilarities.

Shepard (1987) proposed as a universal law that psychological distance and perceived similarity are inversely related via an exponential function

$$s(A,B) = e^{-d(A,B)}, \tag{9.1}$$

where $A$ and $B$ are two stimuli or constructs, $s$ is perceived similarity, and $d$ is perceived dissimilarity or psychological distance. Shepard (1987) further proposed that this exponential function describes the probability that two stimuli fall in a region of stimulus space associated with the same response, which he called a consequential region. Section 9.3.4 compares consequential regions to the GRT notion of a response region.

Another prominent approach that treats similarity as a psychological primitive, which was discussed in chapter 7 (i.e., in section 7.5), is the similarity-choice model (SCM; Luce, 1963; Shepard, 1957). The SCM assumes that in an identification experiment, the probability of responding $R_j$ on trials when stimulus $S_i$ is presented equals

$$P(R_j|S_i) = \frac{\eta_{ij}\beta_j}{\sum_{k=1}^{N_R} \eta_{ik}\beta_k}, \tag{9.2}$$

where $\eta_{ij}$ is the similarity between stimuli $S_i$ and $S_j$ and $\beta_j$ is the bias toward response $R_j$. The denominator of equation 9.2 is just a normalizing constant, so note that this model assumes that, in the absence of any response bias, the probability of confusing stimulus $S_j$ for $S_i$ is simply proportional to the similarity between the two stimuli.

There are many alternative ways to define distance, but to qualify as a distance, certain properties, called the distance axioms, must be satisfied. For example, no matter how distance is defined, the distance from point A to point B must be equal to the distance from point B to point A. If similarity is inversely related to psychological distance, then perceived similarity must also obey these axioms. Considerable empirical effort therefore, has been spent testing whether perceived dissimilarity satisfies the distance axioms. The four distance axioms are

1. *Equal self-similarity*. $d(\text{A, A}) = d(\text{B, B})$ for all points A and B. Therefore, $s(\text{A,A}) = s(\text{B,B})$ for all stimuli A and B.
2. *Minimality*. $d(\text{A,B}) > d(\text{A,A})$ for all points A $\neq$ B. Therefore, $s(\text{A,B}) < s(\text{A,A})$ for all stimuli A $\neq$ B.
3. *Symmetry*. $d(\text{A, B}) = d(\text{B, A})$ for all points A and B. Therefore, $s(\text{A, B}) = s(\text{B, A})$ for all stimuli A and B.
4. *Triangle Inequality*. $d(\text{A, B}) + d(\text{B, C}) \geq d(\text{A, C})$ for all points A, B, and C. Therefore, the dissimilarities among any set of three stimuli should satisfy this same condition. The triangle inequality also implies that if stimuli A and B are similar and stimuli B and C are similar, then stimuli A and C must also be similar.

Much evidence has now been collected that raises questions about the validity of each of these axioms. For example, William James (1890) described an apparent counterexample to the triangle inequality well over a century ago. A flame is similar to the moon because they are both luminous, and the moon is similar to a ball because they are both round, but in contradiction to the triangle inequality, a flame is not similar to a ball. As one example of evidence against symmetry, Tversky (1977) reported that most people judge the similarity of North Korea to China to be greater than the similarity of China to North Korea. In response to examples such as these, many theories were proposed that could account for violations in some or all of the distance axioms. Included in this list were models that retained the assumption that similarity and psychological distance are inversely related, and assumed that violations in the distance axioms occurred because of response bias (Nosofsky, 1991), shifts in selective attention (Nosofsky, 1986), variations in the spatial density of stimulus representations in the psychological space (Krumhansl, 1978), or because percepts are probabilistic rather than deterministic (Ennis & Johnson, 1993). Tversky (1977) accounted for violations of the distance axioms by rejecting the assumption that similarity and distance are closely related and assumed instead that similarity depends on the physical features that define each stimulus, and more specifically on the sets of features that are common and unique to the two stimuli.

### 9.1.1 THE GRT-SIMILARITY MODEL

Similarity is not a fundamental construct in GRT, and an argument can be made that it is not a fundamental psychological construct. For example, human survival never depends on the ability to produce a number that matches perceived similarity, and most humans will live their whole lives without ever producing such a number. On the other hand, confusability is fundamentally important for human survival. Confusing a poisonous mushroom for one that is edible can have fatal consequences. As a result, Ashby and Perrin (1988) argued that in experiments where participants are asked to judge similarity, they have no such construct, so instead they judge the confusability of the two stimuli—that is, they respond with an estimate of how likely the two stimuli would be confused in an identification experiment—perhaps one in which the representations are degraded in some way (e.g., via masking, limited exposure duration, or because of added noise).

Following this logic, Ashby and Perrin (1988) assumed that the similarity of stimulus A to stimulus B is proportional to the probability of responding B on a trial of an identification experiment when stimulus A was presented, where this probability is computed via GRT. According to this model, the similarity of stimulus A to stimulus

B is therefore proportional to the volume of the pdf that describes the perceptual representation of stimulus A that falls in the region of perceptual space associated with response B:

$$s(\mathrm{A,B}) = k \int \cdots \iint_{\mathcal{R}_\mathrm{B}} f_\mathrm{A}(\underline{x})\, d\underline{x}, \tag{9.3}$$

where $k$ is a constant that sets the unit of measurement, $\underline{x} = (x_1, x_2, \ldots, x_r)$ are coordinates in the $r$-dimensional space that describes the observer's perceptual representation of the stimulus, $f_\mathrm{A}(\underline{x})$ is the pdf of the perceptual representation for stimulus A, and $\mathcal{R}_\mathrm{B}$ is the response region for response $\mathrm{R_B}$ (i.e., the set of all $\underline{x}$ that elicit a B response). Setting $k = 1$ fixes the similarity scale to the interval [0,1]. The default assumption is that the response region $\mathcal{R}_\mathrm{B}$ is defined by the optimal classifier—that is, it includes all $\underline{x}$ for which

$$\frac{f_\mathrm{B}(\underline{x})}{f_\mathrm{A}(\underline{x})} > 1. \tag{9.4}$$

Note that considerable knowledge is required to compute this measure of similarity. To begin, one must be able to identify that the subject of this comparison is stimulus A and the object is stimulus B, but in addition, one must also know the pdf that describes all A percepts and the optimal decision bound that separates the A and B response regions. How realistic is this?

The vast majority of studies that collected similarity ratings provided participants with linguistic labels that identified each stimulus. For example, when Tversky (1977) asked participants to judge the similarity of North Korea to China, there was no confusion about the identity of the subject or object of this comparison. The same is true in the William James (1890) counterexample to the triangle inequality, in which participants judge pairwise similarities of a flame, the moon, and a ball. So in most similarity-judgment experiments, the first condition is satisfied. Furthermore, GRT assumes that the observer has access to the $\mathrm{R_A}$ and $\mathrm{R_B}$ response regions. Even so, GRT does not assume that the observer has access to the pdf that describes the stimulus A perceptual representation. GRT assumes that the percept an observer experiences on each encounter with stimulus A can be described by one point in the perceptual space, and that the A pdf describes the set of all possible A percepts. So GRT assumes that the observer has access to the point representations elicited by each A encounter, but it does not assume that the observer has access to the stimulus A perceptual pdf. Therefore, GRT assumes that observers must estimate the equation 9.3 integral by appealing in some way to their past experiences with A and B. As a result, GRT predicts that similarity judgments should be noisy, in the sense that they should be more variable than identification responses, which do not depend on the observer's knowledge of any pdfs.

Note also that GRT predicts that similarity judgments could be qualitatively different if the raw stimuli are presented to observers without linguistic labels. If the stimuli are familiar and highly discriminable then the lack of labels might not matter much because observers would be able to identify the stimuli with high accuracy. If so, then they could use their past experience with the stimuli to estimate confusability and hence similarity. However, if the stimuli are novel or highly confusable, then the absence of labels could have significant effects on judged similarity. This GRT prediction has not been systematically investigated. Even so, the few existing empirical results are promising, with most of these reporting that adding verbal labels to stimuli systematically changes similarity judgments (e.g., Foroni & Rothbart, 2011; Lara et al., 2012; Sloutsky & Fisher, 2004). The one exception that I found was a study with preschool children that reported no effect of labels when the stimuli were discriminable pictures of animals (Noles & Gelman, 2012). More research on this problem is clearly needed. Lacking such research, the remainder of this subsection will assume either that verbal labels for each stimulus are supplied by the experimenter or that the stimuli are familiar and highly discriminable.

**9.1.1.1 Empirical support for the GRT-similarity model** Equation 9.3 assumes that similarity is proportional to unbiased confusability. If this assumption is correct, then the GRT-similarity model should provide a more accurate description of human perceived similarity than MDS-based models or the similarity-choice model because as we saw in chapter 7, GRT provides more accurate accounts of identification confusions than these models—at least in experiments where the stimuli vary on only a few dimensions.

One important arena in which the GRT-similarity model outperforms all distance-based similarity models, is in its ability to account for violations of the distance axioms. As I noted earlier, there is substantial evidence that human similarity judgments violate all of the distance axioms—with the possible exception of minimality. A strength of the GRT-similarity model is that it easily accounts for violations in all of these axioms. Figure 9.1 shows how the model is able to account for violations of equal self-similarity, symmetry, and the triangle inequality. Panel (a) shows a model in which the volume of the stimulus B pdf that lies in the $R_A$ response region is greater than the volume of the stimulus A pdf that lies in the $R_B$ region.

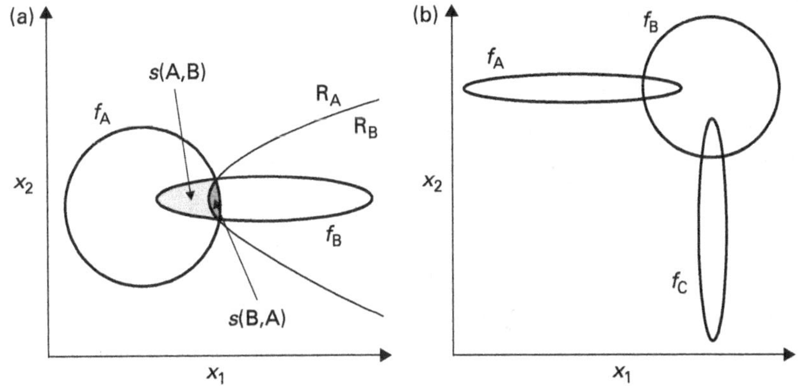


**9.1** Examples of GRT models that violate distance axioms. (a) A model that violates symmetry and equal self-similarity. (b) A model that violates the triangle inequality.

As a result, this model predicts that $s(B,A) > s(A,B)$, which violates the symmetry axiom. Similarly, the model also predicts a violation of the equal self-similarity axiom because it predicts that $s(A,A) > s(B,B)$ [i.e., because $s(B,B) = 1 - s(B,A)$ and $s(A,A) = 1 - s(A,B)$].

Figure 9.1b shows a model that predicts a violation of the triangle inequality. The distributions associated with stimuli A and B have considerable overlap, as do the distributions associated with stimuli B and C. As a result, the dissimilarities between A and B and between B and C are both small. The triangle inequality specifies that the dissimilarity between stimuli A and C must therefore also be small (i.e., less than the sum of the other two dissimilarities). But in the figure 9.1b model, the overlap between the A and C distributions is very small, and therefore the GRT dissimilarity between A and C is very large. Ashby and Perrin (1988) showed that in this model, the dissimilarity between A and C is much larger than allowed by the triangle inequality. Ashby and Perrin (1988) also showed that GRT can account for violations of minimality, but that these violations should be rare.

Because confusability is the psychological primitive in GRT, rather than similarity, note that GRT therefore also predicts that any factor that affects confusability might affect judgments of perceived similarity. Several of these are of illustrated in figure 9.2. First, GRT predicts that similarity judgments should be susceptible to context effects. This phenomenon is illustrated in panels (a) and (b) of figure 9.2. Panel (a) shows GRT predictions of the similarity of stimulus A to B when these are the only two stimuli under consideration, or when all other stimuli are so dissimilar that they do not affect the A to B confusability. Panel (b) shows the same predicted similarity, except after two similar stimuli, C and D, are added to the stimulus ensemble. Note that the addition of C and D decreases the predicted similarity of A to B because some percepts that occur on A trials that would be confused as a B event under the panel (a) scenario would instead be confused as C or D events in the panel (b) scenario. Therefore, there are fewer A, B confusions in the panel (b) scenario, and so GRT predicts reduced A to B similarity.

There is considerable evidence supporting this GRT predicted context effect. Much of this was reviewed by Krumhansl (1978), who also proposed a generalization of the standard MDS distance measure specifically to account for context effects of this type. The resulting model, called the *distance-density model*, predicts that similarity is reduced in regions of stimulus space occupied by multiple stimuli, below what one would expect by measuring the simple interstimulus distances. The distance-density model adds a correction factor to the interpoint distances that adjusts for these stimulus density effects. GRT makes similar predictions, but provides a theoretical interpretation of these context effects.

Second, GRT predicts that base rate differences should systematically alter similarity judgments. This effect is illustrated in panels (a) and (c) of figure 9.2. Panel (c) shows GRT predictions under exactly the same conditions as panel (a), except that the base rate for stimulus A presentations is higher than for B presentations. In other

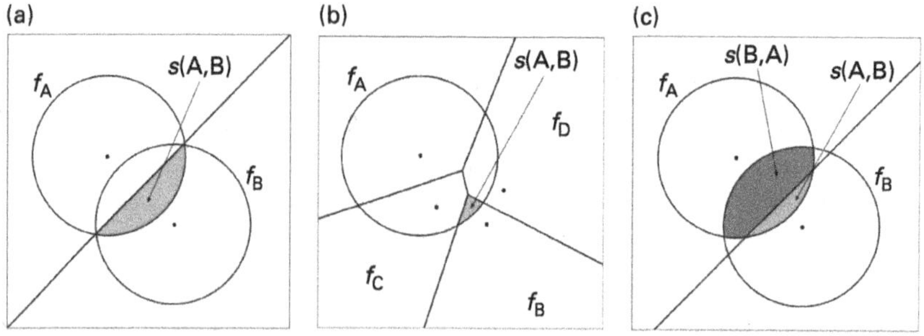


**9.2** Context effects on judged similarity as predicted by GRT. (a) Predicted similarity of A to B when A and B are the only relevant stimuli. (b) Predicted similarity of A to B when the relevant stimuli include A, B, C, and D. To keep this figure simple, a contour of equal likelihood is shown only for the A perceptual distribution. The dots show the locations of each distribution mean. The decision bounds were drawn under the assumption that all distributions have an identical variance-covariance matrix (i.e., equal to $\Sigma = \sigma^2 I$). (c) Same as in (a), except for a task in which the base rate of stimulus A presentations is higher than the base rate of B presentations.

words, in the panel (c) scenario, the observer encounters stimulus A more frequently than stimulus B. This causes the ideal observer to move the decision bound away from the A perceptual mean (i.e., see equation 2.8 of section 2.2.1). In panel (a), the bound is set to all points for which the likelihood ratio $f_A/f_B = 1$. In panel (c) however, the optimal response on trials when the likelihood ratio equals 1 is to respond $R_A$. The percept does not favor either stimulus on these trials, but knowledge that stimulus A occurs more frequently than B means that the optimal strategy is to guess response $R_A$ on these trials. Therefore, GRT predicts that base rate differences will lead to violations of symmetry, with the similarity of the more frequent stimulus to the less frequent being greater than the similarity of the less frequent to the more frequent.

To my knowledge, this base-rate prediction of GRT has not been tested rigorously. Even so, there are some encouraging suggestive results. For example, Tversky (1977) reported results of an experiment in which participants judged the similarity of 21 pairs of countries. The pairs were all selected so that one country was judged to be more prominent than the other. Prominence was not explicitly defined, but presumably, prominence was correlated with familiarity, in the sense that prominence increased with the participant's knowledge and familiarity with the country—that is, with its base rate. For example, participants were all Israeli college students, and they judged the United States to be more prominent than Mexico and Belgium to be more prominent than Luxembourg. The results of this experiment strongly supported the violations of symmetry predicted by GRT—that is, when similarity was queried in three different ways, participants consistently indicated that the less prominent country was more similar to the more prominent country than vice versa.

**9.1.1.2 Relation to the SCM** Note that the GRT assumption that similarity and confusability are proportional differs only slightly from the popular SCM of equation 9.2. In that model, similarity is the fundamental construct, but in the absence of response bias, the SCM assumes that similarity and confusability are monotonically related. To see exactly how the SCM and GRT notions of similarity are related, consider a simple two-stimulus identification task with stimuli A and B. In this case, the SCM includes four similarities: $\eta_{AA}$, $\eta_{AB}$, $\eta_{BA}$, and $\eta_{BB}$. However the SCM assumes that similarity is constrained by the distance axioms. As a result, it assumes that $\eta_{AA} = \eta_{BB} = 1$ and $\eta_{AB} = \eta_{BA} = \eta$, where $\eta$ is constrained to the interval [0,1]. Therefore, in the absence of response bias, the SCM predicts that

$$P(A|B) = \frac{\eta}{1+\eta}. \tag{9.5}$$

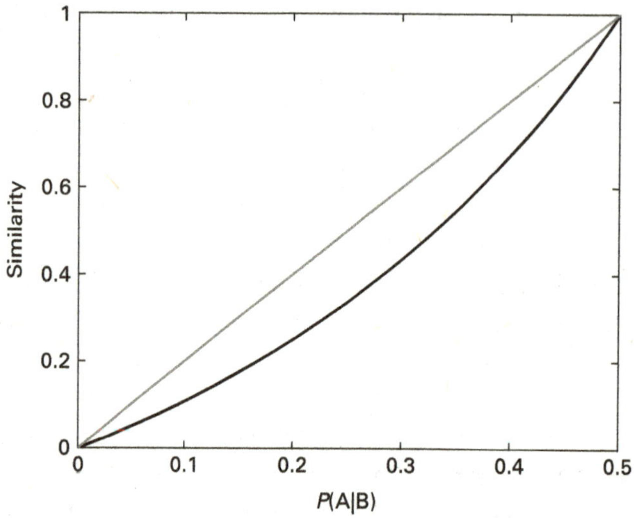


**9.3** Predicted relationships between similarity and the probability of a confusion in a two-stimulus identification task according to the SCM (black curve) and the GRT-similarity model (gray curve; with the scaling constant $k$ from equation 9.3 set to $k=2$).

Solving for $\eta$ produces

$$\eta = \frac{P(\mathrm{A}|\mathrm{B})}{1 - P(\mathrm{A}|\mathrm{B})}. \tag{9.6}$$

$P(\mathrm{A}|\mathrm{B})$ is the probability of a confusion and $\eta$ is the similarity between A and B. In a two-stimulus identification task, the optimal classifier has a maximum confusion probability of 0.5. Therefore, the relevant range for $P(\mathrm{A}|\mathrm{B})$ is [0, 0.5].

Figure 9.3 shows predictions of the SCM and the GRT-similarity model for the similarity between stimuli A and B in this task as a function of the probability of a confusion. GRT predicts these are proportional, and note that the SCM makes nearly identical predictions. So, whereas similarity is the fundamental construct in the SCM and confusability is the fundamental construct in GRT, they both predict nearly the same relationship between these two constructs. Even so, the two models make fundamentally different predictions about what outcomes to expect in this experiment. Note that, in the absence of response bias, the SCM predicts

$$P(\mathrm{A}|\mathrm{A}) = P(\mathrm{B}|\mathrm{B}) = \frac{1}{1+\eta}, \tag{9.7}$$

no matter what A and B stimuli are used in this experiment. Similarly, it also always predicts that

$$P(A|B) = P(B|A) = \frac{\eta}{1+\eta}. \tag{9.8}$$

In summary, the SCM predicts that, no matter what stimuli are used in a two-stimulus identification experiment, the only way an observer can perform differently on each stimulus is because of response bias.

In contrast, GRT can predict any relationship between $P(A|B)$ and $P(B|A)$ (e.g., see figure 9.1)—even in the absence of any response bias. The only constraints on the unbiased version of GRT that assumes optimal responding are that

$$P(A|A) \geq P(B|A) \tag{9.9}$$

and

$$P(B|B) \geq P(A|B). \tag{9.10}$$

But these restrictions are so weak that they allow for considerable differences in how observers respond to stimuli A and B.

**9.1.1.3 Relation to MDS models of similarity** A variety of different MDS models have been proposed to account for human similarity judgments. All of these assume that similarity is inversely related to the psychological distance between the point representations of the relevant pair of stimuli. The most common assumption is that all distances are computed using the Euclidean distance metric. The models differ in their assumptions about whether there are individual differences in how distance is computed and in whether there are perceptual interactions between stimulus dimensions. To keep the discussion in this section as simple as possible, I will focus on the case in which the stimuli differ on only two perceptual dimensions, but all results described in this section generalize to any number of dimensions.

The simplest Euclidean MDS model assumes that all participants compute psychological distance in the same way, and that there are no perceptual interactions between stimulus dimensions (Kruskal, 1964). This model, called the *simple Euclidean MDS model*, assumes that the distance between the point representations of stimuli A and B equals

$$d(A,B) = \sqrt{(x_{1,A} - x_{1,B})^2 + (x_{2,A} - x_{2,B})^2}, \tag{9.11}$$

where $x_{i,K}$ is the coordinate value of stimulus K on perceptual dimension $i$ (where K = A or B and $i = 1$ or 2). Therefore, in this model $d(A,B)$ is just the Euclidean distance between the point representations of A and B.

Carroll and Chang (1970) generalized this model to allow for individual differences. Their idea was that different participants might produce different similarity judgments because of differences in how they allocate attention to the various perceptual dimensions. For example, one participant might allocate more attention to dimension 1, whereas another might attend more to dimension 2. Furthermore, they assumed that attention increases perceptual differences, and therefore also perceptual distance. The resulting model is known as the *weighted Euclidean MDS model* or also as the INDSCAL model (for individual differences in multidimensionsal scaling). We saw this model in chapter 7 (section 7.7) because this is the model that inspired GRT-wIND. The weighted Euclidean MDS model replaces equation 9.11 with a unique distance equation for each participant. The distance between the representations of stimuli A and B is defined for participant $j$ as follows:

$$d_j(\mathrm{A,B}) = \sqrt{w_{1,j}(x_{1,\mathrm{A}} - x_{1,\mathrm{B}})^2 + w_{2,j}(x_{2,\mathrm{A}} - x_{2,\mathrm{B}})^2}, \tag{9.12}$$

where $w_{i,j}$ is the amount of attention that participant $j$ allocates to perceptual dimension $i$ (where in this application $i = 1$ or 2). To appreciate the role that the attention weights play in this model, consider two stimuli that are equally far apart on two perceptual dimensions—that is, for which $x_{1,\mathrm{A}} - x_{1,\mathrm{B}} = x_{2,\mathrm{A}} - x_{2,\mathrm{B}}$. In the simple Euclidean MDS model, the differences between A and B on the two dimensions contribute equally to perceived dissimilarity in this example. Now consider a participant who allocates more attention to dimension $x_1$ than to dimension $x_2$, so that $w_1 > w_2$. Note that the effect of this allocation scheme is that the $x_1$ distance-apart contributes more to $d(\mathrm{A,B})$ than the $x_2$ distance-apart. As a result, perceptual differences are larger on attended dimensions than on nonattended dimensions.

The weighted Euclidean MDS model was generalized to allow for perceptual interactions by Tucker (1972). Tucker used the model of perceptual interactions that was introduced by Tanner (1956), and was described in section 4.3.1 of chapter 4. Briefly, that model accounts for perceptual interactions by allowing the angle between the perceptual dimensions to be oblique. As described in chapter 4, Ashby and Townsend (1986) showed that the Tanner model, in which the angle between dimensions equals $\theta$ is equivalent to a GRT model with orthogonal dimensions, in which all variance-covariance matrices are equal and the common correlation coefficient that characterizes each variance-covariance matrix equals $\rho = -\cos\theta$ (i.e., see theorem 4.1). Tucker (1972) added the assumption that perceptual interactions might cause the angle between dimensions to be oblique to the weighted Euclidean MDS model, thereby producing a Euclidean MDS model that allows each participant to have unique attention weights and a unique angle between dimensions. This model, which is known as the *general Euclidean MDS model*, defines the distance between the representations of stimuli A and B for participant $j$ to be:

$$d_j(\mathrm{A,B}) = \Big[w_{1,j}^2(x_{1,\mathrm{A}} - x_{1,\mathrm{B}})^2 + w_{2,j}^2(x_{2,\mathrm{A}} - x_{2,\mathrm{B}})^2 + 2w_{1,j}w_{2,j}\cos\theta_j(x_{1,\mathrm{A}} - x_{1,\mathrm{B}})(x_{2,\mathrm{A}} - x_{2,\mathrm{B}})\Big]^{1/2}, \qquad (9.13)$$

where $\theta_j$ is the angle between dimensions for participant $j$. Because $\cos 90° = 0$, note that this model reduces to the weighted Euclidean MDS model when the dimensions are orthogonal.

As an illustration of how $\theta$ affects distance in this model, consider the simple example shown in figure 9.4. Panel (a) shows an MDS space with two stimulus representations. Stimulus A has coordinates (0,1), and stimulus B has coordinates (1,0). Note that the Euclidean distance between these two points equals $\sqrt{2}$ (i.e., $\sqrt{(-1)^2 + 1^2}$). Panel (b) shows the same model, except as the angle between the dimensions approaches 0. Note that the points A and B move closer together as the angle $\theta$ moves closer to 0. When $\theta = 0$ exactly, the A and B points will be superimposed, and so their distance-apart will be 0. Equation 9.13 accounts for this change in distance as the angle between the dimensions shrinks because $\cos\theta$ increases from a value of 0 to a value of 1 as $\theta$ decreases from 90° to 0°. As a result, when the attention weights both equal 1, note that when $\theta = 0°$, equation 9.13 reduces to:

$$d(\mathrm{A,B}) = \sqrt{(-1)^2 + 1^2 + [2 \times 1 \times (-1) \times 1]} = 0. \qquad (9.14)$$

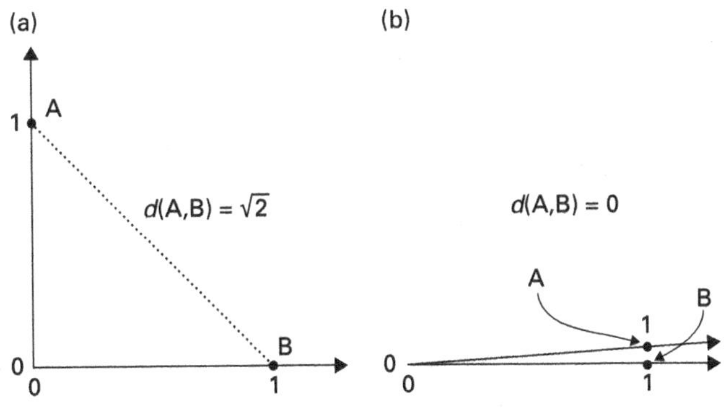


**9.4** (a) A simple MDS model with two stimuli A and B. (b) The same MDS model as in panel (a), except as the angle between dimensions approaches 0.

Ashby and Perrin (1988) showed that all three of these Euclidean MDS models are special cases of the GRT-similarity model—that is, if we restrict the parameters of the GRT-similarity model in certain ways, then the similarity predictions of the GRT model become identical to the predictions of each of these MDS models. As a result, any data that any of these MDS models can fit, can also be fit perfectly by the GRT-similarity model. In contrast, the GRT-similarity model is much more general, in the sense that it can account for many phenomena that are incompatible with all MDS models. We reviewed many of these phenomena in section 9.1.1.1.

The equivalence relations established by Ashby and Perrin (1988) all assume that dissimilarity is inversely related to the GRT similarity measure (i.e., from equation 9.3) via

$$d(\text{A,B}) = -2\Phi^{-1}[s(\text{A,B})], \tag{9.15}$$

where $\Phi$ is the cumulative distribution function of the standard normal z distribution. Note that this equation implies that

$$s(\text{A,B}) = \Phi[-d(\text{A,B})/2]. \tag{9.16}$$

This is not a usual similarity function. Figure 9.5 compares it to the more common exponential similarity function of equation 9.1. Note that the only substantial difference is for small values of dissimilarity.

Ashby and Perrin (1988) used this similarity function to prove the following result.

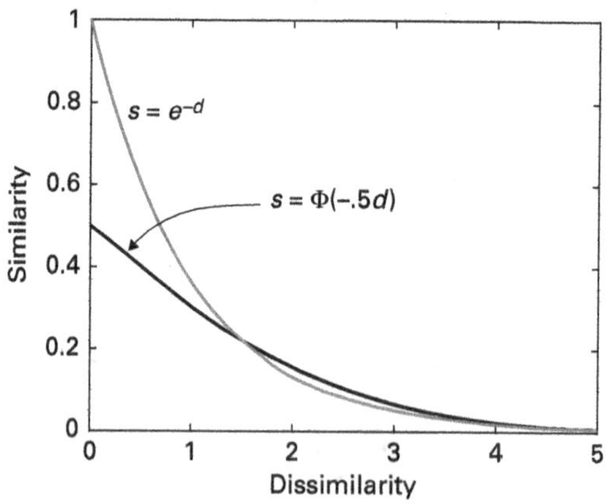


**9.5** The exponential similarity function and the similarity function that makes MDS a special case of GRT.

**Theorem 9.1** Suppose perceived similarity and dissimilarity are related as in equation 9.16. Consider a GRT-similarity model in which all perceptual distributions are multivariate normal with identical variance-covariance matrices equal to

$$\Sigma = \begin{bmatrix} \sigma_1^2 & \rho\sigma_1\sigma_2 \\ \rho\sigma_1\sigma_2 & \sigma_2^2 \end{bmatrix}. \tag{9.17}$$

Then the GRT-similarity model that ignores context predicts that the dissimilarity between stimuli A and B equals:

$$d(\mathrm{A,B}) = \Big[\omega_1^2(\mu_{1,\mathrm{A}} - \mu_{1,\mathrm{B}})^2 + \omega_2^2(\mu_{2,\mathrm{A}} - \mu_{2,\mathrm{B}})^2 - 2\omega_1\,\omega_2\,\rho\,(\mu_{1,\mathrm{A}} - \mu_{1,\mathrm{B}})(\mu_{2,\mathrm{A}} - \mu_{2,\mathrm{B}})\Big]^{1/2}, \tag{9.18}$$

where $\mu_{i,\mathrm{K}}$ is the mean percept of stimulus K (K = A or B) on perceptual dimension $i$ ($i = 1$ or 2), and

$$\omega_i^2 = \frac{1}{\sigma_i^2(1-\rho^2)}. \tag{9.19}$$

*Proof* See Ashby and Perrin (1988). ■

Given theorem 9.1, it is easy to see that the simple Euclidean, weighted Euclidean, and general Euclidean MDS models are all special cases of the GRT-similarity model. In each case, the distance metric predicted by the MDS model can be generated from the GRT-similarity model by setting the relation between similarity and dissimilarity as in equation 9.16, setting all GRT perceptual distributions to multivariate normal, setting the GRT mean percepts to the MDS coordinate values assigned to each stimulus, constraining all variance-covariance matrices to be equal, and ignoring context (i.e., assuming that $d$(A,B) depends only on the stimulus A and B perceptual distributions). Within this special, restricted class of GRT models, the distance metric predicted by the simple Euclidean MDS model is produced if the further constraints are added that all variances equal 1 and the common correlation coefficient equals 0. Note that when these constraints are incorporated into equation 9.18, $\omega_1 = \omega_2 = 1$ and the third term of $d$(A,B) drops out (because $\rho = 0$). After these simplifications, $d$(A,B) in equation 9.18 reduces to simple Euclidean distance (as defined in equation 9.11). Therefore, using the simple Euclidean MDS model is equivalent to assuming a (context insensitive) GRT model in which perceptual independence holds for all stimuli and that all perceptual variances are equal for all stimuli and on all perceptual dimensions. This makes sense because when these distributional assumptions hold, confusability in GRT is inversely related to inter-mean distance, and therefore so is similarity in the GRT-similarity model. Moving any two perceptual distributions apart in this model decreases the confusability of the associated stimuli, and so also their perceived similarity, regardless of the direction of movement (i.e., since perceptual variability is the same in every direction).

The GRT dissimilarity measure described by equation 9.18 of theorem 9.1 reduces to the weighted Euclidean MDS model if no constraints are placed on the variances (i.e., on the $\sigma_i^2$) but if the common correlation coefficient $\rho$ is constrained to equal 0. In this case, $\omega_i^2 = 1/\sigma_i^2$ and the third term of equation 9.18 drops out (again, because $\rho = 0$). Under these conditions, the attention weights $w_{i,j}$ of the weighted Euclidean MDS model (described by equation 9.12) are equal to $w_{i,j} = 1/\sigma_i^2$, and as a result, the more attention allocated to a dimension, the smaller the perceptual variability. Therefore, using the weighted Euclidean MDS model is equivalent to assuming a (context insensitive) GRT model in which perceptual independence holds for all stimuli, and that there are no stimulus-specific differences in perceptual variability on any dimensions.

Finally, if no constraints are placed on any parameters in the variance-covariance matrix $\Sigma$, then the GRT-similarity model described in theorem 9.1 is equivalent to the general Euclidean MDS model. More specifically, the GRT dissimilarity measure defined in equation 9.18 is equivalent to the dissimilarity measure predicted by the general Euclidean MDS model (and described in equation 9.13) when $\omega_i^2 = w_{1,j}^2$ and $\rho = -\cos\theta$. So using the general Euclidean MDS model is equivalent to assuming a (context insensitive) GRT model in which there are no stimulus-specific differences in perceptual variability on any dimensions. Perceptual independence could be violated in this model, but if it is, it must be violated by exactly the same amount in every stimulus. Finally, note that this equivalence mapping establishes exactly the same relationship between the GRT correlation coefficient $\rho$ and the angle between dimensions $\theta$ that Ashby and Townsend (1986) derived when comparing the modern GRT model with Tanner's 1956 model—namely, that equivalence requires that $\rho = -\cos\theta$ (i.e., see theorem 4.1).

### 9.1.2 DISTANCE-BASED MODELS

GRT was developed as a multidimensional generalization of SDT. In SDT, the criterion $X_C$ plays a fundamental role in the decision process. The decision bounds of

GRT are straightforward generalizations of $X_C$ when the perceptual space is expanded from a line to a plane. At about the same time that GRT was being developed, independent efforts were underway to generalize MDS—not to multidimensional spaces, because MDS was always multidimensional. Rather the generalization was from the point representations of percepts in classical MDS to probabilistic perceptual representations that were identical to the perceptual representations of GRT. So this line of work replaced each stimulus point in MDS with a multivariate normal probability distribution. On each trial, the stimulus representation was still a single point, but now that point was a random sample from the multivariate probability distribution that described all possible percepts elicited by that stimulus. In this model, all distances are between random samples from two different distributions, and, as a result, distance is a random variable. There was never an attempt to add decision bounds to this probabilistic version of MDS, so this line of work remained parallel to and independent of GRT.

Consider two stimuli A and B that differ on $r$ perceptual dimensions. Denote the A and B perceived values on dimension $j$ by $x_{j,A}$ and $x_{j,B}$, respectively. Then the most common MDS assumption is that the similarity between A and B is inversely related to their Minkowski distance apart:

$$d(\mathrm{A}, \mathrm{B}) = \left[ \sum_{i=1}^{r} |x_{i,\mathrm{A}} - x_{i,\mathrm{B}}|^{\gamma} \right]^{\frac{1}{\gamma}}, \tag{9.20}$$

where $\gamma$ can take on any fixed value $\gamma \geq 1$. The most common choices are $\gamma = 1$, which results in city-block distance and $\gamma = 2$, which results in familiar Euclidean distance. Note that in either case, $d$(A,B) is not normally distributed when the A and B percepts are sampled from multivariate normal distributions. The nonnormality is because of the absolute values in the city-block case, and the squares and square root in the Euclidean case. Next, these nonnormal random variables are passed through some function that converts the distances to similarities, such as the exponential similarity function of equation 9.1. This latter step changes the distribution again, and as a result of these nonlinear transformations, evaluating this model is a significant computational challenge.

Ennis, Palen, and Mullen (1988) were the first to propose and evaluate this general model. Numerical evaluation showed that, under certain conditions, the relationship of the distance between distribution means and expected similarity approximated a Gaussian function, even when the trial-by-trial similarity function was exponential (i.e., defined by equation 9.1). This resolved an apparent paradox in the literature in which MDS-based models were reported to provide the best account of data characterized by low confusability when the similarity function was exponential, but when confusability was high the best account was provided by a Gaussian similarity function (Ennis, 1988; Nosofsky, 1986; Shepard, 1987). Ennis and Johnson (1993) showed how to simplify the computational challenges associated with this model, and MacKay (2001) derived exact pdfs for the city-block distance version of this model.

The probabilistic MDS model struggles to account for context effects on perceived similarity that come naturally to GRT (e.g., density and base rate effects like those illustrated in figure 9.2), but it seems much better suited than GRT to account for similarity judgments with novel stimuli. If a stimulus is truly novel, then GRT predicts that the entire perceptual history associated with that stimulus is the single point in perceptual space that represents the observer's single exposure to the novel stimulus. The probabilistic MDS model assumes that this single point is all that is needed to judge the similarity of this stimulus to any other stimulus. In contrast, the GRT equation 9.3 similarity measure requires the pdf that describes the effects of all possible exposures to this stimulus—something that is simply not available if the stimulus is novel, even to an ideal observer. Furthermore, note that with novel stimuli, most context effects are not possible. For example, by definition, there can be no base rate differences among novel stimuli (since they are only novel on first presentation). So one intriguing possibility that should be explored in future research is that GRT provides the best account of similarity judgments for familiar stimuli, whereas the probabilistic MDS model provides the best account for novel stimuli.

## 9.2 PREFERENCE

Suppose that on your next visit to Costco, a coffee supplier offers free tastes of a variety of different cups of coffee. How do you decide which one you like best? More than half a century ago, Coombs (1964) proposed that we first imagine our ideal cup of coffee, then we compare the taste of each actual cup to this imagined ideal, and use these results to rank-order the actual cups by preference. The most preferred cup of coffee is the one that is most similar to our ideal cup. Coombs (1964) called this the *unfolding model* because when the stimuli vary on a single psychological dimension, the preference order can be used to unfold the stimulus representations from around the ideal point. This is arguably still the most widely used preference model and it has received support in a wide variety of empirical settings (e.g., Chernev, 2003; Davison, 1979; Dubois, 1975; Roberts et al., 2000; Rousseau et al., 2012).

The original unfolding model assumed a one-dimensional perceptual space (i.e., as in SDT) and was deterministic, in the sense that it assumed point representations for all stimuli and for the ideal. Subsequently, the model was generalized to multidimensional and probabilistic perceptual representations (e.g., De Soete et al., 1986; DeSarbo & Rao, 1984; DeSarbo et al., 1997; Ennis, 1993; Ennis & Johnson, 1994; Mullen & Ennis, 1991; Perrin, 1992; Schönemann & Wang, 1972; Zinnes & Griggs, 1974).

The unfolding model and its generalizations provide an accurate account of preference orderings, but these models are less successful at identifying the sensory characteristics associated with the imagined ideal. Some multidimensional versions of the model produce an MDS solution that situates each of the stimuli and the hypothetical ideal as a single point or probability distribution in a multidimensional space (e.g., De Soete et al., 1986; Zinnes & Griggs, 1974). However, as in traditional MDS, no information is provided about the nature of these dimensions. Sometimes, by noting which stimuli are situated at one extreme on a dimension and which stimuli are situated at the other extreme, it is possible to speculate about the nature of one or more dimensions. For example, if an MDS representation of odors places lemon and lavender at opposite ends of some dimension then one might infer that this dimension measures arousal—both are pleasant, but lemon is stimulating whereas lavender is calming. But with many dimensions, no such obvious ordering will emerge, and whatever inferences are made are generally impossible to test.

One experimental method for estimating the sensory characteristics of a stimulus, which is popular within the field of perception, is called the concurrent-ratings task. In this paradigm, participants rate the magnitude of each stimulus simultaneously on a number of sensory dimensions, and then the observed ratings are used to estimate the participant's sensory, perceptual, or cognitive impressions of the stimulus (Hirsch et al., 1982; Olzak, 1986). For example, consider an experiment in which participants first taste cups of coffee that were prepared using different amounts of ground coffee and different amounts of sugar. In a concurrent-ratings experiment, the participants might be asked to rate each cup on its sweetness and on the richness of its flavor (e.g., both on a 1 to 7 scale).

When stimuli are rated on a single sensory dimension—most commonly sensory magnitude—the resulting data often can be modeled accurately via SDT. In fact, as we saw in section 2.3.1 of chapter 2, the ratings task is a popular experimental method for estimating an ROC curve. When ratings are collected on multiple sensory dimensions, then the percepts are multivariate, rather than univariate, so GRT is the appropriate model. The GRT analysis of concurrent ratings data applies the SDT ratings model separately to each dimension—that is, it assumes that the participant establishes a set of criteria or cut-points on each rated dimension that partitions that dimension into intervals and that a different numerical rating is assigned to each interval, exactly as in figure 2.3 (Ashby, 1988; Wickens, 1992). Like SDT, this model assumes that on each trial, the participant determines which interval the percept is in on each rated dimension and then selects the associated ratings.

Ashby and Ennis (2002) combined the unfolding model and the SDT model of the ratings task to account for simultaneous sensory and liking ratings. This model used the participant's sensory ratings to estimate the sensory representation of the ideal. However, the model was only developed and applied to situations in which the various stimuli all varied on a single sensory dimension. Inglis et al. (2022) extended this model to more complex real-world stimuli that vary on many sensory dimensions. The resulting model, which they called the *GRT-unfolding model*, estimates the distribution of imagined ideals (i.e., across trials and participants) by identifying the ideal mean on each rated sensory dimension and estimating the variance-covariance matrix of the ideal distribution across all rated dimensions.

The GRT-unfolding model is illustrated in figure 9.6 for one hypothetical trial of a coffee-tasting experiment similar to our Costco experience. The main difference is that in this experiment participants are asked to rate: (1) the sweetness of the coffee; (2) the richness of the flavor; and (3) how much they like the coffee—each on a 1 to 4 rating scale. The figure depicts hypothetical events on a trial in which the participant rates sweetness and liking, but not richness. The circle in the left panel is a contour of equal likelihood from the bivariate normal distribution that represents all possible percepts that are elicited by the specific cup of coffee that the participant tastes on this trial. The star labeled $\underline{x}_i$ represents the specific percept experienced by the participant when tasting the current cup of coffee—that is, the specific perceived sweetness and richness of the current cup, which is the $i$th cup of coffee in the experiment. Let $x_1$ denote the perceived sweetness and $x_2$ the richness of the flavor (i.e., so $\underline{x}_i = [x_1, x_2]'$). The percept $\underline{x}_i$ is assumed to be a random sample from the bivariate normal distribution that describes all possible percepts elicited by this cup. Note that the perceived sweetness of this particular cup (i.e., $x_1$) falls in the interval assigned to a rating of 3, so in this hypothetical example, the participant responds with a rating of 3 when asked to judge sweetness.

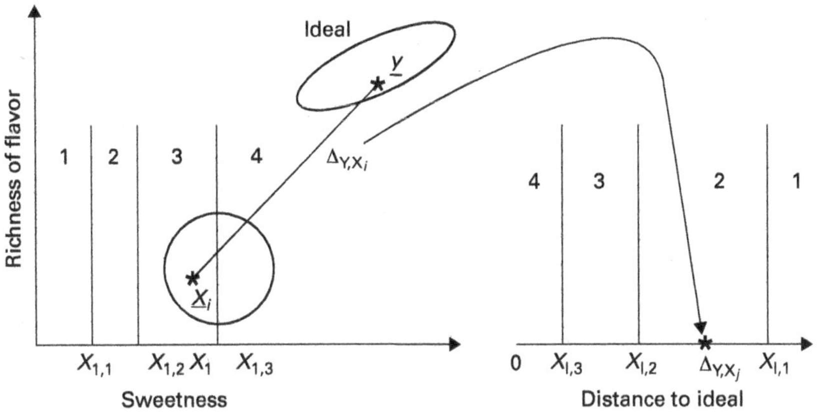


**9.6** A schematic illustrating the GRT-unfolding model for a hypothetical trial in which a participant tastes a cup of coffee and then provides ratings (from 1 to 4) on the coffee's sweetness and on liking. The circle and ellipse in the left panel are contours of equal likelihood from the sensory and ideal distributions, respectively. The participant's responses on this trial are "3" on sweetness and "2" on liking (adapted from Inglis et al., 2022).

The model assumes that if the participant had been asked to rate the richness of the coffee's flavor, rather than its sweetness, then the participant would have evaluated the position of the percept $x_2$ relative to the positions of three criteria established on the richness dimension (i.e., denoted $X_{2,1}, X_{2,2}$ and $X_{2,3}$, respectively). These are not shown in figure 9.6 to keep the figure as simple as possible.

The tilted ellipse in the top of the left panel of figure 9.6 is a contour of equal likelihood from the imagined ideal cup of coffee. Note that on average, the imagined ideal coffee is sweeter and has a richer flavor than the current cup. The star labeled $\underline{y}$ denotes the sensory values of the imagined ideal on this trial, which again is assumed to be a random sample from the bivariate normal distribution that describes all possible imagined ideals. So note that the model predicts that because of a variety of different sources of variability (e.g., in preference and memory), the imagined ideal changes from trial to trial. The model assumes that to respond with a liking rating, the participant imagines the ideal cup of coffee, computes the distance (or similarity) of the current cup to this imagined ideal, and then responds with a rating based on this distance, with greater distances (or lower similarities) eliciting lower levels of liking and therefore smaller ratings. In figure 9.6, the distance, denoted $\Delta_{Y,X_i}$, falls in the interval assigned to a rating of 2 (see the right panel), so the participant responds with a liking rating of 2 on this trial.

The GRT-unfolding model was developed specifically to apply to experiments in which participants rate the stimuli on multiple sensory dimensions. For example, Inglis et al. (2022) applied the model to an experiment in which participants rated the stimuli on six sensory dimensions. In this case, the perceptual representation of each stimulus is a six-dimensional multivariate normal distribution, and so each variance-covariance matrix includes 15 covariances and six variances. If these were all free parameters, then the model would include 27 parameters for each stimulus (15 covariances, six variances, and six means). An enormous amount of data would be required to estimate all these parameters accurately. Furthermore, estimation of the covariances would require simultaneous ratings on all possible pairs of dimensions, plus the assumption that all of these ratings are based on the same sensory sample of the stimulus. Unfortunately, this assumption seems untenable. For example, if a participant is asked to rate a stimulus on six different sensory dimensions, then it seems likely that the participant would reexamine the stimulus one or more times before responding with all six ratings. According to the model, the sensory representation of the stimulus after each examination is represented by a new random sample from the perceptual distribution. If ratings on two dimensions are based on different samples, then the correlation (e.g., across trials) between the ratings would not reflect the correlation between sensory dimensions.

For these reasons, Inglis et al. (2022) only considered applications of the model to experiments in which a single rating is requested on each trial. If there are $r$ sensory dimensions, then this means that each stimulus must be presented to every participant on at least $r+1$ different trials to ensure that all the necessary ratings are collected: $r$ trials are needed to collect a rating on each sensory dimension and one trial is required to collect the hedonic rating. In this case, no information about covariances is available, and as a result, Inglis et al. (2022) assumed that all covariances equal 0 and therefore that the variance-covariance matrix of every perceptual distribution is diagonal. Furthermore, they also assumed, without loss of generality, that all variances equal 1. This assumption just serves to set the arbitrary unit of measurement on each dimension. Collectively, these assumptions mean that, for any given stimulus $i$, the variance-covariance matrix $\Sigma_i = \mathrm{I}$, where I is the identity matrix.

In contrast, no constraints need to be placed on the ideal distribution. Specifically, Inglis et al. (2022) assumed that the imagined ideal stimulus, which is represented by the $r$-dimensional random vector $\underline{y}$, has a multivariate normal distribution with mean vector $\underline{\mu}_Y$ and variance-covariance matrix $\Sigma_Y$, and that no constraints are placed on any parameters in $\underline{\mu}_Y$ or $\Sigma_Y$.

The GRT-unfolding model assumes that the participant compares the current stimulus to the imagined ideal by computing the Mahalanobis distance between the two. This is just regular Euclidean distance, except each direction in perceptual space is weighted by its psychological importance, where importance is measured by the inverse of the variance of the ideal distribution along that direction. This requires some unpacking. To begin, think about why an ideal distribution might have different variances on each dimension. If a participant likes strong coffee but does not care much about sweetness, then the imagined ideal cup will always be strong but may vary considerably on sweetness. If my ideal car is a convertible sports car, then when I imagine it, it will always be a convertible and it will always be a sports car, but sometimes it might be blue and sometimes it might be gray. If a dimension is important for liking, then the imagined ideal will have consistent values on that dimension. In contrast, if a dimension is unimportant for liking, then one would expect the imagined ideals to vary widely on that dimension. In the figure 9.6 example, note that the imagined ideal distribution has greater variance on sweetness than on richness, and that the values on these two dimensions have a slight positive correlation. The greater sweetness variance indicates that sweetness is less critical to liking than richness because when participants imagine their ideal cup of coffee, they are more consistent in their imagined richness than in their imagined ideal level of sweetness.

Mahalanobis distance weights each direction by its psychological importance. To see how this works, consider the simple case in which there are only two stimulus dimensions and that the imagined ideal values on these dimensions are statistically independent. So the variance-covariance matrix of the ideal distribution is diagonal. Let $\sigma^2_{1,\text{I}}$ and $\sigma^2_{2,\text{I}}$ denote the variances of the ideal distribution on dimensions 1 and 2, respectively. Then the GRT-unfolding model assumes that the amount that the participant likes the current stimulus depends on the Mahalanobis distance between the current percept $(x_1, x_2)$ and the imagined ideal $(y_1, y_2)$, which under these distributional assumptions is equal to:

$$\Delta_{X,Y} = \sqrt{\left(\frac{x_1 - y_1}{\sigma_{1,\text{I}}}\right)^2 + \left(\frac{x_2 - y_2}{\sigma_{2,\text{I}}}\right)^2}. \tag{9.21}$$

Note that if both variances equal one then this just reduces to plain Euclidean distance. However, if for example, dimension 2 is much more important to the participant that dimension 1, then $\sigma_{2,\text{I}}$ will be much smaller than $\sigma_{1,\text{I}}$ and as a result, Mahalanobis distance will be much more sensitive to dimension 2 differences between the current stimulus and the imagined ideal than to dimension 1 differences.

In this case, Mahalanobis distance is equivalent to the distance metric used by the weighted Euclidean MDS model (see equation 9.12) and therefore also to the distance metric predicted by the GRT-similarity model of theorem 9.1 when $\rho = 0$ (i.e., see equation 9.18).

This important property of Mahalanobis distance is illustrated in figure 9.7. Note that in this example, the ideal distribution has smaller variance on the richness dimension than on sweetness, indicating that participants treat richness as a more important attribute of their ideal cup of coffee than sweetness. The two circles denote contours of equal likelihood from the perceptual distributions of two actual cups of coffee—$S_i$ and $S_j$. Note that the means of these distributions are both the same Euclidean distance $D$ from the mean of the ideal distribution. The units of Euclidean distance are the same as the units of the sensory space. In contrast, the units of Mahalanobis distance are standard deviations of the ideal distribution. For this reason, percepts elicited by cup $S_i$ will tend to be closer according to Mahalanobis distance to imagined ideal cups of coffee than percepts elicited by cup $S_j$. This is because the $S_i$ percepts are fewer standard deviations from the ideal than the $S_j$ percepts (i.e., the ideal standard deviation is larger on the sweetness dimension than on richness). As a result, the GRT-unfolding model predicts that participants will usually like cup $S_i$ more than cup $S_j$ in this example, despite the fact that both cups are the same Euclidean distance from the ideal mean.

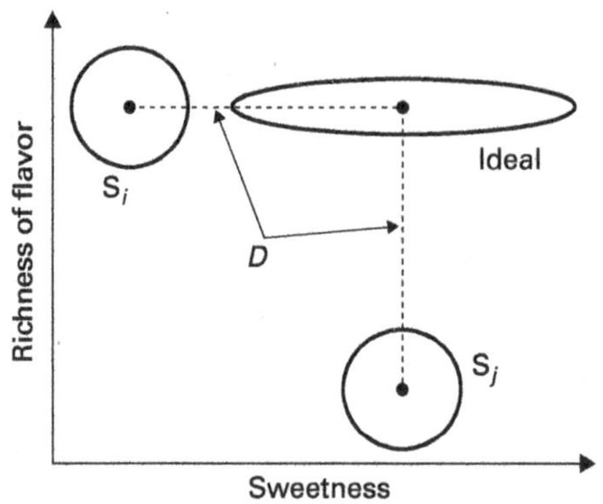


**9.7** An example of how Mahalanobis distance weights the importance of psychological dimensions. Although the mean percepts elicited by cups of coffee $S_i$ and $S_j$ are the same Euclidean distance from the mean of the ideal cup, the $S_i$ mean is closer according to Mahalanobis distance, and, as a result, the GRT-unfolding model predicts that participants will usually prefer cup $S_i$ more than cup $S_j$ (adapted from Inglis et al., 2022).

In the general case in which the stimulus varies on $r$ dimensions, and in which the ideal distribution may be characterized by one or more covariances, the Mahalanobis distance becomes

$$\Delta_{X,Y} = \sqrt{(\underline{x} - \underline{y})' \Sigma_Y^{-1} (\underline{x} - \underline{y})}, \tag{9.22}$$

where $\Sigma_Y$ is the variance-covariance matrix of the ideal distribution (see equation A.6 of appendix A). Because this equation uses the inverse of $\Sigma_Y$, Mahalanobis distance tends to weight each dimension by the inverse of the variability on that dimension. For example, if the correlation coefficient equals zero and there are only two stimulus dimensions then

$$\Sigma_Y = \begin{bmatrix} \sigma_{1,I}^2 & 0 \\ 0 & \sigma_{1,I}^2 \end{bmatrix}.$$

and therefore

$$\Sigma_Y^{-1} = \begin{bmatrix} \frac{1}{\sigma_{1,I}^2} & 0 \\ 0 & \frac{1}{\sigma_{1,I}^2} \end{bmatrix}.$$

The equation 9.22 distance measure is general enough that it allows Mahalanobis distance to preferentially weight a direction through perceptual space that is different from any perceptual dimension. This feature of the metric is illustrated in figure 9.8, which shows three different contours of equal distance predicted by the metric, for three different possible variance-covariance matrices. The GRT-unfolding model predicts that every point on each of these contours would be preferred equally. The black curve shows points that are equal Mahalanobis distance from the point (0,0) when the variance-covariance matrix equals the identity. In this case, as we saw, Mahalanobis distance reduces to simple Euclidean distance, in which case, distance is measured in the same way in every direction. This results in circular contours of equal distance. The dark gray curve was produced when the variance on dimension 1 was large, the variance on dimension 2 was small, and the correlation coefficient was zero. All points on this ellipse are the same Mahalanobis distance from (0,0). So according to this metric, moving a certain Euclidean distance vertically from (0,0) increases Mahalanobis distance more than moving horizontally from (0,0) by the same Euclidean distance.[1]

The light gray curve in figure 9.8 shows that Mahalanobis distance is general enough that it can preferentially weight a direction through perceptual space that is

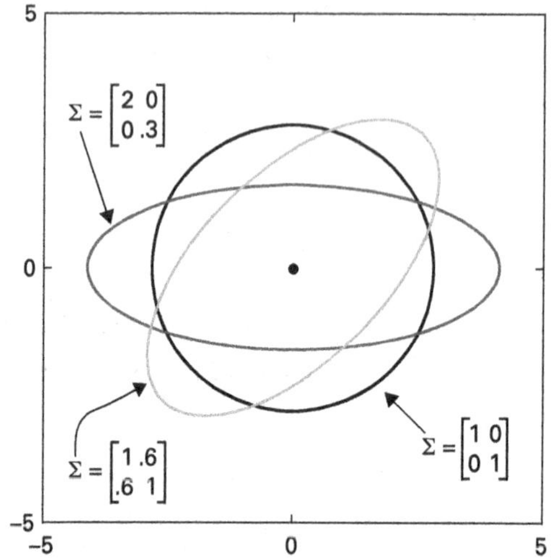


**9.8** Contours of equal distance predicted by three different versions of the Mahalanobis distance metric. The contours are defined by the variance-covariance matrices identified in the figure.

different from any perceptual dimension. This contour was produced by fixing both variances to 1 and setting the correlation coefficient to $\rho = .6$. Note that this version of Mahalanobis distance is most sensitive to changes in the direction from northwest to southeast (i.e., parallel to a line with slope $-1$) and least sensitive to changes in the direction from southwest to northeast (i.e., parallel to a line with slope $+1$). By varying the values of the two variances and the correlation coefficient, any direction through perceptual space can be privileged, although the metric is constrained to always predict that the least privileged direction must be orthogonal to the most privileged.

It is no coincidence that the contours of equal distance predicted by the Mahalanobis distance metric look like contours of equal likelihood predicted by the multivariate normal pdf. In fact, they are identical. This is because the equation of the multivariate normal pdf equals:

$$f(\underline{x}) = \frac{1}{(2\pi)^{r/2}|\Sigma|^{1/2}} \exp\left[-.5(\underline{x}-\underline{\mu})'\Sigma^{-1}(\underline{x}-\underline{\mu})\right], \tag{9.23}$$

where $r$ is the number of perceptual dimensions. Note that the set of all $\underline{x}$ that are associated with the same likelihood (i.e., with the same value of $f(\underline{x})$) satisfy the equation

$$(\underline{x}-\underline{\mu})'\Sigma^{-1}(\underline{x}-\underline{\mu}) = K, \tag{9.24}$$

for some constant $K$. But this is just the set of all $\underline{x}$ that are a Mahalanobis distance from the pdf mean equal to $\sqrt{K}$ (i.e., compare equations 9.22 and 9.24). It should also be noted that the distance metric predicted by the GRT-similarity model described in theorem 9.1 (i.e., in equation 9.18) is also exactly equal to Mahalanobis distance.

Although Mahalanobis distance uses the variances of the ideal distribution to differentially weight the stimulus dimensions, it is worth noting that the variances that characterize the ideal distribution and the variances as weights in Mahalanobis distance have different psychological interpretations. In the ideal distribution, the variances are measures of how much variability there is on each stimulus dimension when the participant repeatedly imagines an ideal stimulus. We would not generally expect these variances to be under the participant's control, just as the observer is assumed to have no control over the perceptual variances in SDT. In contrast, in Mahalanobis distance, the variances are used to measure the importance of each dimension. In general, we expect observers to have control over the psychological importance they assign each dimension—either through conscious deliberation or via the allocation plan that determines how selective attention is distributed across the various stimulus dimensions. For example, this is the typical assumption about the dimensional weights included in the weighted Euclidean MDS model. As a result, the variances that define the ideal distribution and the variances used to compute Mahalanobis distance could logically be different. The GRT-unfolding model uses the same parameter values for both, which is equivalent to assuming that participants estimate the importance of each dimension accurately and with no bias. Of course, a more general version of the GRT-unfolding model could assume that unique parameters define the two types of variance. But these extra parameters should only be added to the model if there is some compelling empirical evidence that they are needed.

The GRT-unfolding model predicts that all sensory ratings are made with bounds that satisfy decisional separability (e.g., as in figure 9.6). Therefore, the predicted probability of each possible sensory rating depends only on the marginal distributions on the rated dimension. These marginal distributions are all normal and as a result, the GRT-unfolding model for sensory ratings reduces to the SDT ratings model. So the predictions for sensory ratings are the same as those described in section 2.3.1 of chapter 2.

The liking ratings depend on the Mahalanobis distance between the current percept and the imagined ideal. This is a random variable since both of these are random samples from their own probability distributions. The good news is that this distance is one dimensional—that is, a single numerical value on each trial—and therefore conceptually the model is the same as the SDT model for sensory ratings. The bad news though is that the distribution of these distances is not normal, and this nonnormality considerably complicates model fitting.

The predicted probability that participants assign stimulus $i$ a liking rating of $j$ equals

$$\begin{aligned} P_L(j|S_i) &= P(X_{I,j} < \Delta_{Y,X_i} \leq X_{I,j-1}) \\ &= P(X_{I,j}^2 < \Delta_{Y,X_i}^2 \leq X_{I,j-1}^2), \end{aligned} \tag{9.25}$$

where $\Delta_{Y,X_i}$ is the Mahalanobis distance between the imagined ideal $\underline{y}$ and the sensory value $\underline{x}_i$ as defined in equation 9.22. Inglis et al. (2022) noted that the random variable $\Delta_{Y,X_i}^2$ has the distribution of a weighted sum of $r$ non-central $\chi^2$ random variables, each with one degree of freedom (Scheffé, 1999). Efficient numerical integration algorithms are available to compute the equation 9.25 probability under these distributional assumptions (e.g., de Micheaux & de Micheaux, 2017). However, in many applications $r$ will be large enough so that this weighted sum will be approximately normally distributed. For example, there were six rated sensory dimensions in the empirical application reported by Inglis et al. (2022), so normality seems a reasonable assumption. If so, then all that is needed to use the SDT ratings model is the mean and variance of $\Delta_{Y,X_i}^2$.

Inglis et al. (2022) showed that when there are $r$ sensory dimensions, $\Delta_{Y,X_i}^2$ has mean

$$\mu_{\Delta^2} = r + \text{trace}(\Sigma_Y^{-1}) + (\underline{\mu}_Y - \underline{\mu}_i)'\Sigma_Y^{-1}(\underline{\mu}_Y - \underline{\mu}_i) \tag{9.26}$$

and variance

$$\sigma_{\Delta^2}^2 = 2r + 4\text{trace}(\Sigma_Y^{-1}) + 2\text{trace}(\Sigma_Y^{-2}) + 4(\underline{\mu}_Y - \underline{\mu}_i)'\Sigma_Y^{-1}(\mathrm{I} + \Sigma_Y^{-1})(\underline{\mu}_Y - \underline{\mu}_i), \tag{9.27}$$

where the trace of a matrix is just the sum of the elements on the main diagonal. Therefore, the normal approximation computes the predicted probability that rating $j$ is assigned to stimulus $i$ on the liking dimension by computing the area between $X_{I,j}^2$ and $X_{I,j-1}^2$ under the pdf of a normal distribution with mean and variance specified by equations 9.26 and 9.27, respectively.

Inglis et al. (2022) reported an application of the GRT-unfolding model to a task in which participants rated pictures of 20 hypothetical planets on six sensory dimensions and on how important each image would be to record and send back to Earth if it happened to be captured by an unmanned deep-space probe. The data were aggregated across participants and then recorded in a 20 (number of planets)

$\times$ 7 (six sensory dimensions plus importance) $\times$ 7 (number of possible ratings) frequency array. These data include 840 degrees of freedom (i.e., 20 $\times$ 7 $\times$ 6), and the GRT-unfolding model that was fit to these data included a total of 183 free parameters.

Overall, the GRT-unfolding model accounted for 95.27% of the variance in the data, which seems impressive given that after parameter estimation there were still 657 degrees of freedom left to test the model (i.e., 840 – 183). Not surprisingly, the model was more successful at accounting for the sensory ratings than the hedonic ratings. In particular, the model accounted for 96.09% of the variance in the sensory ratings and 73.67% of the variance in the importance ratings. The variance-covariance matrix of the ideal distribution from the best-fitting version of the model indicated that the two dimensions that most affected the importance ratings were the prominence of water and the prominence of clouds and that the least influential dimensions were the presence of moons and the prominence of red-yellow colors.

Figure 9.9 shows a contour of equal likelihood from the ideal distribution of the best-fitting GRT-unfolding model when it is projected onto the plane defined by the two most important sensory dimensions—namely, water and clouds. The dots denote the mean of each planet distribution and the ideal mean. Note that water and clouds are negatively correlated in the ideal distribution. This makes sense because as cloud cover increases there is less available surface to display water. Figure 9.9 also shows the planets that were closest to and furthest from the ideal mean, and therefore, according to the model, the planets that were judged to be the most important, and least important, respectively, to photograph and send back to Earth. These planets are displayed in grayscale in figure 9.9, but in the Inglis et al. (2022) experiment, they were presented to participants in color. In those images, the gray areas in the most important (upper-right) planet were blue, whereas the gray areas of the least important (lower-left) planet were brown.

The GRT-unfolding model uses sensory ratings to build a probabilistic, multidimensional representation of the sensory experiences elicited by exposure to each stimulus. If participants rate the stimuli on *r* sensory dimensions then the sensory representations built by the model will be *r* dimensional. And the model will also build a representation of the ideal stimulus in this same space. It then attempts to account for hedonic ratings by measuring differences between the presented stimulus and the imagined ideal on each of these *r* sensory dimensions. This approach can only hope to account for the hedonic responses of participants if the rated sensory dimensions include all stimulus attributes that significantly affect the hedonic response.

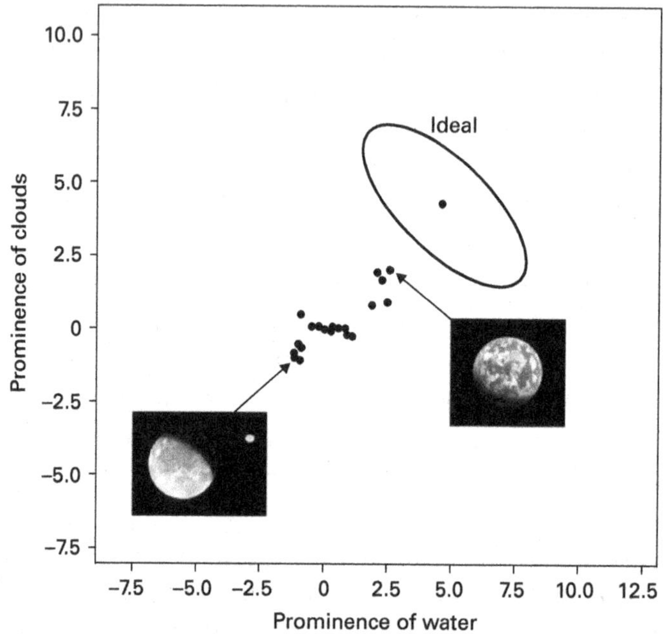


**9.9** A contour of equal likelihood from the ideal distribution of the best-fitting version of the GRT-unfolding model and the sensory means of each planet projected onto the plane defined by the water and cloud dimensions. Note that in the color versions of these planets, which were presented to participants, the dark gray areas in the rightmost planet were blue, whereas the dark gray areas of the leftmost planet were brown. This figure was adapted from Inglis et al. (2022).

To take an extreme example, consider an experiment in which participants rate a set of stimuli on *r* sensory dimensions but that the participants' hedonic responses to those stimuli depend exclusively on some other, unrated sensory dimension. In this case, the hedonic responses will be independent of the stimulus value on any of the rated dimensions, and therefore a comparison of the stimulus to the ideal values on the *r* rated dimensions will not predict the participant's hedonic response. So the efficacy of the GRT-unfolding model depends strongly on the ability of the experimenter to identify all sensory dimensions that could significantly affect the hedonic responses of participants to the selected stimuli.

Given this, the default expectation should be that the model will account for sensory ratings better than it accounts for hedonic ratings. In the experiment described by Inglis et al. (2022), the GRT-unfolding model accounted for 96% of the variance in the sensory ratings and 74% of the variance in the hedonic ratings. Therefore, one plausible account of this difference is that participants based their hedonic responses, at least in part, on some unrated dimension or attribute of the planets. Traditional unfolding models that are based on MDS are applied to data from experiments in which participants make hedonic responses, but in which no sensory ratings are collected. Therefore, as in traditional MDS, these models could just add more unspecified dimensions to the model until goodness-of-fit to the hedonic ratings is maximized. Even so, note that a better fit by such a model would provide only vague information about the sensory qualities of the ideal. Given that the GRT-unfolding model provides precise estimates of the sensory qualities of the ideal on all rated sensory dimensions, the ability to account for 74% of the variance in the hedonic ratings seems impressive, especially since the model was provided no information about how participants might make these judgments.

## 9.3 SAME-DIFFERENT JUDGMENTS

On each trial of a same-different experiment, two stimuli are presented and the observer's task is simply to judge whether the stimuli are the same or different. This is among the simplest judgments that observers can make. For example, evidence suggests that same-different judgments require less visual processing than an identification response (Farah et al., 1991). Another advantage of the same-different task is that it does not require observers to learn arbitrary response labels, unlike, for example, either categorization or identification experiments. Largely for these reasons, the same-different task has become a popular method for studying the visual processing of human faces (e.g., Perry et al., 2008).

Despite its simplicity, the same-different task has a rich empirical and theoretical history (e.g., Farell, 1985; Krueger, 1978). For example, one ubiquitous but perplexing result is that SAME responses are often faster than DIFFERENT responses, especially when the stimuli are fairly simple. As Krueger (1978) noted, finding a single difference between two images is sufficient to know that the correct response is DIFFERENT, whereas an observer can only be confident that a SAME response is correct after exhaustively scanning both images. For this reason, one would expect DIFFERENT responses to be faster than SAME responses. Krueger (1978) proposed that the fast-same effect might be the result of perceptual noise. If all percepts are noisy, as assumed by statistical decision theory, then the percepts elicited by any two stimuli will be different, even on same trials. In this case, finding a single difference between two images is not enough to know that they are different. Instead, a DIFFERENT response would require finding many differences.

A variety of other closely related paradigms use three stimuli, rather than two. These tasks are widely used in food science, where they are often collectively referred to as the *method of triads* (e.g., Ennis, Mullen, & Frijters, 1988). There are a number of variants. In the duo-trio method, a standard stimulus is presented along with two comparison stimuli. One of the comparison stimuli is identical to the standard and one is different. The observer's task is to select the comparison stimulus that is the same as the standard. The triangular method is similar, except the observer's task is to select the comparison stimulus that is different from the standard stimulus.

As was the case with similarity ratings, the appropriate model for same-different judgments depends on whether the observer's response is based entirely on the current percepts or whether it also depends on the observer's history with the stimuli. If it only depends on the percepts, then any noticeable difference in those percepts should be enough to elicit a DIFFERENT response. However, consider a task in which the stimuli are human faces that are arbitrarily labeled as A, B, C, and D. Now consider a trial in which the observer notices a difference in the two percepts but nevertheless identifies both faces as belonging to person A. For example, the amount of facial hair might be different in the two images, or the photographs might be from different perspectives. In this case, one would expect the observer to respond SAME, despite the perceptual difference. R. D. Thomas (1996) called models in which the response depends entirely on the current percepts, *distance-based models* because the decision presumably depends entirely on the distance between the two percepts in psychological space. In contrast, she referred to models that assume observers first identify each stimulus as *subclassification models*.

### 9.3.1 SUBCLASSIFICATION MODELS

Consider subclassification models first. No matter how the subclassifications are made, all such models assume observers will respond SAME if the two stimuli are identified as having the same response label. Therefore, in our hypothetical same-different face task, with faces A, B, C, and D, subclassification models assume a SAME response will be made correctly on same trials if both faces are identified correctly or if both are incorrectly identified but assigned the same incorrect label. To formalize this, let $X_K$ denote the event that the stimulus in position one is identified as stimulus K (for K = A, B, C, or D) and let $Y_K$ denote the event that the stimulus in position two is identified as stimulus K. Then the probability of responding SAME on a trial when the two stimuli are K and M (for any K or M = A, B, C, or D) equals (R. D. Thomas, 1996)

$$P\,[\text{SAME}|(\text{K,M})] = \sum_{i=\text{A}}^{\text{D}} P\,[(X_i, Y_i)|(\text{K,M})]. \tag{9.28}$$

Most approaches assume that the identification of each stimulus occurs independently of the other stimulus. In this case, equation 9.28 reduces to

$$P\,[\text{SAME}|(\text{K,M})] = \sum_{i=\text{A}}^{\text{D}} P\,[X_i|(\text{K,M})]\,P\,[Y_i|(\text{K,M})]. \tag{9.29}$$

Of course, the probability of responding DIFFERENT is just

$$P\,[\text{DIFFERENT}|(\text{K,M})] = 1 - P\,[\text{SAME}|(\text{K,M})]. \tag{9.30}$$

GRT same-different models can be built by computing the predicted probabilities in equations 9.29 and 9.30 from the relevant GRT identification model. In particular,

$$P\,[X_i|(\text{K,M})] = P(\text{R}_i|\text{K}) \tag{9.31}$$

and

$$P\,[Y_i|(\text{K,M})] = P(\text{R}_i|\text{M}). \tag{9.32}$$

Note that the right sides of these two equations specify a specific column of the identification confusion matrix (the column for response $\text{R}_i$) and two specific rows (the rows for stimuli K and M), but the notation is ambiguous with respect to the order of the matrix. For example, consider our hypothetical same-different face task with faces A, B, C, and D. Now consider a trial when faces A and B are presented. Then equation 9.29 includes the term $P[X_\text{B}|(\text{A,B})]$, which we would evaluate by computing $P(\text{R}_\text{B}|\text{A})$ from a GRT identification model. But is this from an identification experiment that only includes stimuli A and B, or from an experiment that includes stimuli A, B, C, and D? The former model is context free and the latter is context sensitive. Either model is logically possible, but conceptually only the context-sensitive model makes sense. By definition, when stimuli A and B are presented, the observer does not know their identities, and therefore could confuse either of them for any of the stimuli in the ensemble. As a result, the probabilities specified in equations 9.31 and 9.32 should be computed from the confusion matrix predicted by some GRT model for the identification experiment that includes all stimuli in the same-different stimulus ensemble—so, in this example, from an experiment that includes all four stimuli: A, B, C, and D.

Of course, if separate identification and same-different experiments are run with the same stimuli and the same participants, and the results from these two experiments are used to estimate the probabilities in equations 9.31 and 9.32, then whether the equalities specified in those equations will hold is an empirical question, even if our assumption that observers are subclassifying each stimulus is correct. This is because it is logically possible that the ability to correctly identify a stimulus is different when that stimulus is presented alone in an identification task as opposed to when it is presented as one of a pair of stimuli in a same-different task. For example, if the stimuli are presented side-by-side in the same-different task, then lateral masking or divided attention could reduce performance relative to a single-stimulus-per-display identification task. To my knowledge, no one has ever run identification and same-different tasks with the same stimuli and participants, so exploring these issues should be a goal of future research.

The critical point here is not that identification performance should necessarily predict same-difference performance, but rather that a GRT subclassification model of same-different performance begins with a GRT identification model and then the relevant entries in the predicted identification confusion matrix are inserted into equations 9.29 and 9.30 to compute predicted probabilities of SAME and DIFFERENT responses. Critically, all parameters of the base identification model should be estimated from the same-different confusion matrix.

Note that in an experiment with $N_S$ stimuli, a same-different task will include $N_S(N_S - 1)/2$ stimulus pairs. As a result, the same-different confusion matrix will have $N_S(N_S - 1)/2$ rows and 2 columns. Column 1 of row *i* will list the proportion of SAME responses on trials when stimulus pair *i* is presented, and column 2 will list the proportion of DIFFERENT responses. Each row of the confusion matrix must add to 1.0, and so the same-different confusion matrix includes only $N_S(N_S - 1)/2$ degrees of freedom. In contrast, the confusion matrix from an identification experiment with the same stimuli includes twice as many degrees of freedom [i.e., $N_S \times (N_S - 1)$], but many fewer stimulus types (i.e., $N_S$ stimuli versus $N_S(N_S - 1)/2$ stimulus pairs). Recall from chapter 7, that the full GRT identification model has $5N_S - 4$ free perceptual parameters in the case when the perceptual space is two-dimensional (see section 7.3). Therefore, the minimum number of stimuli for which the full GRT model is identifiable in the same-different experiment is $N_S = 11$. In this case, there are 55 different pairs of stimuli to present and the same-different confusion matrix has 55 degrees of freedom ($11 \times 10/2$). The full two-dimensional GRT model has 51 free perceptual parameters in this application ($5 \times 11 - 4$), which leaves four degrees of freedom to estimate parameters of the decision process and to test the validity of the model. In contrast, in an identification experiment, the full GRT model is estimable when there are only six stimuli (see section 7.3).

In any experiment, the true response probabilities must be estimated from the sample proportions. At least one-hundred trials per stimulus are required to guarantee that the standard error on any given sample proportion is no more than 0.05. An identification experiment with six stimuli will require 600 trials to meet this standard, which should be possible to collect in a single experimental session. Even so, participants must learn the response label assigned to each of these six stimuli and this learning process should be completed before the 600 trials commence. For this reason, an identification experiment with six stimuli would typically require each participant to complete at least two experimental sessions. In contrast, a same-different experiment with 11 separate stimuli will require 5,500 trials to estimate the 55 response probabilities accurately. The same-different task does not require participants to learn any response labels, but even so, at 600 trials per session, nine or 10 separate sessions would be required from each participant to estimate all of the relevant response probabilities. Given this, it is not surprising that I know of no same-different experiment that collected enough data to fit the full GRT model.

### 9.3.2 DISTANCE-BASED MODELS

Distance-based models of the same-different task assume that each same-different judgment is based purely on the current percepts of the two stimuli. The observer is assumed to respond SAME if those percepts are similar, and DIFFERENT if they are dissimilar. Logically, the decision rule could be stated in two different ways—that is, in terms of either similarity or dissimilarity. In the former case, the decision rule would be to respond SAME if the similarity is large and DIFFERENT if it is small, whereas, in the latter case, the decision rule would be to respond SAME if the dissimilarity is small and DIFFERENT if it is large. The similarity and dissimilarity versions of distance-based models make identical ordinal predictions. However, they can differ in their distributional assumptions, and these distributional differences can greatly affect the computational challenges associated with fitting the model to empirical data.

Let the vectors $\underline{X}_1$ and $\underline{X}_2$ denote the coordinates in perceptual space of randomly selected percepts associated with the stimuli in positions one and two, respectively.

All distance-based same-different models assume responding is determined solely by the location of these two points in perceptual space, and all probabilistic distance-based models assume these two points are random samples from some univariate or multivariate normal distributions that model the observer's perceptual experience with the two stimuli.

The dissimilarity versions of the distance-based models assume that, for any two stimuli K and M,

$$P[\text{SAME} \mid (\text{K,M})] = P\left[d(\underline{X}_1, \underline{X}_2) < X_d \mid (\text{K,M})\right], \tag{9.33}$$

where $d(\underline{X}_1, \underline{X}_2)$ is the distance between $\underline{X}_1$ and $\underline{X}_2$ in psychological space and $X_d$ is a parameter that represents the criterion that separates small from large dissimilarities. There are many different ways to define distance, and as a result, there are many different versions of this model. To my knowledge, the only versions that have been explored formally assume either Euclidian or city-block distance (i.e., see equation 9.20).

The similarity versions of the distance-based models assume that, for any two stimuli K and M,

$$P[\text{SAME} \mid (\text{K,M})] = P\left[s(\underline{X}_1, \underline{X}_2) > X_s \mid (\text{K,M})\right], \tag{9.34}$$

where $s(\underline{X}_1, \underline{X}_2)$ is the similarity between $\underline{X}_1$ and $\underline{X}_2$ and $X_s$ is the criterion that separates small from large similarities. Equation 9.34 seems to imply that similarity is the fundamental construct in these models. However, all current versions define similarity as an inverse function of dissimilarity or distance. Specifically, they all define similarity as

$$s(\underline{X}_1, \underline{X}_2) = \exp\left[-d(\underline{X}_1, \underline{X}_2)^{\alpha}\right], \tag{9.35}$$

where $\alpha$ is always set to either $\alpha = 1$, in which case equation 9.35 is called the exponential similarity function, or to $\alpha = 2$, in which case equation 9.35 is called the Gaussian similarity function. There are even more versions of the similarity-based models than of the dissimilarity-based models because the similarity-based models can set $\alpha$ to either 1 or 2, no matter how distance is defined.

Fitting either the dissimilarity or similarity versions to data is a computational challenge because all versions of both models include nonlinear transformations of the multivariate normal random vectors $\underline{X}_1$ and $\underline{X}_2$. In most cases, these transformations produce a random variable with no known distribution. There are a few exceptions however. One clever trick that minimizes computational difficulties is to choose a similarity function with a value of $\alpha$ that cancels the $1/\gamma$ exponent of the equation 9.20 Minkowski distance metric. The two most popular versions of this approach pair Euclidean distance with a Gaussian similarity function or city-block distance with an exponential similarity function.

Combining Euclidean distance with a Gaussian similarity function simplifies equation 9.34 as follows:

$$\begin{aligned} P\,[\text{SAME} \mid (\text{K,M})] &= P\left[s(\underline{X}_1, \underline{X}_2) > X_s \mid (\text{K,M})\right] \\ &= P\left\{\exp\left[-d(\underline{X}_1, \underline{X}_2)^2\right] > X_s \mid (\text{K,M})\right\} \\ &= P\left\{d(\underline{X}_1, \underline{X}_2)^2 < -\ln(X_s) \mid (\text{K,M})\right\} \\ &= P\left\{(\underline{X}_1 - \underline{X}_2)'(\underline{X}_1 - \underline{X}_2) < -\ln(X_s) \mid (\text{K,M})\right\} \\ &= P\left[\underline{Y}'\underline{Y} < -\ln(X_s) \mid (\text{K,M})\right], \end{aligned} \tag{9.36}$$

where $\underline{Y}$ has a multivariate normal distribution with mean vector $\underline{\mu}_1 - \underline{\mu}_2$ and variance-covariance matrix $\Sigma_1 + \Sigma_2$. As a result, $\underline{Y}'\underline{Y}$ has the distribution of a weighted sum of $r$ noncentral $\chi^2$ random variables, each with one degree of freedom, where $r$ is the number of dimensions of the perceptual space. This is a nonstandard distribution. Even so, there are efficient computational methods for evaluating the equation 9.36 probability (de Micheaux & de Micheaux, 2017).

Combining city-block distance with an exponential similarity function produces

$$P\,[\text{SAME} \mid (\text{K,M})] = P\left[\sum_{i=1}^{r} |Y_i| < -\ln(X_s) \mid (\text{K,M})\right], \tag{9.37}$$

where, as in equation 9.36, the $Y_i$ are the $r$ elements of the random vector $\underline{Y} = \underline{X}_1 - \underline{X}_2$. Each $Y_i$ is normally distributed and the absolute value of a normal random variable has a folded normal distribution, which has a known pdf, mean, and variance. Note that the distance-based version of the model that assumes city-block distance is almost the same since it predicts that

$$P\,[\text{SAME} \mid (\text{K,M})] = P\left[\sum_{i=1}^{r} |Y_i| < X_s \mid (\text{K,M})\right], \tag{9.38}$$

Unidimensional versions of the distance-based same-different models were developed by Frijters (1979) and Ennis, Mullen, and Frijters (1988). Versions that assumed a multidimensional perceptual representation were developed by Ennis and Mullen (1986) and Mullen et al. (1988). These versions all assumed that the decisions were based on dissimilarities, rather than on similarities. However, as noted in section 9.1.2, Ennis, Palen, and Mullen (1988) developed the closely related distance-based, probabilistic multidimensional similarity model that accounted for either an exponential or Gaussian similarity function.

### 9.3.3 FUTURE RESEARCH

There have been almost no attempts to compare the ability of subclassification and distance-based models to account for empirical same-different judgments—and the lack of such attempts is especially striking when compared to the voluminous literature that has examined the ability of competing models to account for the confusions that occur in identification or categorization tasks. One factor that undoubtedly has contributed to the sparse nature of this literature is the major effort required to collect a data set rich enough to test the most general models. As noted earlier, when the stimuli vary on two perceptual dimensions, a same-different experiment would need to include somewhere around 5,500 trials per participant to fit and test the most general GRT models. Collecting all these data would likely require nine or 10 separate experimental sessions. In contrast, an identification experiment rich enough to test the most general GRT models would only require one session for data collection, and perhaps one prior session to ensure that participants learn the appropriate response labels associated with each stimulus.

Despite this methodological challenge, a richer empirical data base is needed to improve our understanding of how people make same-different judgments. One obvious experiment that would not necessarily require all 5,500 trials would be to test for context effects. The GRT subclassification model predicts that same-different accuracy should be lower in regions of perceptual space where stimulus density is high than in regions where there are few stimuli (because there will be more confusions in the denser region). In contrast, the distance-based models predict no effect of stimulus density. Instead, they predict that same-different accuracy should depend only on the discriminability of the presented stimulus pair. Krumhansl (1978) reviewed some evidence that same-different accuracy is lower in regions of perceptual space where stimulus density is high, but more work is needed on this problem.

Another experiment, that is badly needed would include two separate conditions—both would use stimuli that vary on two perceptual dimensions, but one set of stimuli would include obvious response labels, whereas the stimuli in the other set would be difficult or impossible to label. For example, novel stimuli typically have no obvious label, and labels are much more difficult to apply to gustatory or olfactory stimuli than to visual stimuli. Next, same-different judgments would be collected on all possible pairs of stimuli within each stimulus set, and the experiment would include enough pairs that the full GRT models could be tested.

Given such data, a variety of different subclassification and distance-based models could be fit to the resulting same-different confusion matrices. An obvious prediction is that some distance-based model will provide the best fit to the data collected from the difficult-to-classify stimuli. If for some reason, a subclassification model fit these data best then both classes of models should be questioned, since the best-fitting model is assuming a cognitive process that is presumably impossible (i.e., that observers are classifying each stimulus before deciding same or different). Perhaps the more interesting question is which class of models would provide the best fit to the data from the easy-to-classify stimuli. In this case, both classes make plausible assumptions about the decision process. A finding that a distance-based model fits best when the stimuli are difficult to classify, but a subclassification model fits best when the stimuli are easy to classify would be important because this would provide evidence that people have multiple strategies for making same-different judgments.

### 9.3.4 CONSEQUENTIAL REGIONS AND THE EXPONENTIAL SIMILARITY FUNCTION

GRT models of same-different judgments seem closely related to Shepard's (1987) notion of consequential regions. According to Shepard (1987), a consequential region is the set of all points in perceptual space that are associated with the same response. In other words, a consequential region is the same as a response region in GRT. Shepard's main contribution was to show that if consequential regions meet a set of strong mathematical assumptions, then the probability that two stimuli will elicit the same response is an exponentially decreasing function of the distance between their point representations in perceptual space. Shepard (1987) called this the universal law of generalization. The assumptions required for this derivation were that (1) every point in perceptual space is equally likely to be the center of a consequential region; (2) all consequential regions are convex, extend for a finite distance in all directions, and are symmetric around their center; and (3) the probability that a region has volume $v$ has a gamma distribution with shape parameter equal to 2 and finite mean. Shepard (1987) also showed that this third distributional assumption is not critical, in the sense that so long as the first two assumptions hold, the generalization function is approximately exponential for a variety of qualitatively different pdfs on the volume of each consequential region.

Shepard (1987) did not explicitly describe the experimental conditions under which he expected his universal law to hold. Specifically, he did not describe exactly how one would estimate the probability that two different stimuli elicit the same response. There are a variety of plausible alternatives. For example, one estimate could be obtained from the confusion matrix that describes the results of an identification experiment. Specifically, one might estimate the probability that stimulus B elicits the same response as stimulus A with the proportion of A responses on trials when stimulus B is presented [i.e., with $\hat{P}(R_A|B)$]. Another possibility is to use the proportion of SAME responses from a same-different experiment. As empirical support of his law, Shepard (1987) cited data from a variety of different types of experiments, including identification, same-different, and paired-associate learning. Equation 9.29 shows that GRT makes very different predictions about same-different and identification experiments. As a result, GRT predicts that no universal law could describe such different response probabilities. Most experiments that Shepard (1987) cited, however, included pairs of stimuli (e.g., as in same-different and paired-associate learning experiments), and so the most obvious experimental paradigm to test Shepard's universal law is in same-different tasks. Following this approach, one would examine how the proportion of SAME responses on trials when two different stimuli A and B are presented changes as a function of some measure of the distance between the A and B perceptual representations.

Figure 9.10 shows the generalization gradient predicted by GRT in this experiment (black curve), along with the exponential function predicted by Shepard's universal law (solid gray curve), and the Gaussian function $\exp[-d(\mu_A, \mu_B)^2]$ (broken gray curve). The GRT model assumes two bivariate normal perceptual distributions—one for each of stimuli A and B. It assumes the two distributions have identical variance-covariance matrices, and that the same-different responses are made using a subclassification strategy based on the optimal (linear) bound. Distance is defined as the standardized distance between the distribution means (i.e., number of standard deviations between the means). Note that GRT predicts a Gaussian generalization gradient and thus conflicts with Shepard's universal law.

With distance-based models, the probability of a SAME response depends directly on the similarity between the two momentary percepts—that is, on the similarity between the random samples from the two perceptual distributions that are assumed to describe the current percepts of the two stimuli (i.e., see equation 9.34). Ennis (1988) showed that the expected value of this similarity decreases as an approximate Gaussian function of the distance between the means of the two perceptual

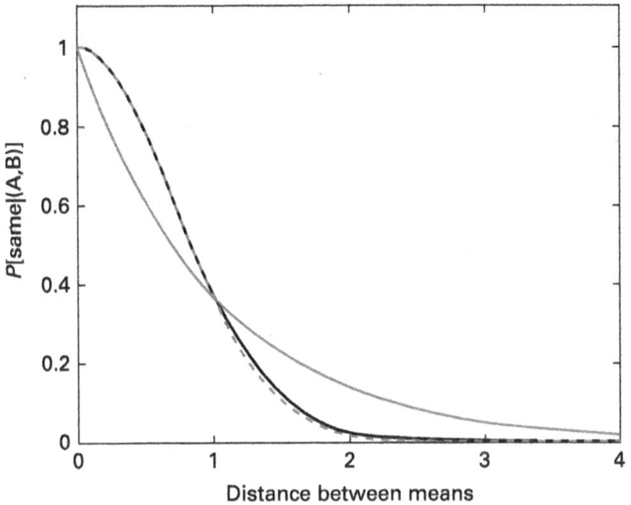


**9.10** The black curve is a generalization gradient predicted by GRT, the solid gray curve is an exponential generalization gradient predicted by Shepard's (1987) universal law of generalization, and the broken gray curve is a Gaussian function.

distributions, regardless of whether similarity is an exponential or Gaussian function of distance on each trial, and regardless of whether distance is computed using a city-block or Euclidean metric. In any empirical application, each summary statistic that would be used for model testing would, at best, be based on the mean similarity across many trials, rather than on any single-trial similarity. As a result, GRT seems to predict a Gaussian generalization gradient, rather than the exponential gradient predicted by Shepard's (1987) universal law of generalization. And this GRT prediction is quite general, in the sense that it holds regardless of whether the decision strategy depends on a subclassification process or is distance based, and in the latter case, on how distance is computed or on the nature of the function that relates trial-by-trial similarity to distance.

So what do the data say? Are empirical generalization gradients exponential or Gaussian functions of psychological distance? Unfortunately, this is a difficult question to answer. The universal law of generalization relates a poorly defined dependent variable (i.e., generalization) to an unobservable independent variable (i.e., psychological distance). On the one hand, defining generalization as the probability that two different stimuli elicit the same response sounds simple and unambiguous. However, there are many different empirical tasks that might be used to estimate such probabilities, including operant conditioning, identification, same-different judgment, paired-associate learning, similarity judgment, and classical conditioning. Many theories—including GRT—predict that observers will use different decision strategies in at least some of these tasks and, as a result, that the estimated generalization probabilities from these different tasks will have different forms. And how does one estimate psychological distance? As we saw in section 9.1, psychological distance is not a primitive in any current theories. The closest are theories, such as MDS and the similarity-choice model, that assume similarity is a fundamental construct. Even these approaches though must add extra assumptions to convert similarity to distance. And as we also saw in section 9.1, they both also assume that the resulting psychological distances obey the distance axioms—all of which appear to be of dubious psychological validity. Furthermore, other theories, including SDT and GRT, do not even acknowledge that distance or similarity are real psychological constructs.

Given all this ambiguity, it is unsurprising that there is no consensus on the true nature of generalization gradients. Shepard (1987) reported a number of empirical curves that were well described by an exponential function, but these came from a variety of qualitatively different tasks and they used MDS to estimate psychological distance. Other researchers have reported that a Gaussian generalization gradient provides a better account of human performance (e.g., Nosofsky, 1986). GRT is agnostic on this debate since it does not recognize psychological distance as a meaningful construct.

# 10

# MODELING RESPONSE TIMES

SDT and GRT were both originally developed to account exclusively for accuracy data. Even so, both versions of statistical decision theory have seen a number of extensions that also attempt to account for response times (RTs). A model that can account for both accuracy and RT data has obvious advantages over the traditional versions of SDT and GRT. For example, consider the simple YES-NO detection task that serves as the workhorse of SDT applications (i.e., as discussed extensively in chapter 2). Recall that the accuracy data from this task have only two degrees of freedom—most typically reported as the proportions of false alarms and hits. This is a meager harvest for the efforts expended to collect the data and allows us to estimate a maximum of only two parameters. For example, most SDT applications would use the two degrees of freedom to estimate $X_C$ and $d'$, with no degrees of freedom left over to test the validity of the model. However, if our model could also account for the RTs collected in this task, without any additional cost we would immediately increase the degrees of freedom to at least six—the proportions of false alarms and hits plus the mean RTs for each cell of the $2 \times 2$ confusion matrix (i.e., mean RTs of hits, misses, false alarms, and correct rejections). This is because, unlike the accuracy data, there are no row constraints on the mean RTs. For example, the proportion of misses plus the proportion of hits must equal 1, but there is no analogous constraint on the mean RT of misses plus the mean RT of hits. And this increase in degrees of freedom would be even more dramatic if we have enough data to estimate RT distributions in each of these four cells. Of course, more degrees of freedom means that we can test more detailed models, which should increase the amount we can learn about the perceptual and cognitive processes that produced our data.

The extensions of statistical decision theory that attempt to account for both accuracies and RTs are generally of two types. One approach is to add assumptions that allow the theory to make RT predictions but that are not detailed enough to account for psychological process. Thus, like the original versions of SDT and GRT, the resulting models are descriptive or, in the language of Marr (1982), computational. The other approach is to add enough structure to the theory to model psychological process—thereby producing models that Marr identified as algorithmic and that are more typically called process models in psychology. This chapter reviews both approaches. It is important to note, however, that this chapter is not intended as a general review of the large RT literature. For a more general treatment, see, for example, Van Zandt (2002) or the admittedly older books by Luce (1986) or Townsend and Ashby (1983). The focus here instead is strictly on extensions of statistical decision theory that can account for RT in addition to accuracy.

As a way to contrast the various models, and to provide a rough assessment of their empirical validity, I will highlight their hazard function predictions. The hazard function, which is widely used in reliability analyses, is defined as

$$h(t) = \frac{f(t)}{1 - F(t)}, \tag{10.1}$$

where $f(t)$ is the probability density function (pdf) and $F(t)$ is the cumulative distribution function (cdf). In reliability theory, $t$ is the time when a component fails or an organism dies, and so $h(t)$ describes the conditional likelihood that a failure or death occurs in the next instant, given that it has not yet occurred. The denominator $1 - F(t)$ is called the survivor function because it describes the proportion of components or organisms that have survived to time $t$. As a model of human lifespan, $h(t)$ is initially high because of the possibility of complications during birth and of birth defects. Following birth, $h(t)$ will tend to decrease as these negative factors become less likely. After reaching a minimal value, $h(t)$ will then tend to slowly increase throughout the lifespan, as normal aging occurs.

Hazard functions are a useful tool for contrasting RT models because there are few constraints on their shape. In contrast, all RT pdfs must increase and then eventually decrease to 0 because of the constraint that they all integrate to 1. Empirical RT pdfs are virtually all unimodal and skewed right, and all models reviewed in this chapter predict RT pdfs with these properties. So contrasting the RT pdf predictions of the various GRT RT models is not illuminating. Focusing on the cdfs is even worse, because all cdfs are nondecreasing functions that rise from 0 to 1. In contrast, hazard functions can have almost any shape—they can increase to infinity, increase to a constant value, decrease monotonically, or increase to a peak and then decrease.

Furthermore, empirical RT hazard functions all seem to have the same distinct qualitative shape—regardless of task. Specifically, for the most part, they are ordered by task difficulty, have generally flat tails, and are nonmonotonic in that they rise to a peak at an intermediate RT and then decrease to their flat tail, and these nonmonotonicities are most pronounced in the easier conditions. This same pattern has been reported in experiments on simple reaction times (Burbeck & Luce, 1982), subitizing (Balakrishnan & Ashby, 1992), memory scanning (Ashby et al., 1993), word-pronunciation times (Holden et al., 2009), and perceptual categorization (Maddox et al., 1998).

An example from perceptual categorization is shown in figure 10.1 (Maddox et al., 1998). Each participant in this experiment completed around 9,000 categorization trials—all with one of three different category structures (one-dimensional RB, linear II, or nonlinear II; see section 5.3). All three conditions included the same 18 stimuli that varied across trials on two perceptual dimensions. The hazard functions shown in figure 10.1 were estimated from the RTs to three different stimuli from one observer in the linear II condition. The three stimuli varied in their distance to

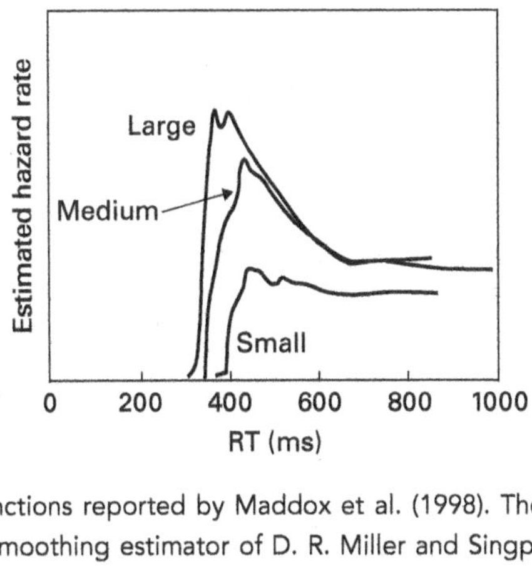


**10.1** Empirical hazard functions reported by Maddox et al. (1998). The three functions were estimated using the random smoothing estimator of D. R. Miller and Singpurwalla (1980) from the RTs of one observer in the linear II condition to three different stimuli. Of the three stimuli, one was closest to the optimal categorization boundary (denoted "Small"), one was an intermediate distance ("Medium"), and one was furthest from the optimal bound ("Large").

the optimal categorization boundary, and therefore also in their difficulty. The labels "Large," "Medium," and "Small" refer to these distances. Note that the hazard functions: (1) are ordered by trial difficulty; (2) all have flat tails; (3) are nonmonotonic with a single obvious peak at an intermediate RT; and (4) the nonmonotonicities are more pronounced the easier the condition. The hazard functions reported by Maddox et al. (1998) had this same qualitative pattern for all observers in all three conditions. Furthermore, the hazard functions reported in the previously mentioned simple reaction time, subitizing, and memory scanning experiments also displayed all these same qualitative properties.

The non-monotonicities that characterize figure 10.1 rule out many alternative RT models. Included in this list are many models that assume processing consists of a series of discrete stages (Ashby et al., 1993). For example, consider a model that assumes a serial process with $n$ discrete stages. Initially, the hazard rate must be low because at the onset of the trial, there are $n$ stages to complete before a response can be given. However, as each successive stage completes, there are fewer stages remaining to be completed, so the hazard rate will tend to rise. If $n$ is large and all stage durations are statistically independent, then by the central limit theorem, the RT will be approximately normally distributed, and, as this argument suggests, the hazard function of a normal distribution is strictly increasing. In fact, R. E. Barlow et al. (1963) showed that if the hazard functions of all stage durations are nondecreasing, then all serial (exhaustive) processing models must predict a nondecreasing RT hazard function.

A special case of this large class of excluded serial processing models is the exGaussian distribution (Hohle, 1965), which is arguably the most popular model of RT pdfs. In fact, Ratcliff (1978) was so confident in the ability of the exGaussian distribution to model RT pdfs accurately that he took the unusual step of first fitting the exGaussian to the RTs he collected and then fitting his drift-diffusion model to the best-fitting exGaussian distributions, rather than to the data he had collected.

The exGaussian is the distribution produced by adding two independent random variables—one with an exponential distribution and one with a normal distribution. Both have nondecreasing hazard functions, and so by the R. E. Barlow et al. (1963) theorem, the exGaussian hazard function must be nondecreasing. This property of the exGaussian distribution is illustrated in figure 10.2, which shows hazard functions predicted by the model for three conditions that roughly match the conditions shown in Figure 10.1. No attempt was made to fit the model to the empirical hazard functions shown in figure 10.1. The focus here is rather on the qualitative properties

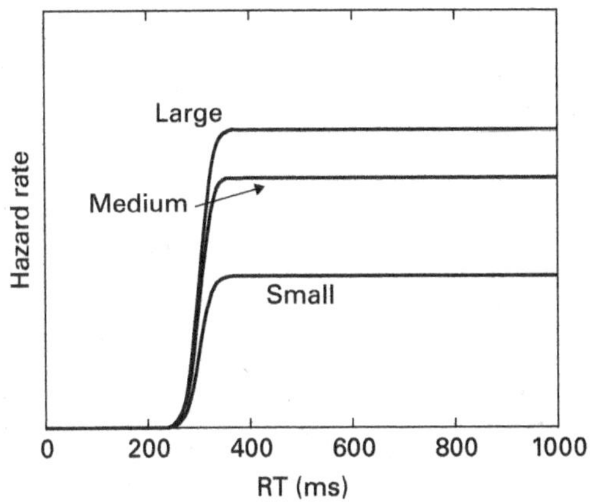


**10.2** Hazard functions predicted by the exGaussian distribution for conditions that are similar to those described in figure 10.1. In each case, the normal distribution had mean 300 ms and standard deviation 20 ms. The mean of the exponential distribution was 230 ms, 140 ms, and 120 ms, respectively for the small, medium, and large conditions.

of the predicted hazard functions. All exGaussian distributions have hazard functions like those shown in figure 10.2—that is, they all rise monotonically to a flat tail. As a result, note that the exGaussian distribution provides a poor model of the data illustrated in figure 10.1.

One obvious worry about this approach—that is, the approach via which we evaluate RT models by their ability to account for the qualitative properties of the empirical hazard functions illustrated in figure 10.1 and in so many other studies—is that some of these properties, most notably the nonmonotonicities, might reflect a bias in the statistical estimators that were used, rather than a property of the RTs themselves. Hazard functions are notoriously difficult to estimate since they are defined as a ratio of two distributional functions, and there is no consensus in the statistical literature on the best estimator (e.g., there is no known uniformly minimum variance unbiased estimator).

In the psychological literature, the most widely used hazard-function estimator is the nonparametric random-smoothing estimator of D. R. Miller and Singpurwalla (1980). This estimator was used by Maddox et al. (1998) to create the hazard function estimates shown in figure 10.1, but it was also used by Burbeck and Luce (1982), Balakrishnan and Ashby (1992), Ashby et al. (1993), and Holden et al. (2009). The random-smoothing estimator smooths a nonparametric estimator of the hazard function in a way that adjusts the amount of smoothing according to how much data are nearby. In time intervals where there are many RTs, the smoothing kernel is narrow, whereas in time intervals where there are few RTs, a wide kernel is used. Despite its attractive properties, both Ashby et al. (1993) and Van Zandt (2002) worried that the estimator might sometimes artificially deflate the estimated hazard rate in the tails, thereby producing a spurious peak (and nonmonotonicity) in the estimated hazard function. To guard against this possibility, Ashby et al. (1993) reported an additional analysis that suggested the nonmonotonicities in their estimated hazard functions were real and not a result of bias in the random smoothing estimator.

To investigate this issue further, I ran a small Monte Carlo simulation study to see if the random smoothing estimator would produce a spurious peak in its estimates of the hazard function predicted by the exGaussian distribution. I used sample sizes and paramter values that were consistent with the empirical hazard functions reported by Maddox et al. (1998) that are shown in figure 10.1. Some sample results are shown in figure 10.3. The gray curve is the hazard function predicted by the exGaussian distribution that I used to draw the random samples. The five black curves show the estimated hazard functions produced by the random smoothing estimator for five separate random samples of 400 RTs each (i.e., the approximate sample size used in the Maddox et al., 1998; study). Note that all five estimators show a consistent bias in that they overestimate the time when the hazard rate starts to rise.[1] More importantly however, note that the five estimates do not show any consistent spurious

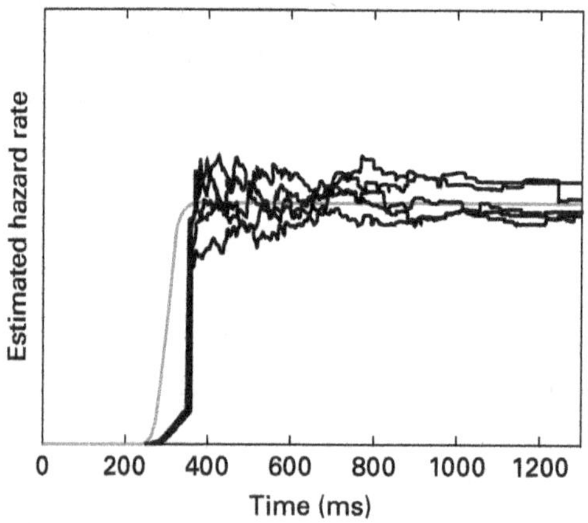


**10.3** Hazard function predicted by the exGaussian distribution (in gray) and five separate random smoothing estimates of this hazard function (in black). Each estimate is based on a separate random sample of 400 RTs drawn from the exGaussian distribution described by the gray hazard function.

peak. Instead, they accurately reflect the flat tail of the exGaussian hazard function. This analysis, together with the additional analysis reported by Ashby et al. (1993), suggests that the nonmonotonic hazard functions reported in so many previous studies reflect some fundamental property of human RTs, rather than a statistical bias in the hazard rate estimator used in those studies.

The remainder of this chapter surveys a variety of different SDT and GRT approaches to modeling RTs. To contrast the various models, and as a first pass at judging their empirical validity, their hazard function predictions will be examined. Specifically, we will examine the predicted hazard functions of each approach under experimental conditions that are similar to those used by Maddox et al. (1998) to estimate the hazard functions shown in figure 10.1. Figure 10.1 is used as the referent for these comparisons, not so much because these data are unusually important, but rather because they seem representative of almost all published hazard function estimates (i.e., regardless of task). Because figure 10.1 was chosen as representative, rather than as some gold standard per se, no attempts were made to fit any of the models to the specific hazard functions shown in figure 10.1. Instead, the goal is simply to examine whether the various models predict hazard functions that share qualitative features with those shown in figure 10.1. We begin with models that Marr (1982) would call computational, and then turn to process-level models of RT.

## 10.1 COMPUTATIONAL-LEVEL ACCOUNTS OF RT

This section reviews models that added assumptions to statistical decision theory that make RT predictions possible, but in which these extra assumptions make no claims about psychological process. The principle example of this approach was to add to either SDT or GRT an assumption that we saw in several earlier chapters called the RT-distance hypothesis. This hypothesis simply assumes that RT decreases with the distance between the percept and the response criterion or decision bound. The idea is that if decisions are made by comparing a percept to a decision bound or criterion, then the greater the distance between the two, the easier, and hence the faster the decision. Empirical support for this assumption was first reported within the context of SDT (Emmerich et al., 1972; Gescheider et al., 1969), and then later in GRT (Ashby et al., 1994). For example, Ashby et al. (1994) reported that categorization RTs decreased monotonically with the distance from the stimulus to the best-fitting decision bound in three separate experiments. Furthermore, they found no evidence that stimulus familiarity or the category prototypes played any special role in determining categorization RT.

Theoretically, the RT-distance hypothesis has been incorporated into statistical decision theory in two different ways. One approach is simply to assume that the RT-distance hypothesis holds, without specifying any functional form for the relationship between distance-from-criterion (or bound) and RT—only that RT decreases monotonically with this distance. This is the form of the RT-distance hypothesis that we saw in earlier chapters. Although this simple monotonicity assumption is not strong enough to make predictions about specific RT distributions, or even about specific mean RTs, it is sufficient to prove theorems 3.2 and 6.3. A second approach, which does make specific RT predictions, is to assume a specific functional form for the RT-distance relationship. The most common assumption is that RT decreases exponentially with distance-from-bound (e.g., Maddox & Ashby, 1996; Murdock, 1985), but models that assume a power function have also been investigated (Maddox & Ashby, 1996). The remainder of this section considers these two latter models in turn.

### 10.1.1 THE EXPONENTIAL RT-DISTANCE MODEL

Murdock (1985) proposed that the relationship between RT and distance-from-bound is exponential—specifically, that RT decreases exponentially with the distance from the percept to the decision bound:

$$RT = \alpha e^{-\beta D} + T_0, \tag{10.2}$$

where $D$ is distance between the percept and the response criterion or decision bound and $\alpha$, $\beta$, and $T_0$ are free parameters. This function is illustrated in figure 10.4, which also shows the predictions of the power-function model considered in the next subsection. Note that $T_0$ is the minimum possible RT, which occurs when a percept is infinitely far from the criterion. In contrast, the maximum possible RT, which occurs when a percept falls exactly on the decision bound is $\alpha + T_0$ (i.e., when $D = 0$). Finally, $\beta$ is the rate at which RT decreases with distance $D$.

A more realistic model would define $T_0$ as a random variable, rather than as a constant. For example, following the lead of the exGaussian model, one might assume that $T_0$ is normally distributed with some mean and variance that are free parameters. I chose not to follow this path because the goal of this chapter is not to build the most realistic model, but rather to understand the predictions of the various statistical decision theory models of RT. Allowing $T_0$ to have its own distribution would

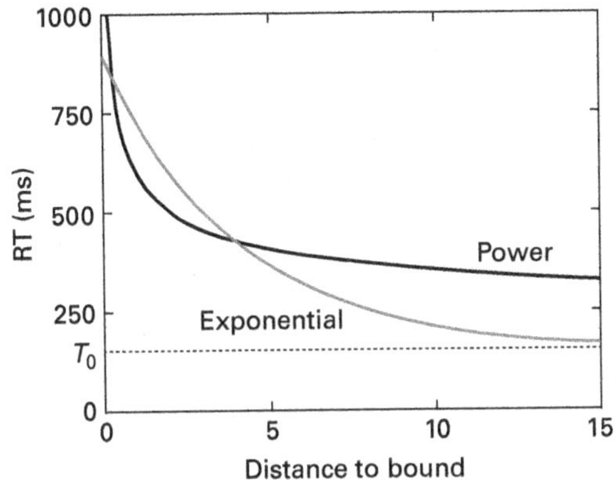


**10.4** The RT-distance functions predicted by the exponential and power-functions versions of the RT-distance hypothesis.

obscure the contributions of the RT-distance function to the model's predictions. For example, versions of the model in which $T_0$ is a random variable make qualitatively different RT hazard function predictions compared to versions in which $T_0$ is a constant.

The exponential RT-distance model is illustrated in figure 10.5. The bottom of figure 10.5 shows two bivariate normal perceptual distributions that might be predicted by some GRT model, along with a linear decision bound that separates the perceptual space into two response regions. The figure depicts events on a trial when stimulus $S_2$ is presented, and the "*" denotes the percept experienced by the observer on this trial [i.e., a random sample from $f_2(x_1, x_2)$]. The parameters $\Delta_1$ and $\Delta_2$ denote the distance from the means of each distribution to the decision bound. Collapsing these distributions onto the dimension orthogonal to the decision bound produces the two univariate normal distributions depicted immediately above the bivariate normals. The distance from the percept to the decision bound is denoted by $D_*$. The figure 10.5 model predicts that on each trial, RT decreases exponentially with $D_*$. This exponential relationship is shown in the top of the figure, which also shows that on this trial, the predicted RT is $RT_*$.

If the GRT perceptual distributions are multivariate normal and the decision bound is linear (e.g., as in figure 10.5), then the marginal distributions that result after collapsing onto the dimension orthogonal to the decision bound (or plane or hyperplane) are always univariate normal. For this reason, applications of this model (or the power-function model that we consider next) are identical in SDT and GRT. The only

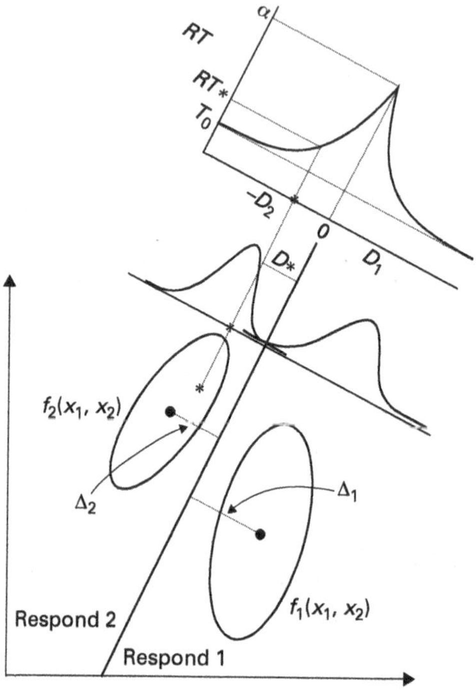


**10.5** An illustration of the RT-distance hypothesis.

difference is in how the means and variances of the univariate normal distributions are computed.

It is straightforward to derive analytic predictions for this model in experiments with two response alternatives and therefore only one decision bound (and hence only one distance-to-bound on each trial). However, analytic predictions are much more problematic in any experiment with more than two response alternatives, because, in these cases, more than one decision bound will be needed to carve the perceptual space into the appropriate number of response regions. As a result, each percept will be associated with separate distances to each different decision bound. This complicates the analysis in two different ways. First, there are multiple alternative models to consider. For example, one model might assume that the RT is completely determined by the minimum of all the different distance-to-bounds, whereas another possibility is that the RT depends on the maximum of these distances. A third model might assume a serial process in which the observer makes independent decisions about which side of each separate decision bound the percept is on, and that these separate decision times are each determined by some exponential

RT-distance model. The penultimate section in this chapter considers an approach that can decide among these alternative model architectures (i.e., systems factorial technology; see section 10.3).

The second complication from applying RT-distance models to tasks with more than two response alternatives is that certain rules for combining different distances can make the model analytically intractable. For example, if RT depends on the minimum of all distance-to-bounds then the distribution of these minimum distances will not be normal. In some cases, analytic predictions will be possible, but they will certainly not be as straightforward as in the case where the perceptual distributions are (multivariate) normal and there is only one decision bound. On the other hand, it it important to note that these complications are not restricted to the exponential RT-distance model but are shared, in one form or another, by all models considered in this chapter.

Murdock (1985) derived the mean, variance, and probability density function for correct RTs in a two-alternative task for the exponential RT-distance model under the assumption that the perceptual distributions are univariate normal. The RT density functions for both correct and incorrect responses are derived at the end of this chapter (see section 10.4). First, note that the model predicts that the minimum possible RT is $T_0$ and the maximum is $\alpha + T_0$, and therefore that the RT pdfs on all trials must equal zero for all $t < T_0$ and for all $t > T_0 + \alpha$. Now consider correct responses. The RT pdf for trials when stimulus $S_i$ is presented and the observer responds correctly equals

$$f_i(t) = \frac{\exp\left\{-\frac{1}{2\beta^2\sigma_i^2}\left[\ln\alpha - \ln(t - T_0) - \beta\Delta_i\right]^2\right\}}{\sqrt{2\pi}\;\beta\;\sigma_i\;\left[\Phi\left(\frac{\Delta_i}{\sigma_i}\right)\right]\;(t - T_0)}, \text{ for } T_0 \le t \le T_0 + \alpha, \tag{10.3}$$

where $\Delta_i$ is the distance from the mean of the $S_i$ perceptual distribution to the criterion or decision bound, $\sigma_i$ is the standard deviation of the $S_i$ (marginal) perceptual distribution along the direction orthogonal to the decision bound, and $\Phi(x)$ is the value of the cumulative Z distribution at $x$ (i.e., $\Phi(x) = P(Z \le x)$, where $Z$ is normally distributed with mean 0 and variance 1). The RT density functions for incorrect responses also equal zero over the same range. For $T_0 \le t \le T_0 + \alpha$, the incorrect RT pdf for stimulus $S_i$ equals

$$f_i(t) = \frac{\exp\left\{-\frac{1}{2\beta^2\sigma_i^2}\left[\ln\alpha - \ln(t - T_0) + \beta\Delta_i\right]^2\right\}}{\sqrt{2\pi}\;\beta\;\sigma_i\;\left[\Phi\left(\frac{-\Delta_i}{\sigma_i}\right)\right]\;(t - T_0)}. \tag{10.4}$$

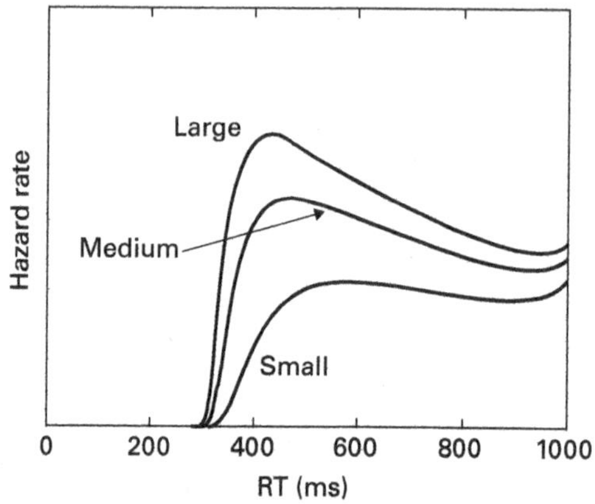


**10.6** Hazard functions predicted by the exponential RT-distance model under experimental conditions that are similar to the conditions used to estimate the hazard functions shown in figure 10.1. This figure was generated with the following parameter settings: $\alpha = 800$, $\beta = .35$, $\sigma_i = 1$, and $T_0 = 280$. The distances from the marginal distribution mean to the decision bound were 5, 6.3, and 7, for the small, medium, and large conditions, respectively.

The hazard function predictions of this model are illustrated in figure 10.6 under conditions that are similar to those used by Maddox et al. (1998) to estimate the hazard functions shown in figure 10.1. Note that overall, the exponential RT-distance model accounts for almost all of the qualitative properties that characterize the empirical hazard functions shown in figure 10.1. In particular, the hazard functions predicted by the model are ordered by trial difficulty, nonmonotonic, the nonmonotonicities are more pronounced on the easier trials, and the hazard rates converge to similar values for longer RTs. On the other hand, the model seems to make two systematic mispredictions. First, the tails are not flat, and second, the tails start to rise for extremely long RTs in all three conditions. In fact, if the RT axis in figure 10.6 was extended by even a few hundred milliseconds, it would be apparent that all three hazard functions eventually become almost vertical. Remember that the model predicts that the maximum possible RT is $\alpha + T_0$, which equals 1,080 ms for the figure 10.6 parameter settings. Therefore, if a response has not yet been given after say, 1,070 ms, the model predicts that the probability it will be given in the next 10 ms is 1. So the increase in the tails that is obvious in figure 10.6 can be attributed to the fact that the model sets a strict upper limit on the maximum possible RT. The fact that empirical hazard functions do not show such an increase suggests that human RTs are not usually constrained by any such upper limit. Allowing $T_0$ to be a random variable mitigates this problem but does not necessarily eliminate it. For example, the hazard function of the normal distribution increases sharply, so if $T_0$ was normally distributed, the hazard functions shown in figure 10.6 would still increase in the tails, albeit not so steeply.

Murdock (1985) concluded that the exponential RT-distance model was able to account for the major accuracy and RT results from both the memory-scanning task (Sternberg, 1966) and the study-test experiment (Murdock & Anderson, 1975). In addition, Maddox and Ashby (1996) reported that the model provided accurate accounts of the RT distributions and the accuracy data collected in a speeded-classification task. Even so, they also reported that the largest RTs they observed were considerably greater than the value predicted by the version of the exponential RT-distance model that provided the best fits to the RT pdfs. Overall then, these results suggest that the exponential RT-distance model provides a reasonably good model of the RTs expected in a variety of different experimental tasks, but that the model makes the apparently false prediction that human RTs are constrained by some fixed upper limit.

### 10.1.2 THE POWER-FUNCTION RT-DISTANCE MODEL

Another closely related RT-distance model assumes that RT decreases with distance-to-bound according to a power function, rather than exponentially (Maddox & Ashby, 1996). Specifically, this model assumes that

$$RT = \alpha D^{-\beta} + T_0. \tag{10.5}$$

This is the function I used to generate the RT pdfs shown in figure 3.3. The function is also illustrated in figure 10.4, where it is compared to the exponential RT-distance function. Note that the power-function model predicts that $T_0$ is again the minimum possible RT, which again occurs when a percept is infinitely far from the criterion. However, in contrast to the exponential model, the maximum possible RT approaches infinity as $D$ approaches 0. Thus, unlike the exponential model, the power-function model assumes there is no upper limit on possible RTs. As noted in the previous subsection, this prediction seems more consistent with the available RT data than the exponential model's prediction of a fixed upper limit.

When the perceptual distributions are normally distributed, the power-function RT-distance model predicts that the RT pdf on trials when the observer responds correctly equals 0 for all $t \leq T_0$, and for $t > T_0$ it predicts that

$$f_i(t) = \frac{\alpha^{\frac{1}{\beta}} \exp\left\{-\frac{1}{2\sigma_i^2}\left[\alpha^{\frac{1}{\beta}}(t-T_0)^{-\frac{1}{\beta}} - \Delta_i\right]^2\right\}}{\sqrt{2\pi}\,\beta\,\sigma_i\left[\Phi\left(\frac{\Delta_i}{\sigma_i}\right)\right](t-T_0)^{\left(\frac{1+\beta}{\beta}\right)}}. \tag{10.6}$$

A derivation is given in section 10.4. On error trials the model predicts that $f_i(t) = 0$ for all $t \leq T_0$, and for $t > T_0$ it predicts

$$f_i(t) = \frac{\alpha^{\frac{1}{\beta}} \exp\left\{-\frac{1}{2\sigma_i^2}\left[\alpha^{\frac{1}{\beta}}(t-T_0)^{-\frac{1}{\beta}} + \Delta_i\right]^2\right\}}{\sqrt{2\pi}\,\beta\,\sigma_i\left[\Phi\left(\frac{-\Delta_i}{\sigma_i}\right)\right](t-T_0)^{\left(\frac{1+\beta}{\beta}\right)}}. \tag{10.7}$$

Hazard function predictions of the model are shown in figure 10.7. Note that these are similar to the hazard functions predicted by the exponential RT-distance model (i.e., compare to figure 10.6). This was expected since the models are identical except for the slightly different functions they assume that relate distance-from-bound to RT (i.e., see figure 10.4). The main difference between the predictions of the two models is in the tails. Rather than the sharp rise predicted by the exponential model, the power-function model predicts a slow decay to an asymptote. This difference is due to the different predictions the models make about percepts that are close to the decision bound and, therefore, that generate the smallest possible distances and the longest possible RTs. The exponential model predicts that there is no distance that can lead to an RT greater than $\alpha + T_0$, whereas the power-function model predicts that percepts close to the decision bound can lead to arbitrarily large RTs.

Maddox and Ashby (1996) reported that the power-function RT-distance model provided accurate fits to the accuracy data and RT distributions collected in a speeded-classification experiment. Even so, they reported that these fits, although good, were slightly worse than the fits derived from the exponential RT-distance model. Overall, the results reported by Maddox and Ashby (1996) suggest that the power-function model might be better than the exponential-function model at accounting for the longest RTs, but that the exponential-function model might be better at accounting for shorter RTs.

## 10.2 PROCESS MODELS OF RT

The RT-distance models provide accurate accounts of accuracy and RT data from a variety of different paradigms. But a complete understanding of a behavior requires

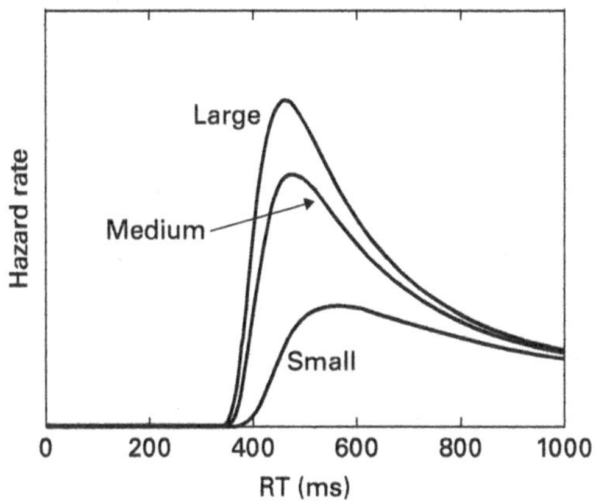


**10.7** Hazard functions predicted by the power-function RT-distance model under experimental conditions that are similar to the conditions used to estimate the hazard functions shown in figure 10.1. This figure was generated with the following parameter settings: $\alpha = 450$, $\beta = .35$, $\sigma_i = 1$, and $T_0 = 100$. The distances from the marginal distribution mean to the decision bound were .5, 1.8, and 2.2, for the small, medium, and large distances, respectively.

more than an ability to predict what response will be made on each trial and how long it will take the observer to make that response. Many researchers want to know *how* the response was selected. What underlying psychological processes mediated the behavior? RT-distance models make no assumptions about underlying psychological process, so they provide little or no insight into this issue. In this sense at least, they were always viewed as incomplete, and, as a result, there is a long history of building process models that are compatible with statistical decision theory.

There are two general approaches. The most popular, which dates back to the 1960s, is to assume that any binary decision can be modeled by a diffusion process or a random walk. An alternative approach, which specifies the underlying processes in more detail, is to assume that the GRT perceptual representation evolves from the operation of a stochastic dynamical system. Although at first glance, these two approaches seem very different, as we will see, they make similar predictions. This section considers each approach in turn.

### 10.2.1 DIFFUSION MODELS

Diffusion models belong to a more general class of sequential-sampling models that assume the observer takes multiple, independent looks at the stimulus on each trial. Each look produces a sample point in perceptual space, and then all these perceptual samples are somehow cumulated to select a response. Different models make different assumptions about how many samples are collected, and about how all these samples are used to select a response. Random-walk models assume a finite number of discrete samples, whereas drift-diffusion models assume continuous sampling. As a result, any given diffusion model can be seen as the result of taking the limit of some random-walk model as the time between samples approaches zero.

**10.2.1.1 Drift-diffusion generalizations of SDT** The first model that could be viewed as a process interpretation of SDT was proposed by Stone in 1960, and was based on Wald's (1947) sequential probability ratio test (SPRT). The SPRT tests between two statistical hypotheses using a sequential process in which a decision is made after each new sample is collected. Depending on the numerical value of the sample: (1) sampling is terminated and the null hypothesis is rejected; (2) sampling is terminated and the null hypothesis is accepted; or (3) another sample is collected and the process is repeated. The SPRT is optimal in the sense that, for any given set of type I and type II error probabilities, no other testing method can achieve a smaller expected sample size.

When used as a model of the YES-NO detection task, Stone's (1960) idea was that the observer makes optimal use of a sequence of sensory samples to test between the null hypothesis that the samples are from the N distribution and the alternative hypothesis that the samples are from the SN distribution. So for example, if the first few sensory values are near zero, the observer might suspect that no signal is present but not yet have sufficient confidence in this decision to terminate sampling and make a response. However, if the next sample is strongly negative, then the likelihood that these sensory values were produced on an SN trial is so low that the observer can confidently terminate sampling and respond NO. If so, then this trial would result in a fast NO response.

The optimality assumption of the SPRT model means that observers are assumed to base their detection decision on the likelihood ratio, rather than on sensory magnitude. Specifically, if the sensory magnitude of the $i$th sensory sample is $X_i$, then the SPRT assumes the observer uses the likelihood ratio $f_{\text{SN}}(X_i)/f_{\text{N}}(X_i)$ to update their decision statistic, rather than the sensory magnitude $X_i$. Although direct tests of whether observers base detection decisions on likelihood ratio or sensory magnitude are challenging, the available evidence strongly favors sensory magnitude (Balakrishnan & Ratcliff, 1996). Furthermore, the SPRT makes one prediction that is easy to disconfirm—namely, that correct and incorrect mean RTs must always be equal. In contrast to this prediction, errors tend to be faster than correct responses when the task is easy and speed is emphasized in the instructions, whereas errors tend to be slower than correct responses when the task is difficult and accuracy is emphasized (e.g., Luce, 1986).

It turns out that the RT-distance models can predict any relationship between RTs on correct versus error trials, and that which one is predicted to be longer depends on the hazard function of the perceptual distributions.[2] Specifically, E.A.C. Thomas (1971) showed that (1) if the hazard function of the perceptual distribution is increasing then the RT-distance models predict that errors are slower than correct responses; (2) if the hazard function of the perceptual distribution is decreasing then the RT-distance models predict that errors are faster than correct responses; and (3) if the hazard function is a constant, which occurs only if the perceptual distributions are exponential, then the RT-distance models predict that correct and incorrect median RTs must always be equal. Normal distributions have increasing hazard functions, so when the RT-distance hypothesis is added to the Gaussian GRT model, the resulting model predicts that errors will be slower than correct responses. This is because the normal distribution has low tails, so outliers that cause errors will tend to be closer to the decision bound than percepts that lead to correct responses.

Link and Heath (1975) generalized the SPRT to a much larger class of random walk models that they called *relative judgment theory* (RJT). RJT models allow the detection decision to be based on sensory magnitude, they can account for any relationship between correct and incorrect mean RTs, and for both optimal and suboptimal decision making. The RJT random-walk models are conceptually similar to the drift-diffusion model described in figure 2.11 in which the accumulation of evidence either drives the random walk generally upward on SN trials or generally downward on N trials. A YES response is given after the first sample on which the evidence exceeds a positive criterion A, and a NO response is given after the first sample on which the evidence falls below a negative criterion –B. There are two fundamental differences between the drift-diffusion model and RJT. First, RJT assumes time is discrete, whereas the drift-diffusion model assumes continuous time. Second, drift-diffusion models assume the drift or step-size distribution is normal, whereas the most general form of RJT makes no assumptions about this distribution. In the drift-diffusion model, the assumption that time is continuous forces normality, because if time is continuous then the accumulation of information is always an infinite sum, and so normally distributed according to the central limit theorem.[3] Since the accumulation sum does not have an infinite number of terms in RJT, the distribution is not necessarily normal, and the flexibility of RJT in setting this distribution is what allows the theory to account for any possible relationship between correct and incorrect mean RTs.

The primary disadvantage of RJT is computational. The derivation of closed form expressions for the predicted mean RTs assumes that on the step when the random walk first crosses either absorbing barrier (either A or –B), the amount that it overshoots that barrier equals zero (e.g., see Townsend & Ashby, 1983). In other words, the derivation assumes that the random walk always eventually takes on the exact value of either A or –B. If A and B are large relative to the mean step size, then the predicted mean RTs should be fairly accurate. However, if this condition is not met, then the predictions derived from these closed-form expressions could be inaccurate.

The drift-diffusion process interpretation of SDT was introduced in section 2.7. The model was first proposed as an account of memory retrieval (Ratcliff, 1978) but has subsequently been applied to a wide variety of different tasks and behaviors (for a review, see, e.g., Ratcliff et al., 2016). As mentioned, the drift-diffusion model can be viewed as a limiting form of RJT, where the limit is taken as the time between samples approaches zero. The basic form of the drift-diffusion model is easily rejected because the normally distributed step sizes that are a consequence of continuous-time sampling force the model to predict equal correct and incorrect mean RTs. Therefore, the basic drift-diffusion model suffers from the same weakness as the SPRT model. Even so, a more general form of the model can account for slow errors by assuming mean drift strength varies probabilistically trial by trial (Ratcliff, 1978), and it can account for fast errors by assuming that the starting value of the diffusion process also varies probabilistically trial by trial (Ratcliff & Rouder, 1998).

This more powerful model, however, comes at considerable cost. First, the model becomes much more difficult to fit. There are closed-form expressions for the RT pdfs predicted by the basic model, but these are complex since they include infinite sums (i.e., see the next subsection). Given this, it is hardly surprising that allowing the mean drift strength and the starting point to vary randomly trial by trial results in a model in which the predictions have no closed-form expression. As a result, fitting this more general (i.e., full) drift-diffusion model requires sophisticated numerical integration over infinite sums (Ratcliff & Tuerlinckx, 2002). Luckily, several computer packages are available that automate this model-fitting process (Vandekerckhove & Tuerlinckx, 2007; Wiecki et al., 2013). This is fortunate because the requisite code is complex enough that writing it successfully from scratch would likely require someone deeply immersed in the subtleties of diffusion theory and with strong coding skills.

The second cost of allowing variability in the mean drift rate and starting point is that the extra parameters required to describe these added probability distributions increase the mathematical flexibility of the model and, therefore, make it more difficult to falsify (Jones & Dzhafarov, 2014a, 2014b). In fact, a common approach is to add even more parameters to the full model by making two extra assumptions (Ratcliff & Tuerlinckx, 2002). First, it is commonly assumed that RT includes a nondecision time that is itself a random variable (similar to $T_0$ in equation 10.2), which is characterized by a probability distribution described by extra parameters that must be estimated from the data. A second common assumption is that observable RTs are contaminated by fast and slow outliers that must be removed before model fitting begins. The algorithm used to discard outliers includes parameters that serve as cutoffs for the discarding process.

In summary, the full drift-diffusion model requires a major computational effort to fit, and the resulting model is so mathematically flexible that a good fit to data should not necessarily be interpreted as strong evidence in support of the model.

**10.2.1.2 Stochastic GRT** Ashby (2000) generalized the basic drift-diffusion model to multiple perceptual dimensions. The model is described in figure 10.8. In this version, the perceptual representations are the same as in classical GRT. Like the drift-diffusion model, application is restricted to tasks with two response alternatives. On each trial, the observer's experience with the stimulus is assumed to produce repeated samples from the relevant perceptual distribution. Each sample is compared to the decision bound and a signed distance is computed, which equals distance-to-bound if the percept is in the A region and minus distance-to-bound if it is in the B region. Note that this computation converts the multidimensional perceptual representation to a single decision dimension, and as a result, from this point onward the model is identical to the drift-diffusion model. As in the diffusion model, the signed distances are cumulated, and sampling continues until the sum crosses an upper or lower barrier (exactly as in figure 2.11, except with an "A" response replacing "YES" and a "B" response replacing "NO").

More formally, the idea is that the observer computes the discriminant value associated with each perceptual sample. Discriminant values, which were introduced in section 5.1.4, have the property that they are positive for all points on one side of a decision bound, negative for all points on the other side, and equal to zero for all

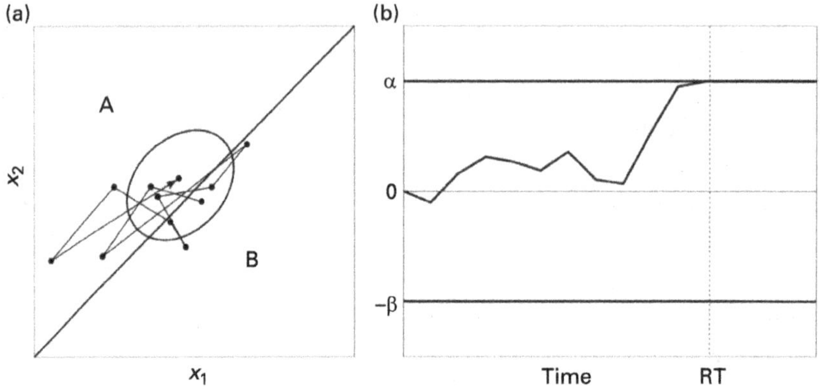


**10.8** The stochastic GRT model proposed by Ashby (2000). The figure depicts hypothetical events that might occur on a trial of a two-alternative forced-choice task on which stimulus A is presented and the observer responds correctly. The perceptual samples are drawn at discrete time for illustrative purposes only. Like the drift-diffusion model, the stochastic GRT model assumes continuous-time perceptual sampling.

points that lie exactly on the bound. For any linear decision bound, the discriminant function evaluated at the point $\underline{x}$, can be written as:

$$h = \underline{b}'\underline{x} + c, \tag{10.8}$$

where the vector $\underline{b}$ and the scalar $c$ are constants chosen so that all points on the bound satisfy $\underline{b}'\underline{x} + c = 0$, and so that $h > 0$ for any $\underline{x}$ in the A region and $h < 0$ for any $\underline{x}$ in the B region.[4]

The stochastic GRT model assumes that response decisions are based on a continuous-time discriminant process computed from continuous-time samples that are drawn from the perceptual distribution that describes the presented stimulus. The discriminant process is a continuous-time version of the equation 10.8 discriminant value. Specifically, consider a trial when stimulus $S_i$ is presented and denote the percept sampled from the $S_i$ perceptual distribution at time $t$ by $\underline{X}_i(t)$. Then the stochastic GRT model assumes that the discriminant process at time $t$ equals

$$h_i(t) = \underline{b}'\underline{X}_i(t) + c + \sigma_{Dc}\epsilon(t). \tag{10.9}$$

The last term models criterial noise; $\epsilon(t)$ is white noise (so zero mean and unit variance at each time $t$) and $\sigma_{Dc}$ is a parameter that models the standard deviation of criterial noise. The subscript "D" indicates that this is a parameter of the dynamic (i.e., stochastic) model. This discriminant process is integrated and compared to criterion values $\alpha$ and $-\beta$ as in figure 10.8b. At each time $t$, the process terminates and response A is given if the integral exceeds $\alpha$, the process terminates and response B is given if the integral is less than $-\beta$, and otherwise the process continues. It is well known that the integral is normally distributed at time $t$ with mean

$$\Delta_i t = (\underline{b}'\underline{\mu}_{Di} + c)\, t \tag{10.10}$$

and variance

$$\eta_i^2 t = (\underline{b}'\Sigma_{Di}\,\underline{b} + \sigma_{Dc}^2)\, t, \tag{10.11}$$

where $\underline{\mu}_{Di}$ is the mean and $\Sigma_{Di}$ is the variance-covariance matrix of the stimulus $S_i$ perceptual distribution. Note that if $t$ is set to 1, then these terms describe the mean and variance of the amount the process moves vertically in unit time. As a result, $\Delta_i$ is called the *drift rate* of the diffusion process and $\eta_i^2$ is often referred to as the variance.

Ashby (2000) provided closed form expressions for the predicted response probabilities, mean RTs, and RT pdfs predicted by this model. These are all straightforward to derive from the predictions of the standard diffusion process with two absorbing barriers (e.g., D. R. Cox & Miller, 1977; Ratcliff, 1978). For completeness, the predicted response probabilities and RT pdfs are given here. The predicted probabilities of responding A or B on a trial when stimulus $S_i$ is presented equal

$$P(A|S_i) = \frac{1 - \exp\left(\frac{-2\Delta_i \beta}{\eta_i^2}\right)}{1 - \exp\left[\frac{-2\Delta_i(\alpha+\beta)}{\eta_i^2}\right]}, \quad P(B|S_i) = 1 - P(A|S_i). \tag{10.12}$$

The predicted RT pdfs equal (i.e., for all $t > T_0$)

$$f_A(t|S_i) = \frac{1}{P(A|S_i)} \frac{\pi\, \eta_i^2}{(\alpha+\beta)^2} \exp\left(\frac{\Delta_i \alpha}{\eta_i^2}\right) \times \sum_{k=1}^{\infty} k \sin\left(\frac{k\pi\alpha}{\alpha+\beta}\right) \exp\left\{-\frac{1}{2}\left[\left(\frac{\Delta_i}{\eta_i}\right)^2 + \left(\frac{k\pi\eta_i}{\alpha+\beta}\right)^2\right](t - T_0)\right\}, \tag{10.13}$$

and

$$f_B(t|S_i) = \frac{1}{P(B|S_i)} \frac{\pi\, \eta_i^2}{(\alpha+\beta)^2} \exp\left(\frac{-\Delta_i \beta}{\eta_i^2}\right) \times \sum_{k=1}^{\infty} k \sin\left(\frac{k\pi\beta}{\alpha+\beta}\right) \exp\left\{-\frac{1}{2}\left[\left(\frac{\Delta_i}{\eta_i}\right)^2 + \left(\frac{k\pi\eta_i}{\alpha+\beta}\right)^2\right](t - T_0)\right\}. \tag{10.14}$$

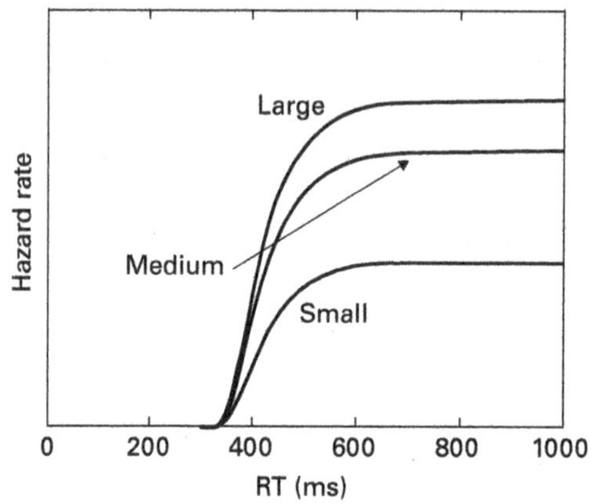


**10.9** Hazard functions predicted by the stochastic GRT model under experimental conditions that are similar to the conditions used to estimate the hazard functions shown in figure 10.1. This figure was generated with the following parameter settings: $\alpha = \beta = 80$, $\eta_i = 3$, and $T_0 = 300$. The drift rates were $\Delta_i = .07$, .17, and .2, for the small, medium, and large distances, respectively.

Note that both RT pdfs include an infinite sum. The terms in each sum converge to zero as $k$ approaches $\infty$, but unfortunately, the convergence is relatively slow. For example, to generate the predicted hazard functions shown in figure 10.9, at least 70 terms in the sum must be computed. But even then, the resulting predictions are unstable farther out in the tails (i.e., for RTs > 1,000 ms). Stable predictions for RTs as long as 5,000 ms required computing at least 120 terms.

Figure 10.9 shows the hazard functions predicted by the stochastic GRT model under experimental conditions that are similar to the conditions used to estimate the hazard functions shown in figure 10.1. Note that unlike the empirical hazard functions, these are all nondecreasing. Ratcliff et al. (1999) showed that if mean drift rate is allowed to vary probabilistically across trials, then the model predicts nonmonotonic hazard functions. Specifically, they rise to a peak and then slowly decay. So it appears that allowing mean drift rate to vary randomly across trials would not only allow the stochastic GRT model to account for slow errors (i.e., slower than correct responses) but also provide a better fit to empirical RT hazard functions.

The stochastic GRT model was constructed specifically as a stochastic generalization of the classical static GRT model. Note that the perceptual representations in both models are described by mean vectors and variance-covariance matrices, and both models postulate a decision bound. Although the stochastic model is the only one to make RT predictions, both models make accuracy predictions, so it is natural to ask whether these accuracy predictions agree or disagree. And if they do agree, what exactly is the relation between the perceptual parameters in the two models? To begin answering these questions, note that the stochastic model has two more free parameters than the static model—namely, the distance to the two absorbing barriers, $\alpha$ and $\beta$. If $\alpha \neq \beta$, then stochastic GRT will exhibit a response bias of a type that cannot be represented in classical GRT. Therefore, it is clear that a necessary condition for equivalence is that $\alpha = \beta$ in the stochastic model. The following result, due to Ashby (2000), provides a complete set of sufficient conditions for the two models to make identical accuracy predictions.

**Theorem 10.1** Consider a two-alternative task in which the observer uses a linear decision bound, and suppose: (1) there is no bias in the placement of the stochastic GRT response barriers (i.e., so that $\alpha = \beta$); and (2) the marginal distributions of perceptual and criterial noise on the dimension orthogonal to the decision bound have a logistic distribution in the static model (rather than the usual assumption of normality). Then the classical and stochastic GRT models predict identical response probabilities if the following three conditions hold.

1. The two models have identical decision bounds.
2. The perceptual distributions in the two models have identical mean vectors (i.e., $\underline{\mu}_{Di} = \underline{\mu}_i$).
3. The variance-covariance matrices that describe the stimulus $S_i$ perceptual distributions in the two models and the criterial noise variances are related as

$$\Sigma_i = \frac{\pi^2}{12(\alpha^*)^2}\Sigma_{Di} \quad \text{and} \quad \sigma_c = \frac{\pi^2}{12(\alpha^*)^2}\sigma_{Dc}, \tag{10.15}$$

where $\Sigma_i$ is the variance-covariance matrix of the stimulus $S_i$ perceptual distribution in the classical (static) GRT model and $\alpha^*$ is the standardized distance in the stochastic model from the origin to the absorbing barriers (i.e., see equation 10.11)—that is,

$$\alpha^* = \frac{\alpha}{\eta_i} = \frac{\alpha}{\sqrt{\underline{b}'\Sigma_{Di}\underline{b} + \sigma_{Dc}^2}}. \tag{10.16}$$

*Proof* See Ashby (2000). ■

This result is important for several reasons. First, it shows that the stochastic GRT model is a true generalization of the classical model. The two versions can make identical accuracy predictions, and in both versions the decision bounds and perceptual means have exactly the same interpretation. The assumption that the relevant marginal distributions in the classical model are logistic, rather than normal, is necessary for exact equivalence, but the logistic and normal distributions are highly similar. For example, the cumulative Z distribution and the most closely approximating cumulative logistic distribution differ by less than 0.01 over the entire Z axis (Bowling et al., 2009). As a result, if the marginal distributions in the classical model are normal, rather than logistic, the static and stochastic models might make slightly different accuracy predictions, but the predictions would likely be so similar that they would be empirically indistinguishable. For example, the Kolmogorov-Smirnov test requires a sample size of over 18,000 to detect a maximum difference in two cumulative distributions functions of 0.01 (i.e., with a type I error rate of 0.05).

Second, the theorem shows that the primary difference between the models is in their interpretation of the noise variance parameters. These variances are not equal, but instead are proportional, and the constant of proportionality is inversely related to the squared standardized distance from the origin to the absorbing barriers in the stochastic model [i.e., to $(\alpha^*)^2$; see equation 10.16]. Since the distance to the barriers in the stochastic model is assumed to be under the observer's control, the theorem therefore formalizes a long-held belief that estimates of perceptual noise obtained from SDT or classical GRT may often be corrupted by decisional influences (e.g., Green & Swets, 1966; Pachella, 2021).

This important conclusion requires a bit of unpacking. A good model should not only provide accurate fits to empirical data but also satisfy certain modeling principles. Among the most important of these is the principle of correspondent change, which states that if a model is correct, then it should be able to account for behavioral changes across experimental conditions by only changing the values of parameters that correspond to the independent variables that were varied to create the various conditions (Townsend & Ashby, 1983). This principle was always seen as a major strength of statistical decision theory models. For example, increasing the penalty for false alarms without changing any other payoffs causes observers to respond YES less frequently in the YES-NO detection task. When fit to the data from two conditions that differ only in the cost of a false alarm, estimates of the SDT criterion $X_C$ will increase with cost, but estimates of $d'$ will remain at roughly the same value across conditions. This makes sense because we didn't change the stimulus in any way, nor did we impair or facilitate the ability of the observer to perceive the stimulus. So we would expect a good model to tell us that the observer's sensitivity to the stimulus has not changed. SDT does exactly this. In contrast, to account for these same data with classical threshold theory, one would have to conclude that sensitivity decreased when the cost of a false alarm increased—because estimates of the threshold would have to increase to account for the reduced tendency of observers to respond YES.

Next consider a somewhat more complex categorization or identification experiment in which there are several different speed-accuracy conditions. Suppose that in one condition, observers are instructed to emphasize accuracy over speed, and in another condition they are encouraged to respond as quickly as possible, even if this extra speed costs them a few errors. Naturally, we expect accuracy to be higher in the accuracy condition than in the speed condition. Therefore, if a static GRT model (or an SDT model) was fit to the data from these conditions, it would have to account for the lower accuracy in the speed condition by assuming that the noise variances are larger than in the accuracy condition. But this represents a violation of the principle of correspondent change. The stimuli, display conditions, and observers are all identical in the two conditions, so the correct model should account for performance differences among the conditions by leaving all parameters that describe the perceptual representations fixed. Classical GRT would fail to meet this criterion, and therefore theorem 10.1 tells us that the the perceptual noise variances estimated from the classical GRT and SDT models will often be corrupted by decisional influences.

According to the stochastic GRT model, however, the observer responds to speed instructions by moving the absorbing barriers closer to the origin—that is, by decreasing $\alpha$. This allows the stochastic model to account for lower accuracy in the speed condition, even if the noise variances are the same in both conditions. In other words, the stochastic GRT model satisfies the principle of correspondent change in this example. Furthermore, the theorem shows how the stochastic GRT model can be used to predict exactly how such estimates will change as a function of speed stress.

A third important consequence of theorem 10.1 follows from the fact that the classical and stochastic versions of GRT make identical accuracy predictions only if the absorbing barriers in the stochastic model are equally distant from the origin (i.e., only if $\alpha = \beta$). In SDT, a response bias occurs if the response criterion (i.e., $X_C$) is set at any point for which the likelihood ratio does not equal one (assuming equal base rates; i.e., see equation 2.5). Classical GRT can represent a response bias in a number of different ways—for example, by using a decision bound that has a different functional form from the optimal bound (e.g., using a linear bound when the optimal bound is quadratic), or using a decision bound of the optimal type, but with suboptimal parameter values. In figure 10.8a, a bias could be introduced by increasing or decreasing the intercept of the decision bound. Both the classical and stochastic versions of GRT model can model response bias in this way, but stochastic GRT also has another completely different option for modeling a response bias—namely, by setting the distances to the two absorbing barriers to different values (i.e., so $\alpha \neq \beta$). If these distances are unequal, then the model will be biased in favor of the response associated with the closer barrier. These two separate sources of response bias are also explicitly included in the drift-diffusion model, as is illustrated in figure 2.11. Furthermore, Ratcliff (1985) showed that they are identifiable, in the sense that manipulating one of these two types of bias leads to different RT and accuracy predictions than if the other type is manipulated.

It is also straightforward to show that the same identifiability holds in the stochastic GRT model—that is, changing the intercept of a linear decision bound has a qualitatively different effect than changing the distance to the two absorbing barriers. To see this, consider the example shown in figure 10.8a. In this case, the A response region is associated with positive values of the discriminant process $h_i(t)$ and the B response region is associated with negative values. So a bias toward response A is created if the constant $c$ in the equation 10.9 discriminant process is increased, which happens when the intercept of the figure 10.8a bound is decreased.[5] Specifically, if the value of $c$ in the equation 10.9 discriminant process changes from $c$ to $c + \Delta_C$, then the discriminant process changes from $h_i(t)$ to $h_i(t) + \Delta_C$. As a result, the decision is to continue sampling if

$$-\beta < \int_0^t [h_i(y) + \Delta_C]dy < \alpha \tag{10.17}$$

or equivalently if

$$-\beta < \Delta_C t + \int_0^t h_i(y)\, dy < \alpha, \tag{10.18}$$

or finally, if

$$-\beta - \Delta_C t < \int_0^t h_i(y)\, dy < \alpha - \Delta_C t. \tag{10.19}$$

In other words, decreasing the intercept of the figure 10.8a decision bound is equivalent to changing the horizontal absorbing barriers to lines with slope $-\Delta_C$ (Ashby, 1983).

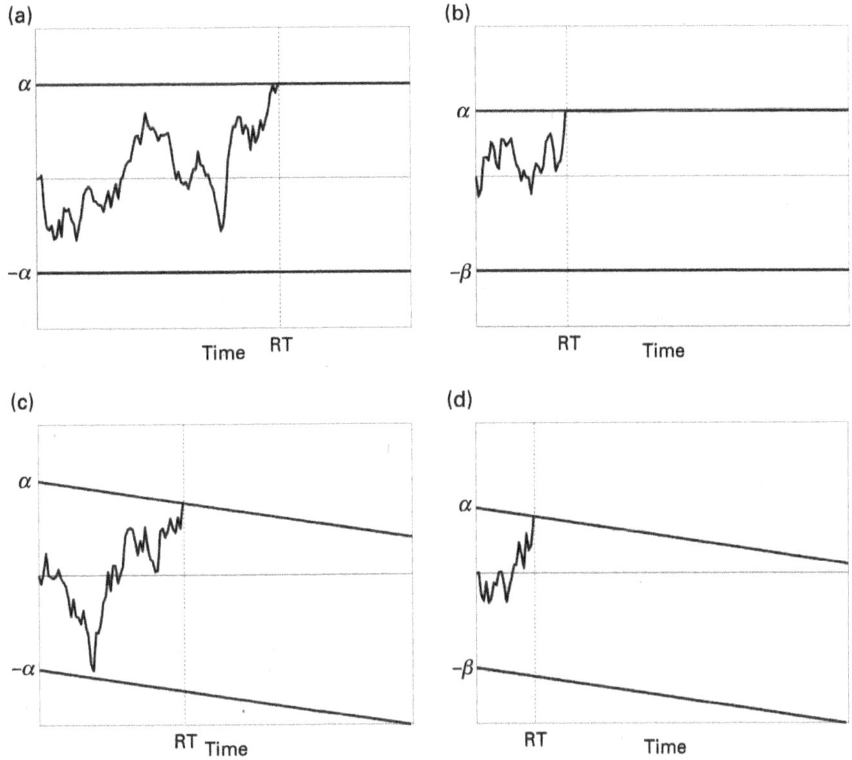


**10.10** Four different versions of the stochastic GRT model. (a) A model in which there is no response bias. (b) A model in which the decision bound is placed in an unbiased position, but the position of the absorbing barriers produces a bias toward response A. (c) A model in which the absorbing barriers are placed in an unbiased position, but the intercept of the decision bound is shifted to produce a bias toward response A. (d) A model in which the position of the absorbing barriers produces a bias toward response A and the intercept of the decision bound also produces a bias toward response A.

The two different ways that the stochastic GRT model can express a response bias are illustrated in figure 10.10. Panel (a) shows the diffusion process of the unbiased model—that is, the model in which the decision bound is set where the likelihood ratio equals one, and the distances to the two absorbing barriers are equal (i.e., $\beta = \alpha$). Panel (b) shows an example in which $\alpha < \beta$, so the model is biased to respond A, but even so, the decision bound is set to an unbiased position. In panel (c) the distances to the two barriers are equal, but the decision bound has been shifted away from the $S_A$ perceptual mean. The model shows a bias to respond A because the criterion to stop and respond A decreases with time. Finally, panel (d) shows an example in which both sources of bias are present—the response A barrier is closer to the origin than the response B barrier, and the decision bound has been shifted away from the $S_A$ perceptual mean. The important point here is that the two different methods of expressing a response bias in the stochastic GRT model clearly have qualitatively different effects on the model's predictions.

Note that the two methods of modeling a response bias also make very different psychological assumptions. If the decision bound is shifted away from the set of points for which the likelihood ratio equals 1, then there are trials where the observer interprets evidence that objectively favors one response as support for the contrasting response (assuming equal base rates). In contrast, if the bound is set where the likelihood ratio equals 1, but the distance to the response A barrier is less than the distance to the B barrier (i.e., if $\alpha < \beta$) then the observer is still biased to respond A, but now each piece of evidence is evaluated objectively. Evidence in favor of stimulus $S_A$ always moves the process toward the A barrier and evidence in favor of stimulus $S_B$ always moves the process toward the B barrier. A response bias occurs only because the observer is willing to stop and respond A on the basis of less overall evidence favoring $S_A$, than they are willing to stop and respond B. Balakrishnan (1999) used a new nonparametric measure of response bias that he developed to test between these two alternatives. Specifically, he used this new measure to reanalyze several datasets that displayed a prominent response bias according to a standard SDT analysis. As mentioned in chapter 2, his results strongly suggested that even when SDT indicated a strong response bias, observers nevertheless still evaluated the sensory evidence objectively. These results suggest that the most appropriate model of a response bias will set $\alpha \neq \beta$, but leave the decision bound in an unbiased position.

**10.2.1.3 The circular GRT diffusion model** One severe limitation of both the drift-diffusion model and stochastic GRT is that these models only apply to tasks in which there are either one or two response alternatives. This is because the models initiate a response when the diffusion process crosses an absorbing barrier. This approach can be used to model simple reaction times by assuming that drift is always nonnegative and there is only one barrier (e.g., Luce, 1986). Or it can be used as in figures 2.11b and 10.8b to model choice RTs in two-alternative tasks by allowing drift to be either positive or negative and assuming there are two absorbing barriers. But until recently, there was no drift-diffusion model that could be applied to tasks with three or more response alternatives. This was not from a lack of interest, but rather from the mathematical challenges associated with generalizing diffusion models to more than two response alternatives. P. L. Smith (2019) proposed a creative solution to this problem that is described in this subsection.

Before proceeding further, a word of caution is in order. Any attempt to model RT data from tasks with three or more response alternatives must be prepared for a more significant contribution of motor processes than in tasks with one or two response options. With three or more alternatives, at least one hand must execute more than one response, and this more complex mapping between response labels and motor actions is likely to cause more motor confusions—especially under speed stress conditions—as well as more variability in the base-time term (e.g., $T_0$ in equation 10.2). Also, partly because of these complications, there is much less published RT data from tasks with three or more response alternatives compared to data from two-alternative tasks, and as a result, testing any such model can be problematic.

With that caveat in mind, note that the ordinates in figures 2.11b and 10.8b denote response evidence. Positive values represent evidence supporting a YES or A response, and negative values represent evidence supporting a NO or B response. Thus, the evidence space is one-dimensional (i.e., the abscissa in figures 2.11b and 10.8b represents time). If we wanted to generalize either model to a task with three or more alternatives, we would need some extra directions to represent evidence supporting these new responses. At the minimum, a two-dimensional evidence space would be required. One obvious approach is to assume that the diffusion process operates inside a cylinder that sits in a three-dimensional space and that a response is initiated as soon as the diffusion process touches the cylinder. The circular cross-sections of the cylinder represent a two-dimensional evidence space and the axis down the center of the cylinder denotes time. To model data from a task with three or more response alternatives, the circular cross-sections of the cylinder could be divided into three or more regions—each one representing a different response. For example, in the unbiased version of this model, each response region in a four-alternative task might be represented by a different 90° arc. The predicted probability of responding A in a task with response alternatives A, B, C, and D would then equal the probability that the diffusion process first touches the cylinder along the part of the circular cross-section assigned to response A.

P. L. Smith (2019) proposed and developed a model of this type to account for response accuracy and RT in the $2 \times 2$ factorial identification experiment that serves as the workhorse for GRT-based investigations of perceptual processes. The model is illustrated in figure 10.11 for a hypothetical trial on which stimulus $A_2B_2$ is presented and the observer responds correctly. The figure depicts the two-dimensional evidence

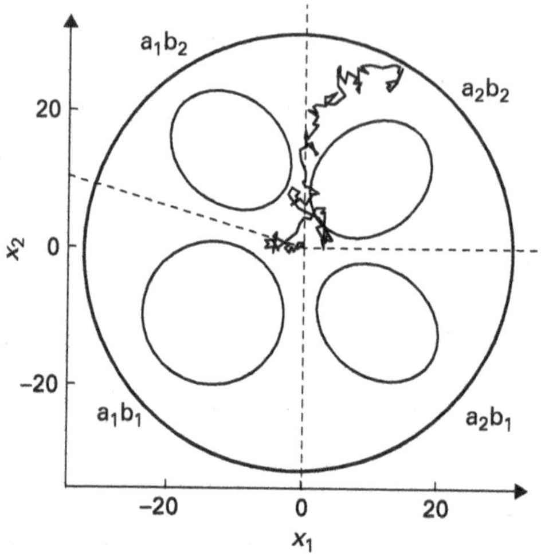


**10.11** The circular GRT diffusion model proposed by P. L. Smith (2019). The figure depicts events on a hypothetical trial of a 2 × 2 factorial identification experiment on which the presented stimulus was $A_2B_2$ and the observer responded correctly. The mean of the $A_2B_2$ perceptual distribution is (10,10), and the random sample from this distribution that was used to generated this simulated diffusion path was (14, 10).

space after the cylinder has been collapsed along the time dimension. The diffusion process of this model is conceptually similar to the diffusion process of Ashby's stochastic GRT model, except for two key differences. First, there is no decision bound in Smith's model. Rather than compare each momentary percept to a decision bound, in the circular-diffusion model the percepts are directly integrated and the cylinder (or circle) represents an absorbing barrier that is placed on the absolute magnitude of this integral.

Second, in Smith's model, perceptual variability is modeled differently within a trial versus across trials. The contours of equal likelihood shown in figure 10.11 describe the distribution of percepts across trials—exactly the same as in the classical, static GRT model. The circular diffusion model assumes that on each trial, one random sample is drawn from the perceptual distribution associated with the presented stimulus—again, exactly the same as in the static GRT model. Next, the circular model assumes that this random sample, which is one point in perceptual space, fixes the mean of a second, within-trial multivariate normal distribution that defines the diffusion process. The variance-covariance matrix of this second, within-trial distribution equals $\sigma^2$I, where $\sigma^2$ is an independent parameter that is not logically related to any parameters that define the perceptual distributions shown in figure 10.11. Finally, the model assumes that repeated random samples are drawn from this second, within-trial distribution, and these are then integrated until the diffusion process touches the circular absorbing barrier.

Figure 10.11 depicts hypothetical events on a trial when stimulus $A_2B_2$ is presented. The sample path depicted in figure 10.11 was created by taking one random sample from the $A_2B_2$ perceptual distribution, which turned out to equal (14, 10), and then taking repeated random samples from a bivariate normal diffusion distribution with mean (14, 10) and variance-covariance matrix $\sigma^2$I. Finally, these samples were integrated (or added) to create the sample path shown in the figure.

Note that the diffusion distribution on this trial had a positive mean on both perceptual dimensions. Next, note that the perceptual dimensions have been set (without loss of generality) so that the point (0, 0) lies at the center of the four distributions. The value of the diffusion process at time $t$ equals the value of the integral at time $t$. Therefore, the starting point of the diffusion process is the origin of the perceptual space and the integration continues until the process first touches the circular cross-section of the cylinder. On the trial depicted in figure 10.11, this absorption occurs along the section of the circle associated with response $a_2b_2$, so the model predicts that the observer responds correctly on this trial.

Although the single random sample drawn from the $A_2B_2$ perceptual distribution had positive values on both perceptual dimensions, and therefore the mean of the diffusion distribution was positive on both dimensions, variability in the diffusion distribution could cause some diffusion samples to have negative values on one or both perceptual dimensions. Even so, the positive means of the diffusion distribution (i.e., 14 and 10) suggest that most samples will have positive values on both dimensions, and as a result, the integral of these mostly positive values should eventually approach $(\infty, \infty)$ (perhaps though, after first crossing the circular barrier in some quadrant associated with an incorrect response). Similarly, the mean of the $A_1B_1$ perceptual distribution shown in figure 10.11 equals (–10, –8), so on most $A_1B_1$ trials, the single random sample drawn from this distribution will have a negative value on both dimensions, and, as a result, most samples from the resulting diffusion distribution will have negative values on both dimensions, in which case we expect the resulting integral to eventually approach $(-\infty, -\infty)$. It is via these properties that the model predicts accurate responding in this task.

Although this model has intuitive appeal, it is not surprising that deriving analytic predictions poses significant mathematical challenges. P. L. Smith (2016) overcame these challenges in a creative way. First, he derived analytic predictions for the zero-drift version of the model, and then he used a change-of-measure theorem from the stochastic processes literature, called the Girsanov theorem, to derive predictions for the nonzero-drift model in terms of the zero-drift predictions.

The zero-drift model assumes that the mean of the diffusion distribution equals (0, 0) for every momentary sample and on every trial. Under these conditions, note that the diffusion is equally likely to proceed in any direction because of the assumption that the variance-covariance matrix of the diffusion distribution equals $\sigma^2$I. Therefore, because the absorbing barrier is a circle at each point in time, all predictions from this model depend only on the distance of the process from the origin. Direction of travel is irrelevant. In other words, if the coordinates of the process at time $t$ are $[x_1(t), x_2(t)]$, then the predictions of the model depend only on the radius (i.e., A) of the circular barrier and on the distance $D(t)$ of the process from the origin at time $t$, namely on

$$D(t) = \sqrt{x_1^2(t) + x_2^2(t)}. \tag{10.20}$$

This is an enormous simplification for two reasons. First, it converts the problem from a diffusion process in multiple dimensions to a one-dimensional problem—namely, to determining how quickly $D(t)$ will equal A. Second, the behavior of the equation 10.20 model is well understood. Equation 10.20 describes a Bessel process and the pdf of the time when $D(t)$ first crosses a barrier at A is well known (i.e., the so-called first-passage pdf; see P. L. Smith, 2016; for details). P. L. Smith (2016) then used the Girsanov theorem to derive the pdf for the nonzero-drift model from this zero-drift pdf. In the nonzero-drift model, direction does matter. For example, if we pass a line through the origin and through the mean of the diffusion distribution, then drift along this line is more likely in the general direction from the origin to the mean than in the opposite direction. As a result, the pdf of the nonzero-drift model changes with the angle of drift as well as the radius of the circular barrier.

In the 2016 article, Smith derived predictions for this model under the assumption that the mean of the diffusion distribution had the same constant value on all trials for each stimulus. P. L. Smith (2019) then generalized the model to the case where the mean of the within-trial diffusion distribution is itself a random vector that has its own across-trial multivariate normal distribution. In other words, in the 2019 version, the mean of the diffusion distribution on each trial is a random sample from some other bivariate normal distribution that describes trial-by-trial variability in the mean percepts. The contours of equal likelihood depicted in figure 10.11 describe these across-trial distributions. This version of the model provides a stochastic interpretation of GRT because the across-trial distributions have the same interpretation as the perceptual distributions in static GRT. In both models, the observer's perceptual experience with a stimulus is represented by a single sample from the relevant across-trial perceptual distribution. P. L. Smith et al. (2023) simplified the computational methods for fitting this more general model, and P. L. Smith and Corbett (2019) generalized the 2016 model to a four-dimensional evidence space.

From a psychological perspective, it seems troubling that the diffusion is not driven by the GRT perceptual distributions, but rather by a separate diffusion distribution that assumes perceptual independence and equal variability on all dimensions. In this model, the stimulus affects the mean drift, but has no effect on any other distributional properties of the diffusion. In particular, the variances and covariances of the diffusion distribution do not depend in any way on the stimulus. GRT assumes that a perceptual dependence means that the perceived value of one stimulus component affects the perception of some other component. If such a dependence exists, one would expect it to characterize all sensory samples, not just the one sample that determines the mean of the diffusion distribution. For example, in Ashby's stochastic GRT model that we considered in the previous subsection, all within-trial sensory samples are drawn from the perceptual distribution. So if the perceptual distribution is characterized by a perceptual dependence, then all sensory samples will exhibit that dependence. P. L. Smith (2019) made the assumption that there are separate within-trial and across-trial perceptual distributions for mathematical, rather than psychological reasons. P. L. Smith (2016) used the Girsanov change-of-measure theorem to derive predictions for the nonzero-drift model from predictions for the zero-drift model. These latter predictions depended on the assumptions that the zero-drift predictions are radially symmetric and therefore depend only on the distance of the diffusion from the origin (i.e., so that equation 10.20 defines a Bessel process). Radial symmetry requires a zero mean, but it also requires that the variance-covariance matrix of the diffusion distribution equals $\sigma^2\text{I}$. The Girsanov theorem makes it possible to generalize this zero-drift model to one with nonzero mean drift, but it still requires that the diffusion around this nonzero mean satisfies radial symmetry, and therefore it requires that the variance-covariance matrix of the diffusion distribution equals $\sigma^2\text{I}$.

The assumption that the diffusion distribution always satisfies perceptual independence and homogeneity of variance is therefore necessary to derive analytic predictions from the model. From a psychological perspective however, it seems like a weakness. Even so, it is not clear whether this weakness has any substantive effect on the model's predictions. In other words, it is unknown how much the predictions would change if the within-trial sensory samples were drawn from the across-trial perceptual distributions, rather than from circular diffusion distributions. This is an important question for future research.

Another fundamental difference between Ashby's stochastic GRT model and the circular GRT diffusion model is that the circular-diffusion model does not include any decision bounds. The dotted lines in figure 10.11 are not decision bounds. Instead, these lines mark the locations of the borders between sections of the absorbing circle that are associated with the different alternative responses. The circular-diffusion model accounts for the speed-accuracy trade-off by assuming that the observer adjusts the diameter of the circular absorbing barrier according to task demands. Increasing the diameter increases accuracy and RT, whereas decreasing the diameter has the opposite effect. The dotted lines in figure 10.11 therefore show the locations of the new borders under any increased level of speed stress, assuming no change in response bias.

Although the dotted lines in figure 10.11 look like decision bounds, they serve a very different role from GRT decision bounds. Two examples that highlight these differences are shown in figure 10.12. Panel (a) shows a sample path that was generated on a trial when a single random sample equal to (1, 10) was drawn from the stimulus $A_2B_2$ perceptual distribution. Note that the diffusion terminates on an arc of the circle associated with response $a_2b_2$, so the model responds correctly on this trial. Even so, note that most of the sample path lies in the shaded region. If the dotted lines were decision bounds, then the shaded region would define the $a_1b_2$ response region. In other words, according to classical GRT, most of the evidence on this trial favors response $a_1b_2$ and the diffusion veers over to the section of the circle associated with response $a_2b_2$ only at the last instant before a response is initiated. In the stochastic versions of SDT and GRT considered previously, each point in the shaded region would generate a discriminant value (e.g., a signed distance to the vertical decision bound) that would favor response $a_1b_2$ and the integral of these $a_1b_2$-favoring discriminant values would overwhelm the meager evidence on this trial that favors response $a_2b_2$. So the stochastic models that include a decision bound (or a response criterion) would either respond $a_1b_2$ on this trial or continue sampling for a much longer period of time.

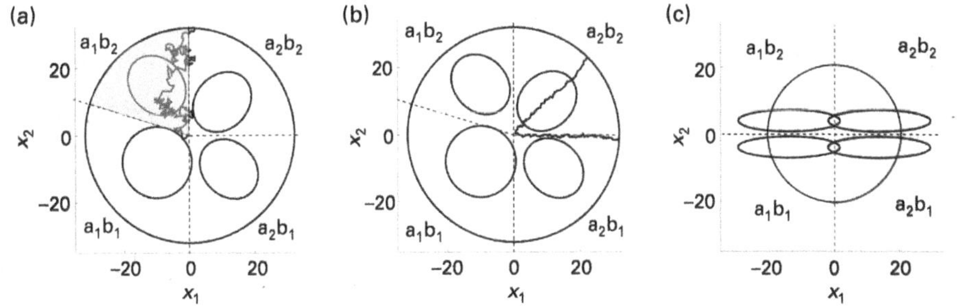


**10.12** Three examples in which the assumptions of the circular GRT diffusion model may be problematic.

The figure 10.12a example illustrates a property of the circular GRT diffusion model that might be called perceptual path independence. By this I mean that the model predicts that although longer paths mean longer RTs, it also makes the strong prediction that the direction a path takes before striking the circular barrier is irrelevant to any model predictions. In other words, the model predicts that all sensory evidence is ignored before the diffusion process strikes the barrier.

A different example of a possibly problematic prediction is shown in figure 10.12b. This panel shows simulated paths on two different trials. The northeast trending path was generated on a trial when a single random sample equal to (20, 20) was drawn from the stimulus $A_2B_2$ perceptual distribution, and the east trending path was generated when the single random sample from the stimulus $A_2B_2$ distribution was equal to (20, 0).[6] The former case, generated from a random sample of (20, 20), is meant to illustrate a trial when the sensory evidence strongly indicates that stimulus components A and B are both at level 2 on this trial. In contrast, the latter case, where the random sample equals (20, 0), illustrates a trial in which the sensory evidence strongly indicates that component A is at level 2, but is completely equivocal about the level of component B. This might occur because the observer simply failed to perceive component B accurately on this trial, or perhaps because the experimenter physically altered component B to be intermediate between the previously seen levels 1 and 2. Whatever the reason, the important point is that the model predicts roughly equal RTs on these two trials. The component B fidelity difference between the two trials affects the direction of diffusion, and therefore the predicted accuracy, but has little or no effect on predicted RT. I don't know of any $2 \times 2$ factorial identification data that could be used to test this prediction of the circular diffusion model, but if an experimenter presented a stimulus $A_2B_\bullet$, where $B_\bullet$ is a stimulus component intermediate between $B_1$ and $B_2$, then to my mind at least, it seems highly likely that RT would be considerably slower than on $A_2B_1$ or $A_2B_2$ trials.

Assuming this intuition is correct, then note that this is not a problem of perceptual path independence because the east-trending path in figure 10.12b is always intermediate between responses $a_2b_1$ and $a_2b_2$. Instead the problem seems to be with the assumption that the absorbing barrier is a circle. Since the radius of a circle is the same in all directions, this means that the criterion to stop sampling and respond is the same on trials when the path is highly diagnostic of one single response (as is the northeast-trending path in figure 10.12b) or completely equivocal about the validity of two different responses (as is the figure 10.12b east-trending path). It seems more psychologically plausible that observers would use a more lenient criterion in the former case than in the latter case. In the figure 10.12b example, this would mean replacing the circular barrier with one that is scalloped inward or star shaped, with the points of the star located where the dotted lines meet the figure 10.12b circle. In such a model, observers would terminate sampling and make a response earlier on trials when the sensory evidence strongly favors one response than on trials when the evidence was equivocal between two responses. The problem, of course, with this generalization of the model is that it is no longer analytically tractable. Because this star-shaped model is not radially symmetric, the computational approach used by P. L. Smith (2016) is no longer valid. As a result, predictions would most likely have to be derived via Monte Carlo simulation.

Finally, figure 10.12c illustrates another problem with the assumption that the absorbing barrier is always a circle. Note that in this example, the four perceptual distributions have much more variability along dimension $x_1$ than along dimension $x_2$. As a result, a circle seems a poor choice for an absorbing barrier. An ellipse that is much wider on dimension $x_1$ than on dimension $x_2$ seems a better choice. Again though, the problem with generalizing the model to allow elliptical barriers is that such a model would no longer be analytically tractable.[7]

**10.2.1.4 A generalization of the stochastic GRT model to the 2 × 2 identification task** Conceptually at least, it is straightforward to generalize the circular GRT diffusion model to include decision bounds or, alternatively, to use the P. L. Smith (2019) approach to generalize Ashby's stochastic GRT model to the $2 \times 2$ factorial identification task. For example, consider the GRT perceptual distributions and decision bounds shown in figure 10.13a. Note that the two bounds have been parameterized so

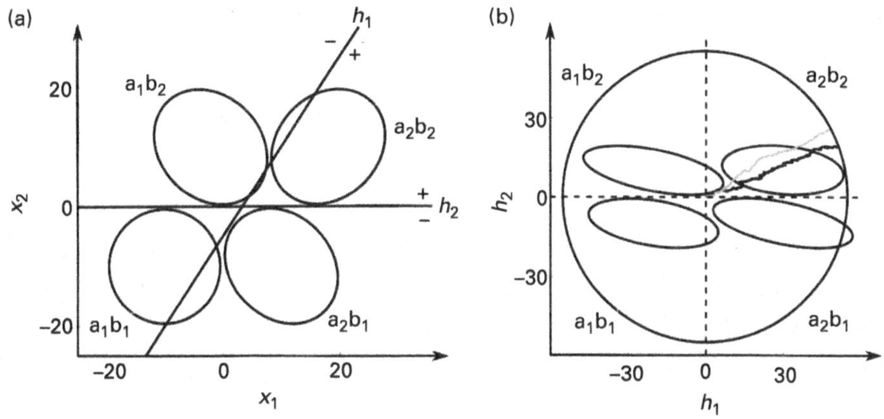


**10.13** (a) A GRT model of the 2 × 2 factorial identification experiment in which the perceptual distributions satisfy perceptual separability on dimension $x_2$ but not on dimension $x_1$. (b) The four contours of equal likelihood describe the joint distributions of the two discriminant values ($h_1$, $h_2$) on each of the four trial types. The black curve is a sample diffusion path of this generalized stochastic GRT model on a trial when the presented stimulus is $A_2B_2$. The gray curve is a sample diffusion path of a generalized version of the circular GRT diffusion model.

that the $a_2b_2$ response region is defined as the set of all $(x_1, x_2)$ such that both discriminant functions are positive—that is, as the set of all points for which $h_1(x_1, x_2) > 0$ and $h_2(x_1, x_2) > 0$. Similarly, the $a_1b_1$ response region is defined as the set of all points such that both discriminant functions are negative.

Consider a trial of the 2 × 2 factorial identification task on which stimulus $A_2B_2$ is presented. Suppose, as in Ashby's stochastic GRT model, that repeated samples are drawn from the $A_2B_2$ perceptual distribution, and that discriminant values are used to select a response. Ashby's stochastic GRT model was developed specifically for two-alternative tasks, and therefore for tasks with only one decision bound and, hence, only one discriminant function. As a result, in this model, the evidence space is one dimensional. In the figure 10.13a model, however, there are two decision bounds and, therefore, two discriminant functions. So each sample from the $A_2B_2$ perceptual distribution generates two discriminant values $h_1$ and $h_2$, and if these are integrated across time, a diffusion process is defined through a two-dimensional evidence space. A sample path from this generalized model is shown in 10.13b.

More formally, consider a trial when stimulus $A_iB_j$ is presented and denote the percept sampled from the $A_iB_j$ perceptual distribution at time $t$ by $\underline{x}_{ij}(t)$. Then this generalized stochastic GRT model assumes that two discriminant processes are defined at each time $t$—one associated with each decision bound—and that these equal

$$h_1(t) = \underline{b}_1' \underline{x}_{ij}(t) + c_1, \tag{10.21}$$

and

$$h_2(t) = \underline{b}_2' \underline{x}_{ij}(t) + c_2. \tag{10.22}$$

Note that $h_1$ and $h_2$ are both linear transformations of a random vector that has a bivariate normal distribution. As a result, $h_1$ and $h_2$ both have univariate normal distributions and the pair $(h_1, h_2)$ has a bivariate normal distribution. Denote the entries in $\underline{b}_1'$ by $\underline{b}_1' = [b_{11}\ b_{12}]$ and the entries in $\underline{b}_2'$ by $\underline{b}_2' = [b_{21}\ b_{22}]$. Next define the matrix

$$\mathrm{B} = \begin{bmatrix} b_{11} & b_{12} \\ b_{21} & b_{22} \end{bmatrix} \tag{10.23}$$

and the vector

$$\underline{c} = \begin{bmatrix} c_1 \\ c_2 \end{bmatrix}. \tag{10.24}$$

Then on trials when the presented stimulus is $\mathrm{A}_i\mathrm{B}_j$, the mean vector and variance-covariance matrix of $(h_1, h_2)$ are

$$\underline{\mu}_{h,ij} = \mathrm{B}\,\underline{\mu}_{ij} + \underline{c} \tag{10.25}$$

and

$$\Sigma_{h,ij} = \mathrm{B}\,\Sigma_{ij}\,\mathrm{B}', \tag{10.26}$$

respectively, where $\underline{\mu}_{ij}$ is the mean of the $\mathrm{A}_i\mathrm{B}_j$ perceptual distribution and $\Sigma_{ij}$ is its variance-covariance matrix.

Figure 10.13b shows the joint distributions of $h_1$ and $h_2$ from panel (a) on trials when each of the four possible stimuli are presented. Note the strong negative correlation between $h_1$ and $h_2$ that is present on all trials. This is because the bounds in figure 10.13a are not orthogonal. If the perceptual distribution satisfies perceptual independence, then $h_1$ and $h_2$ will be statistically independent for any set of orthogonal bounds.

The black curve in figure 10.13b is a sample path from this generalization of the stochastic GRT model. This path was generated by integrating repeated $(h_1, h_2)$ samples on trials when stimulus $\mathrm{A}_2\mathrm{B}_2$ was presented. The circle represents an absorbing barrier on this two-dimensional evidence space—exactly as in the circular GRT diffusion model. Note that this sample path terminates on a section of the circle associated with response $a_2b_2$, so the model predicts that the observer responds correctly on this trial. The problem with this model is that analytic predictions are unavailable, for precisely the same reason that the circular diffusion model was unable to assume that all samples are drawn from the perceptual distributions (i.e., because the diffusion distribution is not radially symmetric, and therefore the Girsanov theorem does not apply).

Recall that in the circular diffusion model, analytic predictions are possible under the assumption that on each trial, exactly one random sample is drawn from the perceptual distribution, and that this sampled value then becomes the mean of a diffusion distribution that is bivariate normal with variance-covariance matrix equal to $\sigma^2$I. If these same assumptions are incorporated here, then the figure 10.13 model becomes analytically tractable. In fact, the model then becomes identical to the circular GRT diffusion model, except the $(h_1, h_2)$ distributions described by equations 10.25 and 10.26 replace the perceptual distributions. The gray curve in figure 10.13b is a sample path from this analytically tractable model. This path was generated from a diffusion distribution in which the variance $\sigma^2$ was set equal to the mean of the $h_1$ and $h_2$ variances.

One advantage of this generalized stochastic GRT model is that it does not suffer from perceptual path independence. Therefore, in the figure 10.12a example, the model would likely respond $a_1b_2$, rather than $a_2b_2$. On the other hand, because the assumption that the absorbing barrier is a circle is necessary to derive analytic predictions, this generalized stochastic GRT model is characterized by the same possible weaknesses that are described in panels (b) and (c) of figure 10.12.

### 10.2.2 PARALLEL DYNAMICAL-SYSTEM MODELS

Visual cortex contains many different neurons with firing properties that are described by very specific tuning functions. Some neurons respond to a narrow range of spatial frequencies, some to a narrow range of orientations, some are color sensitive, and some respond to a specific direction of motion. Two similar stochastic interpretations of GRT were inspired by this architecture (Ashby, 1989; Townsend & Wenger, 2015; Townsend et al., 2012; Townsend & Wenger, 2004; Wenger & Townsend, 2006). In these approaches, a set of parallel processing channels are each tuned to a different stimulus component. For example, to

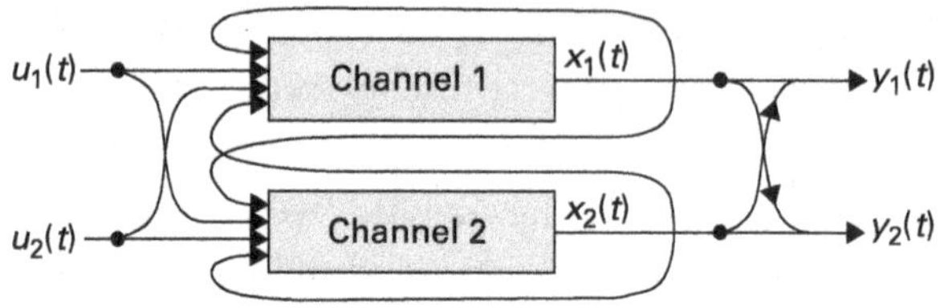


**10.14** The architecture of the linear-system GRT model proposed by Ashby (1989) when it is used to model results from a 2 × 2 factorial identification experiment.

model results of the 2 × 2 factorial identification experiment described in figure 4.2a, such a model would include two or more channels—at least one sensitive to spatial frequency (or bar width) and at least one sensitive to bar orientation. The channels may or may not interact, and processing would continue on each trial until sufficient evidence has been collected to emit an overall identification response.

**10.2.2.1 A linear-system GRT model** Ashby (1989) proposed that the channels build the GRT perceptual representation. Specifically, in this account, the output of the channels at time $t$ defines the location of the percept at time $t$ in perceptual space. Furthermore, Ashby (1989) proposed that the functioning of the channels could be modeled by linear-system theory (e.g., C. T. Chen, 1970). The architecture of this model is described in figure 10.14 for the case where there are two stimulus components that vary across trials. Each channel is a perfect integrator (i.e., the output is simply the integral of the input). For example, consider a 2 × 2 factorial identification experiment where the four stimuli are the Gabor patches shown in figure 4.2a, so component A is bar orientation and component B is spatial frequency. Consider a trial when stimulus $A_2B_2$ is presented. The linear-system GRT model assumes that all lower-level visual neurons that respond to orientation $A_2$ feed into channel 1, which then integrates this input and outputs a value that equals the perceived orientation of the presented stimulus. Likewise, all low-level visual neurons that are sensitive to spatial frequency $B_2$ feed into channel 2, the inputs are all integrated, and the output is the perceived spatial frequency or bar width of the presented stimulus. The linear-system GRT model allows the inputs, percepts, and outputs to all be time varying. In figure 10.14, the inputs at time $t$ are denoted by $u_1(t)$ and $u_2(t)$, and the percepts at time $t$ are denoted by $x_1(t)$ and $x_2(t)$. The outputs $y_1(t)$ and $y_2(t)$ are discriminant values, so there is one output for each decision bound (more on this below).

Note that the model shows three different kinds of possible interactions. The input mixing on the left side of the figure allows each component to influence processing in both channels. The channel interactions in the center of the figure allow each channel to either inhibit or excite both itself and the opposite channel. Finally, output mixing on the right side of the figure allows both channels to affect each discriminant value.

Predictions of the linear-system GRT model are derived from a system of straightforward linear ordinary differential equations[8]

$$\dot{\underline{x}}(t) = A\underline{x}(t) + \sigma_p \underline{\epsilon}(t) + B[\underline{u}(t) + \sigma_s \underline{\epsilon}(t)] \quad (10.27)$$

$$\underline{y}(t) = C\underline{x}(t) + \underline{d} + \sigma_c \underline{\epsilon}(t), \quad (10.28)$$

where $\dot{\underline{x}}(t) = d\underline{x}(t)/dt$, $\underline{x}(t)' = [x_1(t), x_2(t)]$ is a vector that describes the activation in channels 1 and 2 at time $t$, $\underline{u}(t)' = [u_1(t), u_2(t)]$ is a vector that describes the stimulus values of components A and B at time $t$, $\underline{y}(t)' = [y_1(t), y_2(t)]$ is a vector that describes the discriminant values of two different decision bounds at time $t$, and $\underline{\epsilon}(t)$ is white noise. A, B, and C are all $2 \times 2$ matrices that contain the various mixing coefficients, and $\underline{d}$ is a vector containing the intercepts of the various decision bounds (not shown in figure 10.14). So, for example, $b_{12}$ is a coefficient that measures the amount of stimulus component 2 that is input to channel 1, $a_{12}$ measures the amount of channel 2 activation that is input to channel 1, and $c_{12}$ measures the amount of channel 2 activation that contributes to the discriminant value of decision bound 1.

The linear-system GRT model includes three types of noise—each of which is modeled as a white noise process. The first noise term in equation 10.27 models perceptual noise. At any time $t$, perceptual noise is normally distributed with mean $\underline{0}$ and variance-covariance matrix equal to $\sigma_p^2 I$. The second noise term in equation 10.27 models all noise that is generated before perceptual processing begins, which would include noise in the stimulus (e.g., the Poisson nature of light), and in the case of visual stimuli, any noise generated in the retina or lateral geniculate nucleus. At any time $t$, stimulus noise is normally distributed with mean $\underline{0}$ and variance-covariance matrix equal to $\sigma_s^2 I$. Finally, the noise term in equation 10.28 represents criterial noise, which at time $t$ is normally distributed with mean $\underline{0}$ and variance-covariance matrix equal to $\sigma_c^2 I$.

Equation 10.28 computes a discriminant value for each decision bound that might appear in the static version of GRT. For example, consider a two-alternative task where the decision bound is a diagonal line that satisfies $x_2 = x_1$. In this case, the discriminant value equals $h(x_1, x_2) = x_2 - x_1$, which is implemented in equation 10.28 by setting

$$C = \begin{bmatrix} -1 & 1 \\ 0 & 0 \end{bmatrix}, \tag{10.29}$$

and

$$\underline{d} = \begin{bmatrix} 0 \\ 0 \end{bmatrix}. \tag{10.30}$$

In a $2 \times 2$ factorial identification experiment, we would expect two decision bounds, which would be implemented by setting at least one term in the second row of C to some nonzero value.

The final step is to convert these discriminant values to a response. Ashby (1989) discussed several different possibilities. For example, in two-alternative experiments, the decision process can be modeled as in Ashby's (2000) stochastic GRT model in which the single discriminant value is integrated until it crosses a positive barrier at $+\alpha$ or a negative barrier at $-\beta$. In the $2 \times 2$ factorial identification experiment, an obvious choice is to model the decision process as in the figure 10.13 2D generalization of the stochastic GRT model. One significant advantage to the linearity assumptions inherent in equations 10.27 and 10.28 is that they make these equations analytically tractable. For example, closed-form solutions are readily obtained in the noise-free version of the model (see, e.g., C. T. Chen, 1970).

Equations 10.27 and 10.28 describe how the percept changes in response to changes in the stimulus—for example, if the stimulus is presented tachistoscopically or is masked in some way. However, in many identification experiments, the stimulus is unmasked, displayed at constant intensity, and is response terminated. Under these conditions $\underline{u}(t) = \underline{u}$ for all $t$ (at least for all $t$ that matters to our modeling efforts). Let $\underline{u}_{ij}$ denote this constant input on a trial of a $2 \times 2$ identification task when the presented stimulus is $A_iB_j$. Assuming the system is stable, then under these conditions, activity in the channels will converge on an equilibrium state and then perform a random walk around this point until a response is made.[9] If the various noise sources are not time dependent—in other words, if each noise source takes one specific value on each trial, then channel activity will converge to a fixed equilibrium point on each trial, but these points will vary trial by trial depending on the specific values of each type of noise. It is straightforward to characterize all of these equilibrium states. On each trial, equilibrium occurs when $\dot{\underline{x}}(t) = \underline{0}$, and therefore when $\underline{x}(t)$ is a constant. Since the stimulus on this trial was $A_iB_j$, denote this constant percept as $\underline{x}_{ij}$.

Substituting these values into equation 10.27 and solving for $\underline{x}_{ij}$ produces

$$\underline{x}_{ij} = A^{-1}(\sigma_S B\underline{\epsilon}_S + \sigma_P\underline{\epsilon}_P - B\underline{u}_{ij}), \tag{10.31}$$

where $\underline{\epsilon}_s$ and $\underline{\epsilon}_p$ represent two different random samples from a bivariate normal Z distribution.[10]

Equation 10.31 is a linear combination of bivariate normally distributed vectors, and so $\underline{x}_{ij}$ is itself bivariate normally distributed, with mean vector

$$\underline{\mu}_{ij} = -A^{-1}B\underline{u}_{ij}, \tag{10.32}$$

and variance-covariance matrix equal to

$$\Sigma_{ij} = \sigma_S^2 A^{-1}B\,B'(A^{-1})' + \sigma_P^2 A^{-1}(A^{-1})', \tag{10.33}$$

where $'$ denotes matrix transpose. This equilibrium analysis shows that the static, classical version of GRT overlaps with this dynamic version. Specifically, the perceptual distributions in the static version can be viewed as equilibrium-level perceptual distributions in the special case in which some matrices A and B exist such that the mean vectors and variance-covariance matrices of the static model can be described by equations 10.32 and 10.33. Even under such conditions, though, note that the linear-system GRT model is considerably more general because it also makes predictions about the path and timecourse the percept takes to reach this equilibrium state. So theoretically, for example, this model could be used to predict the confusions that an observer would make if processing was interrupted for some reason before an equilibrium-level percept is registered. And the model also makes predictions about how any form of time-varying stimulus should affect the percept.

Note that equations 10.32 and 10.33 describe how input mixing and channel interactions affect the perceived value of the stimulus. Therefore, these equations provide an opportunity to study the core GRT perceptual properties of perceptual independence and perceptual separability at a deeper level. What do these properties tell us about channel interactions? Ashby (1989) showed that in the absence of stimulus noise, perceptual independence holds if A is a diagonal matrix—that is, if there are no lateral interactions between channels. This agrees with intuition. If the channels don't interact, then processing in one channel should be statistically independent of processing in the other. Ashby (1989) also showed that in the absence of stimulus noise, a sufficient but not necessary condition for perceptual separability to hold is that matrices A and B are both diagonal. The condition that B is diagonal makes sense because under these conditions, there is no input mixing, so the level of component B does not directly affect processing of component A, and vice versa. However, the condition that A is also diagonal is somewhat surprising since it suggests that a violation of perceptual separability can also occur because of direct channel interaction—that is, because of a perceptual dependence. So note that the linear-system GRT model makes several strong predictions here. First, a finding that perceptual separability holds should increase the probability that perceptual independence also holds, because the conditions that are sufficient for perceptual separability (A and B are both diagonal) guarantee perceptual independence (A is diagonal). Second, if perceptual independence is violated, then violations of perceptual separability are also likely, whereas violations of perceptual separability say little about whether or not perceptual independence is satisfied. To my knowledge, there have been no attempts to test these predictions.

This dynamical version of GRT interprets perceptual properties, such as perceptual independence and perceptual separability, in terms of network architecture, but of course, this exercise is valuable only if the predictions of the linear-system GRT model are reasonably accurate. The obvious problem is that the dynamical model assumes changes in the percept can be modeled by a linear dynamical system and the evidence is overwhelming that human perception is nonlinear. This doesn't mean that the model isn't useful. A variety of approaches have shown that linear approximations provide good accounts of neural responses over a wide range of conditions (e.g., Boynton et al., 1996; Vazquez & Noll, 1998). Even so, the well documented nonlinearities in perception mean that even the best linear-system model must eventually fail—for example, under boundary-level conditions.

Let's look at this issue more closely. A dynamical system is linear if and only if it satisfies the *superposition principle*, which is defined as follows.

**Superposition Principle** Consider the following experiment. To prepare for each condition, we wait until the system is at rest—that is, until the outputs are all zero. The experiment then includes the following three conditions.

*Condition 1*: Input any $\underline{u}_1(t)$ and record the output, which we can denote as $\underline{y}_1(t)$.
*Condition 2*: Input any $\underline{u}_2(t)$. Call the output $\underline{y}_2(t)$.
*Condition 3*: Input $\underline{u}_3(t) = a_1\underline{u}_1(t) + a_2\underline{u}_2(t)$. Call the output $\underline{y}_3(t)$.

Then the superposition principle holds and the system is linear if and only if the output $\underline{y}_3(t)$ equals

$$\underline{y}_3(t) = a_1\underline{y}_1(t) + a_2\underline{y}_2(t), \qquad (10.34)$$

for any constants $a_1$ and $a_2$ and any inputs $\underline{u}_1(t)$ and $\underline{u}_2(t)$ (e.g., C. T. Chen, 1970).

Note that any system that exhibits saturation will violate superposition. For example, suppose a channel has some maximum activation value of say 1.0, and assume $u_1$ is some input that causes an output of 0.9. Then superposition requires that if we double the input, then the output must also double to 1.8. But this is clearly impossible since the maximum output of the system is 1.0. This system therefore violates the superposition principle, and as a result, is a nonlinear system. Ashby (1989) proposed a partial workaround to this problem. Specifically, Ashby (1989) proposed using Stevens' Law as a way to model a nonlinear transformation that converts stimulus measurements to system inputs (Stevens, 1957). For example, the Stevens' exponent on visual area is 0.7 (e.g., see the Wikipedia page titled "Stevens's power law"). Therefore, if $u_1$ is the system input for an object with area $w_1$ cm$^2$, then $u_1 = w_1^{0.7}$. So if we double the area of the input from $w_1$ cm$^2$ to $2w_1$ cm$^2$. Then note that $u_1$ changes from $w_1^{0.7}$ to

$$(2w_1)^{0.7} = 2^{0.7}w_1^{0.7} = 1.62w_1^{0.7}.$$

In other words, although we doubled the stimulus area, the input to the network less than doubles. Using Stevens' Law in this way should improve the validity of the model, but the linearity assumption should be viewed skeptically in any experiment where the stimulus levels are near either extreme, or that uses a rapid serial visual presentation (RSVP) that allows successive stimuli to interact in some nonlinear fashion.

**10.2.2.2 A parallel-processing GRT model** Consider again a $2 \times 2$ factorial identification task that uses the four Gabor patches shown in figure 4.2a as stimuli. The linear-system GRT model would account for data from this experiment with a single spatial frequency channel and a single orientation channel. But neurons in the early visual cortex that are sensitive to spatial frequency (so-called simple or complex cells) each respond to a fairly narrow range of spatial frequencies (approximately $\pm 1$ octave), and orientation-sensitive neurons each respond to a narrow range of orientations ($\pm 12° - 15°$, e.g., DeValois & DeValois, 1988). The evidence is good that these neurons are organized into a limited number of channels in which all neurons in the same channel have similar tuning functions (e.g., Graham, 1989). For example, in the case of spatial frequency, the evidence suggests humans have somewhere around eight different frequency-sensitive channels (e.g., Hess & Plant, 1985; Watson & Robson, 1981). The important point here is that this model of spatial vision predicts that multiple channels should be expected to respond to each stimulus presentation because the tuning curves of the various channels overlap. So the assumption of the linear-system GRT model that only one spatial frequency channel is active on each trial oversimplifies the neuroscience.

Townsend et al. (2012) proposed a stochastic, parallel-processing interpretation of GRT that more realistically includes multiple channels for the processing of each stimulus component—specifically, they proposed one channel for each level of the stimulus component. So in a $2 \times 2$ identification task, their model would include two spatial frequency channels that are active on each trial—the one tuned most closely to spatial frequency $B_1$ and the one tuned most closely to spatial frequency $B_2$. Similarly, the model assumes there will also be two orientation channels active on each trial—the one most closely tuned to orientation $A_1$ and the one most closely tuned to orientation $A_2$. Clearly, given the $\{A_1B_1, A_1B_2, A_2B_1, A_2B_2\}$ stimulus ensemble, these are the most relevant sensory channels. Next, the model assumes the $A_1$- and $A_2$-sensitive orientation channels race against each other, and the first one to reach criterion determines the identified level of component A. Similarly, the $B_1$- and $B_2$-sensitive spatial frequency channels also race against each other, and the first one to reach criterion determines the identified level of component B. So the model assumes that processing of component A is parallel and self-terminating, as is the processing of component B, but that processing of the winning A and B channels is parallel and exhaustive—in other words, an identification response must wait for the slower of the two winning races to finish.

The most general version of the Townsend et al. (2012) model is distribution free and makes only weak assumptions about the nature of processing within each channel. Specifically, it makes no assumptions about the possible linearity or nonlinearity of the parallel channels. This version was used to develop tests of GRT perceptual and decisional properties that are based on various RT summary statistics. Those tests are reviewed in section 6.3 of chapter 6. Wenger and Townsend (2006) also used the model to account for configurality effects in the perception of faces and words.

At first glance, the parallel-processing model of Townsend et al. (2012) and the linear-system GRT model seem qualitatively similar. Both models postulate a set of parallel-processing channels, and both models interpret perceptual properties, such as perceptual independence and perceptual separability, in terms of similar features of the network architecture—specifically, in terms of interactions between parallel processing channels. Even so, these apparent similarities are somewhat illusory because the models make very different assumptions about exactly what the channels are processing. The channels in the linear-system model process the stimulus components directly. At any time $t$, current activity in the channels—that is, $[x_1(t), x_2(t)]$—defines the percept at time $t$. So the primary goal of the linear-system model is to describe how the percept changes dynamically over time. In contrast, in the parallel-processing model of Townsend et al. (2012), activation in each channel is assumed to be proportional to the evidence that the relevant stimulus component is at the level to which the channel is tuned. So activation in the parallel channels is not perceptual, per se, but rather evidence in favor of a decision about the level of a stimulus component. The model is general enough that evidence could be interpreted in different ways—for example, as a discriminant value or as a likelihood ratio. Either way though, the Townsend et al. (2012) model makes no predictions about how the percept might evolve over time.

The linear-system GRT model converts the momentary percept to decisional evidence by computing momentary discriminant values via the matrix C and the vector $\underline{d}$ (e.g., see equations 10.27 and 10.28). So in the linear-system GRT model, $\underline{x}(t)$ in equation 10.27 describes the percept at time $t$ and $\underline{y}(t)$ in equation 10.28 describes the evidence at time $t$ that the percept was generated by some specific stimulus (or class of stimuli). Therefore, the activity in the channels of the Townsend et al. (2012) model corresponds most closely to the output values of the linear-system GRT model—that is, to $\underline{y}(t)$—rather than to the channel activities.

This difference also explains why the two models seem to model decision processes so differently. In the parallel-processing model, the decision about the level of a component is determined by which component channel reaches criterion first. A decision bound or discriminant value is never explicitly computed. The greater the activity in the channels, the faster the decision. In the linear-system model, high levels of activity in the channels signal a high-intensity percept, but not necessarily a quick decision. The output equation of the model (i.e., equation 10.28) converts these high-intensity percepts to discriminant values. These discriminant values may or may not be extreme. Whether they are depends on differences across the entire stimulus ensemble. For example, if high-intensity percepts are expected from all stimuli, RTs might be long, despite the high-levels of activity in each channel.

Note also that the default decision strategy of the parallel-processing model assumes decisional separability on both dimensions because a decision about whether component A (or B) is at level 1 or 2 depends only on which component A (or B) channel reaches criterion first. The model can accommodate violations of decisional separability because it allows lateral interactions between the A channels and the B channels, but note that whether or not such interactions exist, the model always assumes selective attention—that is, it assumes that to decide whether component A is at level 1 or 2, all relevant attention is allocated to the A channels and activity on the B channels is ignored (and vice versa). For these reasons, this model seems especially well suited to account for rule-guided behavior of the type that is common, for example, in RB categorization tasks (see section 5.3).

In fact, exactly the same architecture was proposed in a recent neuropsychological model of how extended practice eventually causes rule-guided behaviors to become automatic. Much evidence suggests that rule-guided behaviors are initially controlled by a distributed neural network centered in prefrontal cortex. Kovacs et al. (2021) proposed that in addition to initiating behavior, this network also trains a faster and more direct network that includes projections from sensory association cortex directly to rule-sensitive neurons in premotor cortex. After much practice, the direct network is sufficient to control the behavior, without any contribution from prefrontal cortex. The model, called CARM (Cortex Automatizes Rules Model), was implemented as a biologically detailed neural network constructed from spiking neurons and displaying a biologically plausible form of Hebbian learning. This model is relevant because it could provide a neurobiological implementation of the Townsend et al. (2012) parallel-processing GRT model.

The architecture proposed by CARM is shown in figure 10.15 for application to a task in which the observer is implementing a one-dimensional rule (identified there as "Rule 1"). For example, given the four Gabor patches shown in figure 4.2a, the rule might be "respond $A_2$ if the patch has bars with a steep orientation and $A_1$ if the bars have a shallow orientation." The portion of the model that includes the visual cortex and prefrontal cortex could be interpreted as a neurobiological implementation of the Townsend et al. (2012) parallel-processing GRT model. The portion that includes the premotor and primary-motor cortices mediates the transition to automatic responding, so this portion of CARM suggests how predictions of the parallel-processing GRT model might be generalized from novel to automatic behaviors.

The prefrontal-cortex units that implement a one-dimensional orientation rule in CARM are qualitatively identical to the orientation channels in the Townsend et al. (2012) parallel-processing GRT model. Specifically, CARM assumes that two simulated prefrontal cortex neurons implement this rule—one that signals if the

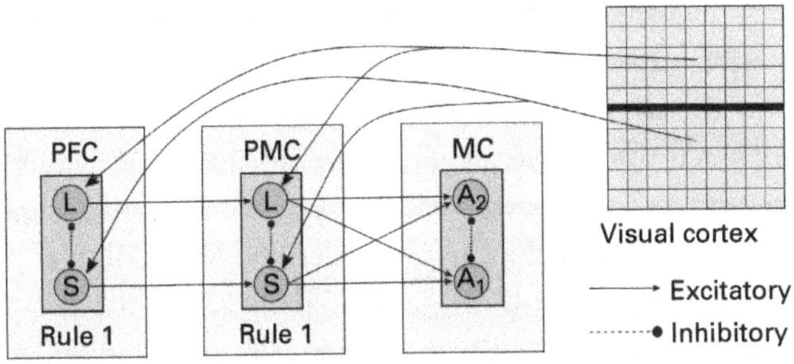


**10.15** Architecture of CARM—a model of how rule-guided behaviors become automatized that was proposed by Kovacs et al. (2021). PFC = prefrontal cortex; PMC = premotor cortex; MC = primary motor cortex.

stimulus has a small (or shallow) orientation that operates like the $A_1$ channel in the parallel-processing GRT model (the S units in figure 10.15) and one that signals if the orientation is large (or steep) that operates like the $A_2$ channel in the parallel-processing model (the L units in figure 10.15). The idea is that orientation-sensitive units in visual cortex that respond to shallow orientations project to the PFC-S neuron, whereas visual cortical units that respond to steep orientations project to the PFC-L neuron. In this way, the S neuron responds to any shallow orientation and the L neuron responds to any steep orientation. The model assumes that these PFC rule units develop as a result of lifelong practice with a rule. Despite this lifelong practice, initial responding in this task is not automatic because experimental participants presumably have no prior experience pressing an $A_1$ or $A_2$ button to signal whether they judge the orientation to be steep or shallow. In CARM, the pathways from prefrontal cortex to primary motor cortex (i.e., through premotor cortex) translate the $A_1$ versus $A_2$ decision (or shallow versus steep) to a motor command (press the $A_1$ button versus press the $A_2$ button).

As a neural interpretation of the Townsend et al. (2012) parallel-processing GRT model, the idea is that on a trial when say, stimulus $A_2B_2$ is presented, the prefrontal L and S units will both respond to stimulus component $A_2$. We expect the L unit to respond more strongly though because its visual cortical inputs are tuned to steep orientations. Even so, CARM assumes the L and S units race against each other (in parallel), it assumes there is lateral inhibition between these two units, and it assumes that processing continues until a criterion level of activation occurs that signals whether the response should be $A_1$ or $A_2$—all exactly the same as in the parallel-processing GRT model.

The structures depicted in the premotor and primary-motor cortices could be viewed as a possible extension of the parallel-processing GRT model. During early training sessions, the stimulus activates the appropriate prefrontal rule unit (i.e., S or L), which then activates the analogous premotor rule unit, which then activates the appropriate unit in motor cortex that causes the model to respond $A_1$ on trials when the orientation is shallow and $A_2$ when the orientation is steep. The same visual cortical units that project to the S and L prefrontal units also project to the S and L premotor units. Initially, however, these visual cortex-to-PMC synapses are not strong enough to activate the appropriate unit in motor cortex. This is because there is no prior association between shallow versus steep orientations and the appropriate button presses required to signal this decision. So instead, during initial training the motor response is completely determined by the parallel, self-terminating processing that occurs between the PFC-S and PFC-L units—again, exactly as in the parallel-processing GRT model.

As this task is practiced, repeatedly pressing the $A_1$ button on trials when the orientation is shallow and the $A_2$ button on trials when the orientation is steep will cause the synapses between the visual cortex and the premotor S and L units to strengthen via Hebbian learning. This is because the prefrontal input to these units will cause more activation in the correct premotor unit than in the incorrect unit, and therefore more synaptic strengthening.[11] Eventually the visual cortex-to-premotor cortex synapses will be strong enough so that visual input alone will be enough to cause the premotor unit to activate the appropriate target in motor cortex. The pathway through prefrontal cortex is still active, but because it is longer, it no longer controls behavior. At this point, the behavior has become automatic. Kovacs et al. (2021) reviewed considerable empirical evidence in support of this model.

The most general version of the parallel-processing GRT model was used to develop summary-statistics tests of GRT perceptual properties, but this model lacked sufficient detail to make specific trial-by-trial RT predictions. To investigate these predictions, Townsend et al. (2012) proposed a special case of the general model that included more assumptions. Specifically, they derived trial-by-trial RT predictions from a version of the model in which the channels were modeled as a linear system, using a set of equations that were mathematically identical to equation 10.27 of the linear-system GRT model. Even so, this linear-system version of the parallel-processing model was not equivalent to the linear-system GRT model because of the difference in how the models define channel activity. Channel activity in this special case of the parallel-processing model is still defined as evidence in favor of a decision about the level of a stimulus component, in contrast to the linear-system GRT model where channel activity defines the current percept.

Unfortunately, analytic predictions are unavailable for either model. Ashby (1989) made no attempt to generate trial-by-trial RT predictions, and Townsend et al. (2012) derived predictions via Monte Carlo simulation (as did Wenger & Townsend, 2006). Kovacs et al. (2021) also used Monte Carlo simulation to derive trial-by-trial RT predictions from CARM. Rather than model each channel as a linear system, Kovacs et al. (2021) modeled each channel as a spiking neuron (using the Izhikevich, 2003; model of the regular spiking neuron).

One reason that simulations are required to make RT predictions from these parallel processing interpretations of GRT is because of the lateral inhibition they typically include, which considerably complicates any attempts to derive analytic predictions. Despite its analytic intractability, lateral inhibition is an almost necessary feature because of its well-known computational benefits and its prevalence in biological networks. Lateral inhibition has long been known as a key feature of winner-take-all networks. If the output of each channel is an input to the other competing channels, then the channel with the greatest activation will have the greatest inhibitory effect. For example, in a two-channel model, like the one illustrated in figure 10.14, the more active channel will inhibit activation in the less active channel more than vice versa. This imbalance magnifies the activation difference between the two channels, thereby increasing response accuracy—that is, assuming the stimulus-driven input tends to be greatest in the channel associated with the correct response.

This property of lateral inhibition is illustrated in figure 10.16, which plots proportion correct in a two-stimulus, two-response task as a function of the strength of the lateral inhibition between two channels. These results were generated via Monte Carlo simulation from a simplified version of the linear-system special case of the parallel-processing GRT model. On each trial, processing continued until activation in one channel exceeded a common threshold, at which point the model emitted the response associated with that channel. I assumed that the stimulus would activate both channels because of overlap in the tuning curves, but that the input to the channel leading to the correct response would be greater than the input to the incorrect channel. Even so, errors occurred because of perceptual noise (i.e., because of $\sigma_p$ in equation 10.27). The strength of lateral inhibition was defined as the numerical value of the off-diagonal entries in the (symmetric) connectivity matrix A from equation 10.27. This value was varied from 0 (so no lateral inhibition) to 0.9. Results are shown for two signal-to-noise ratios, which is defined here as the difference in

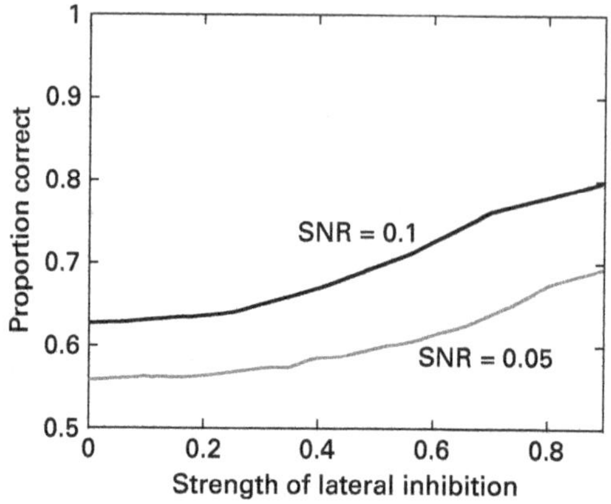


**10.16** Accuracy in a simplified version of the figure 10.14 model as a function of the strength of the lateral inhibition between the two channels. Results are shown for two different signal-to-noise ratios (SNRs), where SNR is defined as the difference in the strength of the inputs to the two channels divided by the standard deviation of the white noise input.

the strength of the stimulus-driven inputs divided by the common standard deviation of the white noise input (i.e., by $\sigma_p$ in equation 10.27). Note that, as expected, in both cases accuracy increases monotonically with the strength of the lateral inhibition.

In summary, lateral inhibition is prevalent in biological systems, and it can significantly improve response accuracy, but it also is almost always associated with analytic intractability. Monte Carlo simulations are always possible, but are there any general RT predictions we can expect from a parallel race model that includes lateral inhibition among competing channels? Fortunately, this is a well-studied problem. The classic treatment is by Bogacz et al. (2006), who showed that a variety of different two-channel parallel race models, which implement lateral inhibition in different ways, all reduce to the drift-diffusion model. The basic idea is that in two-channel models with lateral inhibition, strong activation in one channel causes strong inhibition in the other, which causes the model to display the kind of push-pull dynamics that characterize the drift-diffusion model. As a result of this equivalence, we should expect the parallel processing interpretations of GRT to make hazard function predictions that are similar to those of the stochastic GRT model shown in figure 10.8. Simple race models without any lateral inhibition, also known as counting models or accumulator models, lack these push-pull dynamics because an increase in activation in one channel has no effect on activity in the other channel. Also, as this argument suggests, Bogacz et al. (2006) showed that there is no version of these simple race models that reduces to the drift-diffusion model.[12]

### 10.2.3 RELATIONSHIPS BETWEEN THE PROCESS MODELS OF RT AND THE RT-DISTANCE HYPOTHESIS

It is also worth noting that all of the process models of RT considered in this chapter make predictions that are generally consistent with the RT-distance hypothesis. This is most easily seen in applications to tasks with two response alternatives. First, consider the stochastic GRT and the linear-system GRT models. Both of these compute a discriminant value for each momentary percept, integrate these momentary discriminant values, and then respond when this integral first exceeds a positive threshold or first falls below a negative threshold. Especially with linear decision bounds, the discriminant value of a percept is closely related to the distance from that percept to the decision bound (e.g., Ashby & Maddox, 1992). In general, the farther a percept is from the decision bound, the more extreme its discriminant value. Therefore, the stochastic GRT and the linear-system GRT models both predict faster RTs for stimuli farther from the decision bound because these stimuli will generate more extreme discriminant values, and therefore the resulting diffusion processes will reach an absorbing response barrier more quickly. In fact, these predictions are so stable that the stochastic GRT and the linear-system GRT models could be viewed as providing process interpretations of the RT-distance hypothesis.

The relationship between the RT-distance hypothesis and the Townsend et al. (2012) parallel-processing GRT model is only slightly weaker. Recall that in the most general version of this model, each channel processes evidence favoring a decision about the level of some stimulus component, where evidence is never defined explicitly. Neither SDT nor GRT typically define evidence explicitly either, but in a task where the two response alternatives are A and B, in both theories a natural definition of the strength of evidence associated with the percept $\underline{x}$ would be

$$e(\underline{x}) = |\log L(\underline{x})| = \left|\log \frac{f_A(\underline{x})}{f_B(\underline{x})}\right| = |\log f_A(\underline{x}) - \log f_B(\underline{x})|. \quad (10.35)$$

According to this definition, there is no evidence that favors one response over the other when the likelihood ratio equals 1. Both responses are equally likely to be correct. As the likelihood ratio moves away from 1—in either direction—the evidence about which response is correct increases. According to this reasonable definition of evidence, note that in SDT, if the likelihood ratio is monotonic with increases in sensory magnitude, then evidence increases with the distance from the percept to the response criterion $X_C$. The same is true in GRT if the likelihood ratio is monotonic in the marginal distributions that are created when the multivariate perceptual distributions are projected onto the axis that is orthogonal to the decision bound (e.g., see figure 10.5). So under these conditions, channel activity in the parallel-processing GRT model will increase with the distance from the percept to the criterion or decision bound, and since the model responds as soon as this evidence reaches a positive criterion, predicted RT will decrease with the distance of the percept to the response criterion or decision bound. In other words, under these conditions, the RT predictions of the parallel-processing GRT model will satisfy the RT-distance hypothesis.

## 10.3 COMBINING GRT WITH SYSTEMS FACTORIAL TECHNOLOGY

Systems factorial technology (SFT; Townsend & Nozawa, 1995) provides a set of methods for studying fundamental properties of cognitive processing, including architecture (parallel versus serial), stopping rule (self-terminating versus exhaustive), and cognitive capacity (limited, unlimited, or super). Liu et al. (2023) adapted SFT to an experimental paradigm that is conducive to GRT applications. The result enhances a traditional GRT analysis because besides the traditional GRT inferences about perceptual separability, perceptual independence, and the nature of the decision bounds that observers appear to use, combining GRT and SFT also allows inferences about whether the stimulus dimensions are processed serially or in parallel, and with limited, unlimited, or super capacity.

The adapted experimental paradigm is described in figure 10.17. Note that the stimuli form a $3 \times 3$ factorial structure, in which three levels of stimulus dimension 1 are crossed with three levels of stimulus dimension 2. The experiment includes separate identification and categorization phases. The identification phase is run first. This phase includes several identification tasks that use the stimuli identified as the target category in figure 10.17—that is, stimuli $A_1B_1$, $A_1B_2$, $A_2B_1$, and $A_2B_2$. The main condition here is a standard $2 \times 2$ factorial identification task that includes all four stimuli. The confusion matrix from this condition is analyzed via the standard GRT methods that were discussed in detail in earlier chapters. Liu et al. (2023) also included all six possible two-stimulus identification conditions that are possible with these four stimuli (i.e., $A_1B_1$ versus $A_2B_1$; $A_1B_1$ versus $A_2B_2$, etc.). These were included

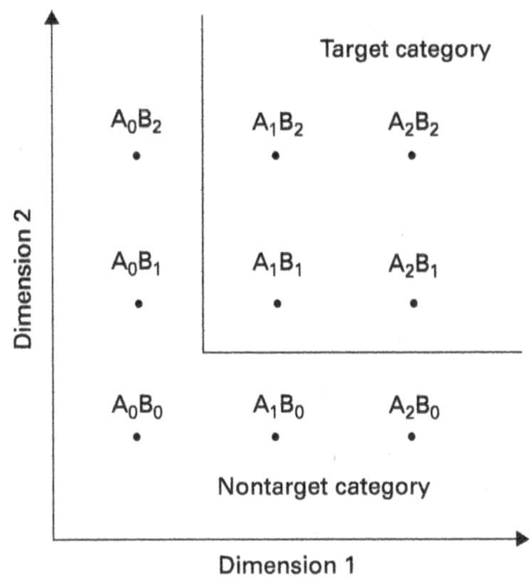


**10.17** The double factorial design necessary to apply systems factorial technology.

for two reasons—first to facilitate the signal-detection-based summary statistics tests described in chapter 6, and second, to serve as low-load control conditions for calculations of cognitive capacity (more on this below). The two bidimensional conditions (i.e., $A_1B_1$ versus $A_2B_2$ and $A_1B_2$ versus $A_2B_1$) are used only for the signal-detection-based summary-statistics analyses and are not needed for the capacity analysis, and therefore could be omitted to shorten the experiment.

After the identification conditions are complete, the categorization phase begins. The categorization task includes all stimuli shown in figure 10.17 divided into the target and nontarget categories illustrated there. From the observer's perspective, note that the stimuli in the target category are all familiar because these are exactly the stimuli that were used in the identification tasks, whereas the stimuli in the nontarget category are all novel. The observer's task is to assign each stimulus to its correct category. The data collected during the categorization phase form the repository for an SFT analysis. These analyses focus on the RTs produced by the observers to the target stimuli. The nontarget RTs are generally not included in any SFT analyses. Therefore, the familiarity difference between target and nontarget stimuli does not affect SFT analyses since all target stimuli are equally familiar.

Note that accurate responding to any stimulus in the target category of the figure 10.17 categorization task requires exhaustive processing of both stimulus dimensions.[13] The best accuracy that is possible with any strategy that is based on the perceived value of a single stimulus dimension is 67% correct. Typical applications that use the figure 10.17 design will require accuracy that is considerably higher than this. For example, in the application reported by Liu et al. (2023), the accuracy of all participants was above 85% correct. Accuracy levels this high rule out self-terminating search strategies. As a result, an SFT analysis of the RTs from the figure 10.17 categorization task focuses on the question of whether the stimulus components are processed serially or in parallel, and on the cognitive capacity that drives that processing.

The question of whether processing is serial or parallel is addressed by estimating the survivor interaction contrast (SIC). The survivor function is defined as

$$S(t) = P(T > t) = 1 - P(T \leq t) = 1 - F(t). \tag{10.36}$$

The name comes from survival analysis because if $T$ defines the time of system failure or death, then $S(t)$ defines the probability of surviving to time $t$. The survivor function $S(t)$ necessarily equals 1 when $t = 0$ and 0 when $t$ is infinite. In between it must be nonincreasing.

The SIC was motivated by the familiar mean interaction contrast illustrated in figure 10.18. If A and B are two independent variables, each with two levels, then in a $2 \times 2$ factorial design the parallel mean RT profiles shown in figure 10.18 suggest that A and B do not interact, and instead that their effects are additive. Mean RT profiles, like those shown in figure 10.18, are parallel, and therefore show no evidence of an interaction, if and only if

$$\overline{\mathrm{RT}}_{22} - \overline{\mathrm{RT}}_{21} = \overline{\mathrm{RT}}_{12} - \overline{\mathrm{RT}}_{11}. \tag{10.37}$$

As a result, the mean interaction contrast is defined as

$$MIC = [\overline{\mathrm{RT}}_{22} - \overline{\mathrm{RT}}_{21}] - [\overline{\mathrm{RT}}_{12} - \overline{\mathrm{RT}}_{11}]. \tag{10.38}$$

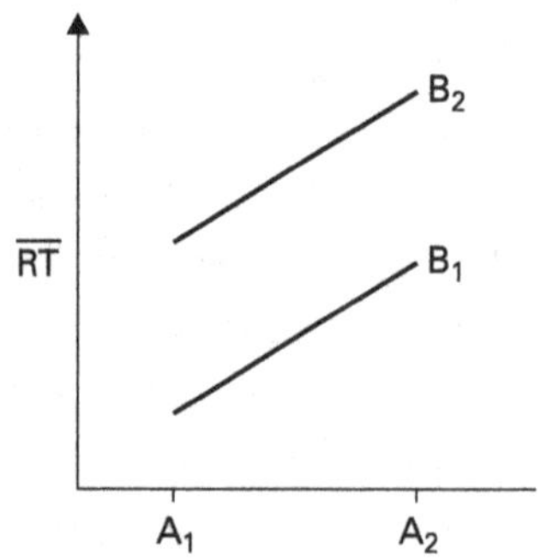


**10.18** Mean RTs from a hypothetical 2 × 2 factorial identification experiment.

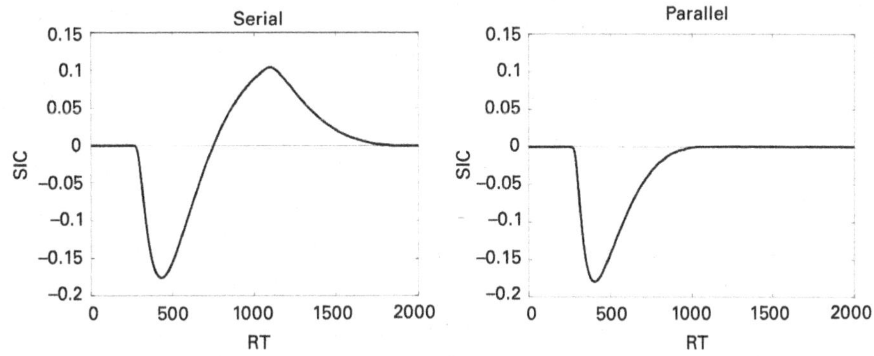


**10.19** Survivor interaction contrasts (SICs) predicted by serial exhaustive and parallel exhaustive models that assume the decision time on the level of each stimulus component is determined by the exponential RT-distance model.

Note that $MIC = 0$ if and only if there is no A versus B interaction, and therefore, any value of $MIC \neq 0$ is evidence that factors A and B interact in some way.

The SIC has the same form as the MIC. Let $S_{ij}(t)$ denote the survivor function of the RTs on trials of the figure 10.17 categorization task when stimulus $A_iB_j$ is presented. Then the SIC is defined as (Liu et al., 2023)

$$SIC(t) = [S_{22}(t) - S_{21}(t)] - [S_{12}(t) - S_{11}(t)]. \tag{10.39}$$

Townsend and Nozawa (1995) showed that serial exhaustive, serial self-terminating, parallel exhaustive, and parallel self-terminating models each predict distinct and qualitatively different SICs. Figure 10.19 shows the SICs predicted for the figure 10.17 task by parallel and serial exhaustive models that assume the time required to judge the level of each stimulus component is determined by the exponential RT-distance model. Note that the serial model predicts a prominent positive component for longer RTs that is absent from the parallel model predictions. Townsend and Nozawa (1995) showed that this qualitative difference is characteristic of an enormous class of serial and parallel models.

Liu et al. (2023) estimated SICs for an experiment that used the figure 10.17 experimental design. The stimuli were rectangles that varied in height and width. The SICs were all qualitatively identical to SICs characteristic of parallel exhaustive search—that is, they all had an early negative dip and then equaled zero for all longer RTs—exactly like the parallel exhaustive predictions shown in figure 10.17. Note that this result supports the assumptions of the linear-system and parallel-dynamical GRT models we reviewed in section 10.2.2.

SFT also provides estimates of the cognitive capacity that fuels stimulus processing. The definitions of cognitive capacity are based on the notion of parallel processing (e.g., Townsend & Ashby, 1978). Suppose only one item is displayed, so only a single channel is in operation. Let $\lambda$ denote the rate of processing on this single channel. Now consider a scenario in which the display includes two items, and that two channels therefore operate in parallel. If the average rate of processing on the two channels is again $\lambda$, then capacity is said to be unlimited. Adding an extra channel has no effect on processing rate, as if the system has an unlimited supply of energy to drive each channel. If the average rate of processing drops to a value less than $\lambda$, then capacity is said to be limited. Now there is a cost to adding an extra channel since, on average, processing is slowed on every channel when the load is increased. Finally, if the average rate of processing increases to a value greater than $\lambda$, then capacity is said to be super. Super capacity is a sign of holistic or configural processing. For example, identifying a whole face is likely to be easier and faster than identifying isolated components of the face (e.g., the eyes, nose, or mouth).

For experimental designs like the one illustrated in figure 10.17, in which accurate responding to stimuli in the target category requires exhaustive processing, Townsend and Wenger (2004) proposed defining the cognitive capacity for stimulus $A_iB_j$ at time $t$ as

$$C_{ij}(t) = \frac{\log[F_{i.}(t)] + \log[F_{.j}(t)]}{\log[F_{ij}(t)]}, \tag{10.40}$$

where $F_{i.}(t)$, for example, is the cumulative distribution function describing the RTs on trials when the stimulus is $A_iB_j$ and the observer responds correctly in the two-stimulus identification task that uses stimuli $A_1B_j$ and $A_2B_j$—that is, in a task where the observer can ignore component B and therefore is required only to make a decision about the level of component A.[14] Note that this condition was included in the identification phase of the experiment.

The idea is as follows. Suppose there are two parallel channels—one that processes component A and one that processes component B. Suppose further that processing on each channel is independent of activity on the other channel. Let $T_i$ denote the completion time on channel $i$ (for $i =$ A or B). Then because processing is exhaustive

$$\begin{aligned} F_{AB}(t) &= P(RT \le t) \\ &= P(T_A \le t \text{ and } T_B \le t) \\ &= P(T_A \le t)\, P(T_B \le t) \\ &= F_A(t)\, F_B(t). \end{aligned} \tag{10.41}$$

Of course, $F_{AB}(t)$ can be estimated directly from the observable RTs on trials when both stimulus components must be processed, but $F_A(t)$ and $F_B(t)$ are both unobservable. Even so, note that if capacity is unlimited, then $F_A(t)$ and $F_B(t)$ could be estimated in separate conditions in which the observer's task is to just identify the level of component A and ignore B or vice versa—that is, in conditions in which only one channel is operating in isolation. If capacity is unlimited, then equation 10.41 should still hold under these conditions because, for example, $F_A(t)$ would equal the observable RT cumulative distribution function on trials of this two-stimulus identification task. For this reason, $F_{i\cdot}(t)$ and $F_{\cdot j}(t)$ in equation 10.40 are estimated from separate conditions in which only one of the two stimulus components needs to be identified—specifically, from the RTs collected in the two-stimulus identification tasks that were included in the identification phase of the experiment. If capacity is limited however, then estimating $F_{i\cdot}(t)$ and $F_{\cdot j}(t)$ in this way will cause the left side to be less than the right side because the limited capacity will slow down the processing of both components simultaneously, relative to conditions under which each component is processed in isolation. In this way, equation 10.41 can be used to estimate capacity. For example, taking the log of both sides and dividing both sides by $\log[F_{AB}(t)]$ leads to the capacity coefficient defined in equation 10.40.[15]

The capacity coefficient described by equation 10.40 assumes error-free responding. Townsend and Altieri (2012) generalized the measure to tasks where errors are common. The generalized capacity coefficient is defined as

$$C_{ij}(t) = \frac{\log[F_{i\cdot}(t)P(C|i\cdot)] + \log[F_{\cdot j}(t)P(C|\cdot j)]}{\log[F_{ij}(t)P(C|ij)]}, \tag{10.42}$$

where $F_{i\cdot}(t)$, $F_{\cdot j}(t)$, and $F_{ij}(t)$ are exactly the same as in equation 10.40—that is, they are the cumulative distribution functions on correct response trials of the three tasks described there, and $P(C|i\cdot)$, $P(C|\cdot j)$, and $P(C|ij)$ are the probabilities of a correct response in each task.

Liu et al. (2023) reported that for all observers the capacity coefficient estimates (i.e., the estimates of $C_{ij}(t)$ in equation 10.42) were significantly greater than 1.0—that is significantly greater than the value predicted by unlimited capacity. In other words, all observers were processing the stimulus dimensions of the rectangle stimuli with super capacity, which suggests holistic or configural processing. The standard GRT analysis of the accuracy data from the identification conditions revealed strong violations of perceptual separability, and these SFT-driven RT analyses reinforce that conclusion. In addition, the SFT analyses also strongly suggested that processing of the components of the rectangles occurred in parallel—a conclusion that would be impossible with any standard application of GRT.

In summary, augmenting the standard $2 \times 2$ factorial identification experiment with the extra conditions required for an SFT analysis provides considerable additional inferential power. Most impressively, adding the categorization condition to the end of the experiment should allow strong inferences to be made about whether processing of the stimulus dimensions was serial or parallel. And including the two-stimulus identification conditions makes it possible to estimate cognitive capacity, which could significantly enhance any GRT-based conclusions about, for example, configural processing.

## 10.4 DERIVATION OF EQUATIONS 10.3–10.7

The exponential and power-function versions of the RT-distance hypothesis both predict that

$$RT = T + T_0, \tag{10.43}$$

where $T_0$ is a constant and

$$T = g(D), \tag{10.44}$$

where $g$ is a differentiable strictly decreasing function and $D$ is the distance from the current percept to the response criterion or decision bound (see figure 10.5). Under the assumption that the percepts are normally distributed, $T$ is a random variable. By the change of variables theorem,

$$f_T(t) = f_D\left[g^{-1}(t)\right]\left|\frac{dg^{-1}(t)}{dt}\right|. \tag{10.45}$$

In the exponential model

$$T = g(D) = \alpha e^{-\beta D}, \tag{10.46}$$

and so

$$D = g^{-1}(T) = -\frac{1}{\beta}\ln(T/\alpha). \tag{10.47}$$

Therefore,

$$\left|\frac{dg^{-1}(t)}{dt}\right| = \left|-\frac{1}{\beta T}\right| = \frac{1}{\beta T}. \tag{10.48}$$

In the power-function model

$$T = g(D) = \alpha D^{-\beta}, \tag{10.49}$$

and so

$$D = g^{-1}(T) = \alpha^{\frac{1}{\beta}} T^{-\frac{1}{\beta}}. \tag{10.50}$$

Therefore,

$$\left|\frac{dg^{-1}(t)}{dt}\right| = \left|-\frac{\alpha^{\frac{1}{\beta}}}{\beta} T^{-\left(\frac{1+\beta}{\beta}\right)}\right| = \frac{\alpha^{\frac{1}{\beta}}}{\beta} T^{-\left(\frac{1+\beta}{\beta}\right)}. \tag{10.51}$$

If the percepts are normally distributed, then in both models, distance-to-bound on correct trials will have a truncated normal distribution. For example, if the distance from the decision bound (or response criterion) to the mean of the stimulus $S_i$ perceptual distribution is $\Delta_i$, then the distance-to-bound pdf on $S_i$ trials when the observer responds correctly equals 0 for $d < 0$, and

$$f(d) = \frac{\exp\left[-\frac{(d-\Delta_i)^2}{2\sigma_i^2}\right]}{\sqrt{2\pi}\sigma_i \Phi\left(\frac{\Delta_i}{\sigma_i}\right)}, \quad \text{for } d \geq 0, \tag{10.52}$$

where $\sigma_i$ is the standard deviation of the $S_i$ perceptual distribution in the direction orthogonal to the decision bound (see figure 10.5) and $\Phi(x)$ is the cumulative Z distribution function (i.e., the area under the standard normal distribution to the left of $x$). Note that the $\Phi\left(\frac{\Delta_i}{\sigma_i}\right)$ term in the denominator, which is also equal to $1 - \Phi\left(\frac{-\Delta_i}{\sigma_i}\right)$, is the probability of a correct response on stimulus $S_i$ trials. Dividing by this term is necessary to ensure that $f(d)$ is a true pdf (i.e., integrates to 1).

Substituting equations 10.46, 10.48, and 10.52 into equation 10.45 yields the pdfs of the exponential model for the correct classification times (i.e., for $T$ in equation 10.43). The constant $T_0$ in equation 10.43 simply translates the pdf to the right by the amount $T_0$ (so $t$ in the pdf is replaced by $t - T_0$). Similarly, substituting equations 10.49, 10.51, and 10.52 into equation 10.45 and replacing $t$ by $t - T_0$ yields the pdfs of the power-function model for correct RTs. The pdfs for error trials are derived in a similar manner.

# 11

# STATISTICAL DECISION THEORY AND NEUROSCIENCE

SDT and GRT were both developed before the cognitive neuroscience revolution that began in the 1990s. As a result, for their first several decades of existence, both theories were purely perceptual and cognitive. But during the past several decades, there has been progress on two fronts. First, much has been learned about the architecture and functioning of the neural circuits that implement the perceptual and decisional processes hypothesized by SDT and GRT. And second, GRT analyses have recently been extended to neuroscience data, in particular to data from neuroimaging experiments.

This chapter reviews this neuroscience literature, with a focus on three separate questions.[1] First, do recent neuroscience discoveries support or disconfirm the assumptions of statistical decision theory? Second, what is known about the neural circuits that mediate the perceptual and cognitive processes hypothesized by statistical decision theory? In other words, is it possible to extend statistical decision theory to Marr's (1982) implementational level? Third, can GRT analyses be extended to neuroscience data—for example, to data from neuroimaging experiments? If so, then this raises the exciting possibility that it may be possible to test for perceptual independence and perceptual separability in different brain regions, and if either of these properties is violated in the behavioral data, to trace these violations through the neural circuits that mediate behavior in the task under study.

## 11.1 NEUROSCIENCE TESTS OF STATISTICAL DECISION THEORY ASSUMPTIONS

Statistical decision theory provides an extremely general account of perception and decision-making that has been applied to a wide variety of different tasks and behaviors. Nevertheless, as we saw in chapter 1, it makes some core assumptions that are assumed to hold in all applications. Section 1.1 focused on two of these—namely, the axioms of perceptual variability and universal decisions. But statistical decision theory also assumes that there are two separate stages of processing—a sensory/perceptual stage that must necessarily precede a second decision stage. The end result of the first stage is a percept, which the second stage then uses to select an appropriate response.

When SDT was first proposed, there were virtually no data that could be used to test any of these assumptions. By the time GRT was developed, the situation was better, but not by much. There was good evidence that sensory representations are noisy, but there was little or no evidence relevant to the other GRT assumptions. As a result, these other assumptions were justified almost exclusively by their intuitive appeal and by the fact that they produced a model that could account for a wide variety of behavior. The past few decades, however, are often referred to as the golden age of neuroscience. For example, in 2014, Michio Kaku wrote, "we have learned more about the thinking brain in the last 10-15 years than in all of human history." This enormous bank of new knowledge includes strong tests of all statistical decision theory assumptions. This section reviews those tests. As we will see, the assumptions all have solid neuroscience support, and as a result, the foundations of statistical decision theory are considerably stronger than when SDT and GRT were first developed.

### 11.1.1 SEPARATE SENSORY AND DECISION PROCESSES

Statistical decision theory assumes that decision processes act on a fully formed percept. This percept may depend on the nature of the task, but it does not depend on the response that is eventually made. For example, if we denote the pdf of perceptual effects elicited by presentation of stimulus $A_iB_j$ by $f_{ij}(x_1, x_2)$, then GRT assumes that this pdf is the same on every trial that stimulus $A_iB_j$ is presented, regardless of what response is made on each trial. In other words, statistical decision theory assumes that a sensory representation is formed first, and then decision processes use this representation to select a response. As a result, the theory predicts that the neural networks that form the perceptual representation of the stimulus are both upstream and relatively separated from the networks that use this representation to select a response.

It has been known for decades that the flow of information passes through many synapses as it makes its way from sensory receptors to the motor effectors that initiate a motor response elicited by the stimulus. For example, consider a trial of a visual identification experiment. Presentation of a stimulus excites rods and cones in the retina, which excite retinal ganglion cells that project to the lateral geniculate nucleus of the thalamus, which projects to V1 (primary visual cortex), then V2, V4, and many other regions before the M1 (primary motor cortex) neurons are stimulated that cause the agent to press one response key or the other. So the general neuroanatomy has long been known to be consistent with the statistical decision theory assumption of separate sensory and decision processes. Even so, this anatomy is also generally consistent with a theory in which response selection co-occurs with the processes that form the perceptual representation. This type of intermingling of sensory and decision processes violates the statistical decision theory assumption that sensory/perceptual and decision processes are relatively separate.

Some early, much-publicized evidence seemed to suggest that decisions *are* made in the visual cortex. From the perspective of statistical decision theory, the most problematic evidence came from widespread reports of category-specific agnosias, in which a patient responds normally to most visual stimuli but is impaired in their ability to identify, recognize, or classify objects from some specific category. Especially troubling is that the impaired category can be quite specific. For example, category-specific agnosias have been reported for animals, for fruits and vegetables, for tools, and for conspecifics (for a review, see, e.g., Mahon & Caramazza, 2009). The most sensational of such agnosias, called prosopagnosia, is for human faces. Some prosopagnosia patients are even unable to recognize their own faces. Category-specific agnosias are known to result from lesions in inferotemporal cortex and other high-level visual areas. For example, prosopagnosia is associated with lesions to the fusiform gyrus in inferotemporal cortex (Kanwisher & Yovel, 2006). These results seem to suggest that visual cortex is learning the categories, rather than just forming the perceptual representations, as assumed by statistical decision theory. In GRT, a category is defined by a response region in perceptual space, not by a perceptual distribution. Any percept from any stimulus that falls in a response region will elicit the motor response associated with that region, even if most of the associated perceptual distribution falls in some other response region. The discovery of category-specific agnosias therefore seemed to suggest that the visual areas are also learning the decision bounds that partition perceptual space into contrasting response regions.

Despite this seemingly damning evidence, there are theoretical interpretations of category-specific agnosia that are consistent with statistical decision theory. Specifically, category-specific agnosias are also generally consistent with the hypothesis that visually similar objects are represented in nearby areas of visual cortex. If so, then lesions to a connected region of visual cortex should cause an impairment in the ability of visual cortex to form perceptual representations for a set of visually similar stimuli. GRT predicts that this perceptual deficit will then cause deficits in any task that depends on these perceptual representations. This hypothesis is supported by the many well-known reports that neighboring neurons in virtually all visual cortical areas tend to fire to similar stimuli.

If visual cortex is learning categories, then an obvious place to find them is within area TE, which is the most anterior region in inferotemporal cortex, and also the final stage of purely visual processing in the mammalian brain. If area TE is learning categories, then the perceptual representations stored in TE should be of category exemplars. However, in contrast to this prediction, TE neurons are maximally activated by shapes or object parts that are more complex than simple edges or textures, but not complex enough to represent a natural object (for a review, see, e.g., Tanaka, 1996). Therefore, the combined activation of at least several TE neurons are required to represent an exemplar of a natural category, such as tools or fruits and vegetables (Rolls, 2009; E. Thomas et al., 2001). Furthermore, the evidence suggests that: (1) TE neurons that are tuned to similar features cluster in columns (Fujita et al., 1992); (2) a relevant visual stimulus activates neurons in several columns (Wang et al., 1998; Yamane et al., 2006); and (3) two similar visual stimuli activate columns that contain neurons sensitive to features the two stimuli share in common (Tsunoda et al., 2001). As a result, a lesion to this area will corrupt the perceptual representation of similar visual objects and therefore cause performance deficits that are characteristic of a category-specific agnosia.

There is also other strong evidence that decision processes are not mediated in visual cortex. For example, many years ago, Rolls et al. (1977) recorded from neurons in inferotemporal cortex of monkeys during a simple visual-learning task. In these studies, one visual stimulus was associated with reward and another with a mildly aversive taste. After training, the associations were reversed. If the associations were learned in visual cortex, then the firing properties of visual cortical neurons should have changed when the associations reversed. However, Rolls et al. found no change in the response of any of these cortical neurons, although other similar studies found changes in the responses of neurons in other downstream brain areas (e.g., orbitofrontal cortex).

Several subsequent category-learning studies reported similar results (de Beeck et al., 2001; Freedman et al., 2003; Sigala, 2004; E. Thomas et al., 2001; Vogels, 1999). In each of these experiments, monkeys were trained to categorize visual objects—for example, as a tree or non-tree, or as belonging to one of two categories of arbitrary complex shapes. During category learning, recordings were collected in inferotemporal cortex from single neurons that were activated by the stimuli used in the experiment. The most important result was that categorization training did not make these neurons more likely to respond to other stimuli in the same category or less likely to respond to stimuli belonging to the contrasting category. In other words, there was no evidence that the structure of the categories caused any learning-related changes in the firing properties of the neurons.

Even so, many studies have reported that in similar experiments, neurons in inferotemporal cortex do often change their firing properties during categorization training. For example, Sigala and Logothetis (2002) trained two monkeys to classify faces into one of two categories and then in a separate experiment to classify fish. In both experiments, some stimulus features were diagnostic to the categorization decision and some were irrelevant. Categorization training caused many neurons in inferotemporal cortex to increase their sensitivity to the diagnostic features much more than to the irrelevant features. Similar results were reported by De Baene et al. (2008) and Sigala (2004). So categorization training causes neurons to become more sensitive to relevant stimulus dimensions, but not to the category membership of the stimulus that causes them to fire. These changes are consistent with the widely held view that categorization training causes changes in the allocation of perceptual attention (Nosofsky, 1986). As we saw in several earlier chapters, statistical decision theory models changes in perceptual attention via changes in the variances of the perceptual distributions (Maddox et al., 2002; Soto et al., 2015). For example, GRT-wIND accounts for two different types of attentional effects. First, it allows for individual differences in the total amount of perceptual attention allocated to the task, with more overall attention reducing all perceptual variances by the same proportion. Second, it allows for categorization training to change the proportion of perceptual attention allocated to each stimulus dimension, with more attention reducing the perceptual variance on that dimension by the same amount in all perceptual representations (see section 7.7 for details). So in GRT-wIND and other statistical decision theory models, perceptual attention changes the perceptual representations, but these changes are the same in all stimuli, regardless of their category membership, which seems to agree perfectly with the single-unit recording data.

But how does perceptual attention change the firing properties of neurons in inferotemporal cortex? The current consensus is that these changes are mediated by top-down projections—for example from prefrontal cortex—back to visual areas (e.g., Gilbert & Sigman, 2007; Kastner & Ungerleider, 2000; Salin & Bullier, 1995). Note though that the existence of these back-projections could be interpreted as evidence that decision processes do influence perceptual representations. Perhaps, but such projections do not violate the assumptions of statistical decision theory if their effects can be modeled as changes in perceptual variance. Furthermore, the fast RTs in many visual tasks suggest that information conveyed by bottom-up, feedforward processing is often sufficient for object recognition and classification (e.g., Hung et al., 2005; Thorpe & Fabre-Thorpe, 2001). These fast RTs support a model in which the top-down attentional effects occur primarily between trials and that the response to each presented stimulus is mediated by a simple feedforward two-stage model of perceptual decision making of the type assumed by statistical decision theory.

An alternative approach to testing whether there are separate sensory and decision stages is by appealing to the principle of correspondent change (Townsend & Ashby, 1983). This principle states that if a model is correct, then it should be able to account for behavioral changes across experimental conditions by only changing the values of parameters that correspond to the independent variables that were varied to create the various conditions. So, for example, if SDT is correct, then varying the brightness of a target stimulus across different conditions should produce data that SDT can fit simply by varying the value of $d'$ (i.e., and holding either $X_C$ or $\beta$ constant).

The principle of correspondent change predicts that if there are separate sensory and decision stages, then changing experimental variables such as verbal instruction and reward history should only affect decision-related processes. Neuroscience experiments generally support this prediction. In particular, conditions that encourage changes in decision processes predominantly induce activation changes in frontal and motor areas, rather than in sensory association areas. For example, shifts in response criteria that are induced either through instruction or by changes in feedback are correlated with changes in activity in non-sensory areas, such as prefrontal cortex and the ventral striatum (M.-Y. Chen et al., 2015; M. B. Miller et al., 2001). Other examples come from studies of the speed-accuracy trade-off, which is usually attributed to shifts in the response criterion (e.g., Heitz, 2014). In conditions that emphasize speed, high levels of activity are found in the striatum and in the presupplementary motor area in both humans (Forstmann et al., 2008) and monkeys (Lauwereyns et al., 2002), and individual variability in levels of response caution are correlated with changes in striatal activity (Forstmann et al., 2008) as well as with the strength of white matter tracts that connect the presupplementary motor area to the striatum (Forstmann et al., 2010). Overall then, as predicted by statistical decision theory, the evidence strongly suggests that experimental manipulations that should alter decision processes predominantly affect neural processing in nonvisual areas that are downstream from the visual cortex.

In summary, the current neuroscience evidence strongly supports the statistical decision theory assumption of separate sensory and decision processes. The best current evidence suggests that decisions about what response to make are not learned or formed in the visual cortex. Instead, the evidence supports the statistical decision theory assumption that the visual cortex builds the final perceptual representation of the visual stimulus as activation moves up from the primary visual cortex (i.e., V1) to the inferotemporal cortex. If so, then it follows that decision processes should be mediated by structures receiving direct projections from these visual areas. There are three obvious candidates: the prefrontal cortex, the medial temporal lobes (i.e., the hippocampal system), and the basal ganglia. As we will see in the next section, there is evidence that all three of these targets contribute to decision-making.

### 11.1.2 ALL SENSORY REPRESENTATIONS ARE NOISY

This is the axiom of perceptual variability that was introduced in section 1.1.1 of chapter 1. See that section for a review of the overwhelming evidence in support of this assumption.

A related, but somewhat different question is whether this inherent variability affects performance. Because behavior depends on activity in many neurons, an alternative view that contradicts the predictions of statistical decision theory is that the noise is somehow averaged out and, as a result, has little or no effect on human behavior. As it happens, though, there is considerable evidence against this averaging-out hypothesis. First, neural responses are highly nonlinear and therefore could amplify small perturbations that are caused by noise (Faisal, 2012). For example, channel noise can significantly affect spike timing in the axons of cortical neurons (Faisal & Laughlin, 2007; Shadlen & Newsome, 1998). Furthermore, there is strong evidence that variability in the responses of sensory neurons affects sensory discrimination. For example, electrical stimulation of motion-sensitive neurons in area MT of temporal cortex can bias perceptual judgments of depth (DeAngelis et al., 1998), and several studies have found that trial-by-trial variability in the activity of MT neurons has predictable effects on judgments of depth (e.g., Britten et al., 1996; Purushothaman & Bradley, 2005; Uka & DeAngelis, 2004). Similar results have been reported with fMRI. For example, a number of fMRI studies of YES-NO detection tasks have reported stronger BOLD responses in visual cortex on YES trials than on NO trials, even on noise-only trials when no signal was present (Pajani et al., 2015; Ress & Heeger, 2003; Watkins et al., 2006). Collectively, these results strongly suggest that noise at the neural level influences behavioral decisions (Gold & Ding, 2013).

In summary, noise is a ubiquitous component of all percepts. This not only validates a core axiom of statistical decision theory, but note that it also implies that decision processes must be flexible enough to deal with noise.

### 11.1.3 EVERY MOTOR RESPONSE REQUIRES A DECISION

This is the axiom of universal decisions that was introduced in section 1.1.2. That section situated the axiom in its historical context but did not discuss supporting neural evidence. Fortunately for statistical decision theory, there is strong neural evidence that even the simplest behaviors require a decision (e.g., see Gold & Shadlen, 2007; for a review). But what constitutes neural evidence of a decision? And how could a decision ever be attributed to a single neuron, or even to a group of neurons?

The neuroscience field has agreed on a number of criteria that single-neuron recordings must satisfy before concluding that the associated neurons are contributing to the decision-making process. A complete discussion is beyond the scope of this chapter, so only a brief overview will be given here. For more details, see Gold and Shadlen (2007), for example. First, to rule out the possibility that the neurons are responding to the stimulus, it is sufficient to show that their activity is correlated with the response more than with the stimulus. For example, consider two identical trials of some two-alternative perceptual task in which the observer makes contrasting responses. If the neuron in question is responding to the stimulus, then the activity on these two trials should be highly similar, whereas if the neuron is contributing to the decision, then its activity on the two trials should be dramatically different. Second, a more challenging problem is to rule out the possibility that the neural activity is strictly motor. For example, neurons in the spinal cord that initiate finger movements are simply responding to inputs from some upstream region, and would not typically be classified as playing a key role in any decision-making process. Several different approaches are often used to settle this question (Shadlen & Newsome, 2001). One is to include a delay period during which observers must withhold their response until after a go signal is presented. If the neuron is participating in the decision, then it should be active during the delay period, whereas if its activity is strictly motor-related, then the neuron should be silent until the go signal is presented. Another criterion is that activity in decision-related, but not motor-related neurons should be modulated by the difficulty of the decision. For example, consider two signal-present trials in a YES-NO detection task. Suppose the observer responds YES on both trials but the two signals are of different intensities. The decision is easier on the trial when the signal is more intense, but the motor response is the same on both trials. As a result, a decision-related neuron should respond more strongly to the more intense signal, whereas activity in a motor-related neuron should be similar on the two trials.

Using criteria of this type, neurons that appear to encode a decision have been reported in a variety of different perceptual tasks. For example, consider a simple vibrotactile frequency-discrimination task, in which participants must discriminate between two mechanical vibrations applied to their fingertips that differ only in vibration frequency. Single-unit recordings in monkeys show that neurons in somatosensory cortex respond in this task as if they encode a sensory representation, with increasing firing rate as vibration frequency increases, and in a way that does not depend on the animal's final decision (Hernández et al., 2000). In contrast, neurons in prefrontal cortex (e.g., Brody et al., 2003) and premotor cortex (Romo et al., 2004) respond in this task as if they encode the actual judgment leading to a behavioral decision.

Perhaps the most widely used task to study the role of decision processes in simple sensory discriminations is the random-dot motion discrimination task. In this task, participants view a cloud of moving dots, some of which move in a coherent direction while the others move randomly. In most such experiments, the proportion of coherently moving dots varies across trials. The observer's task is to report the direction of motion of the coherently moving dots—for example, left versus right. Single-unit recording studies in monkeys have implicated visual area MT in the sensory representation of the stimulus and a variety of other areas—including the lateral intraparietal cortex (LIP), the dorsolateral prefrontal cortex, and the superior colliculus—in the decision of how to respond. For example, the average response rate of neurons in visual area MT increases with motion coherence (Gold & Shadlen, 2007), and stimulating these neurons has similar effects on behavior as varying actual motion coherence (Hanks et al., 2006). In contrast, the average response of LIP neurons increases with time following motion onset, and at a faster rate the greater the motion coherence. More importantly, the average firing rate of LIP neurons reaches the same threshold for all levels of motion coherence right before the behavioral response is made and shows a steep drop afterward (Roitman & Shadlen, 2002). Furthermore, any motion in the dots that occurs after the response threshold has been reached is ignored in the final choice (Kiani et al., 2008). Evidence for similar sorts of evidence accumulation during decision-making has also been found in human experiments, using both fMRI (Heekeren et al., 2004) and EEG (Kelly & O'Connell, 2013; O'Connell et al., 2012; van Vugt et al., 2012).

The bottom line is that in virtually every task where a search was made, neurons have been found that appear to implement a decision to act. As a result, the assumption of statistical decision theory that every motor response requires a decision has considerable neural support.[2]

## 11.2 THE NEURAL BASIS OF GRT

When SDT and GRT were first proposed, the field of neuroscience was still in its infancy. Basic neuroanatomy was still being worked out, and almost nothing was known about the functions of many of the neural circuits that had been identified. The last half century has seen great progress on both of these problems, and, as a result, a functional neuroanatomy of statistical decision theory is beginning to emerge. Even so, this area of research is complex and highly fluid. The best current neuroanatomical theory of GRT changes with almost every new issue of the *Journal of Neuroscience*. This section provides a snapshot of this evolving work. A book-length treatment would be required for a complete review and justification of the models described in this section.

Chapter 5 reviewed considerable evidence that decisional separability and independent decisions strategies are mediated by different neural networks than similarity-based strategies, such as those implemented by the GLC and SPC. Therefore, this section is divided into three subsections. The first describes the neural systems that build the perceptual representations of visual stimuli, the second considers the systems that mediate independent decisions strategies, and the third focuses on similarity-based strategies.

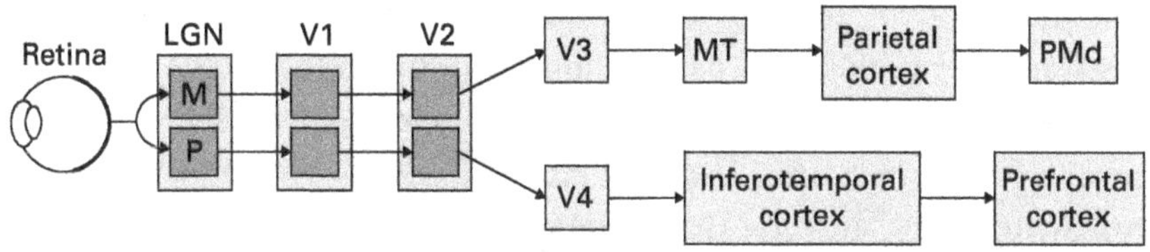


**11.1** A highly simplified schematic showing some of the major brain regions in the visual system (LGN = lateral geniculate nucleus of the thalamus; M = magnocellular; P = parvocellular; MT = middle temporal area of extrastriate visual cortex (also called area V5), PMd = dorsal premotor cortex).

### 11.2.1 THE PERCEPTUAL REPRESENTATION OF VISUAL STIMULI

Any stimulus presentation has its first central-nervous system effects on peripheral sensory receptors—including on rods and cones embedded in the retina in the case of visual stimuli, and on hair cells in the cochlea in the case of auditory stimuli, and on olfactory receptor cells embedded in the olfactory epithelium in the case of odors. Stimulation of the sensory receptors triggers a cascade of activity that then progresses through the brain until eventually reaching prefrontal cortex, at least if the sensation is salient enough to elicit conscious awareness. At each successive stage on this journey, new pathways tend to emerge as brain regions send projections to multiple downstream areas. Any functional activity observed in a neuron in primary visual cortex (area V1) is most likely driven by activity in the lateral geniculate nucleus of the thalamus. In contrast, functional activity in a visually sensitive neuron in prefrontal cortex could be driven by activity in many different areas of visual cortex. Back-projections from high-level to low-level areas cause even more complications. As a result, understanding the functional role of a brain region or projection becomes increasingly difficult with the increase in the number of synapses between that region and the sensory receptors that stimulate it. For this reason, the neuroscience that mediates the perceptual representation of stimuli is better understood than the neuroscience that mediates the decision processes assumed by GRT. And even within this former domain, the accuracy and detail of models of initial visual encoding (e.g., Wandell & Brainard, 2023) far surpasses our understanding of the neural representation of more complex real-world stimuli that depend, at least in part, on activation in sensory association areas.

Figure 11.1 is a highly simplified schematic showing some key areas that mediate visual perception along with some of their most prominent downstream projections. The dorsal stream that begins with magnocellular neurons in the lateral geniculate and progresses through parietal cortex is often referred to as the "where" pathway because of its sensitivity to the spatial location of attended objects. In contrast, the ventral stream that begins with parvocellular neurons in the lateral geniculate and progresses through inferotemporal cortex is often referred to as the "what" pathway because of its sensitivity to the identity of attended objects. Figure 11.1 oversimplifies the neuroanatomy by showing the dorsal and ventral streams as mostly separate, whereas many neuroanatomical studies have identified a variety of interconnections between the two streams that are not shown in figure 11.1 (e.g., Milner, 2017).

Many studies have confirmed that the ascending visual pathways shown in figure 11.1 are arranged hierarchically, in the sense that neurons in earlier areas tend to respond to simpler visual features than neurons in later areas. For example, Hubel and Wiesel (1962) were awarded a Nobel prize for discovering that neurons in V1, which they called simple and complex cells, respond to simple light or dark bars that have a specific location, width, and orientation. In contrast, inferotemporal neurons are tuned to complex objects, such as faces, that appear almost anywhere in the visual field (Tanaka, 1996).

A variety of models of object perception have been proposed that mimic this hierarchical structure. One prominent example from this class is the HMAX model (P. H. Cox & Riesenhuber, 2015; Jiang et al., 2006; Riesenhuber & Poggio, 1999; Serre, Wolf, et al., 2007). In the first, input stage of this model, images are densely sampled by arrays of two-dimensional Gabor filters. These units were inspired by the Hubel and Wiesel (1962) simple cells and, in homage, are called S1 units. Each S1 unit responds preferentially to a bar of a certain orientation, spatial frequency, and spatial location. Units in the second layer were inspired by the Hubel and Wiesel (1962) complex cells, and are therefore called C1 units. Each of these receives input from sets of S1 units that all have the same preferred orientation and similar, but not identical, preferred spatial frequencies and receptive field locations. The output of each C1 unit equals the maximum of all its S1 inputs, which is why the model is called HMAX. C1 units respond best to a bar of the same orientation as the S1 units that feed into it, but over a broader range of spatial frequencies and spatial positions than any single S1 unit. To increase feature complexity, neighboring C1 units that are tuned to similar spatial frequencies are then pooled to provide input to a third layer of S2 units, which roughly correspond to neurons that have been studied in cortical regions V2 and V4. Intermediate S2 features can either be hardwired (Riesenhuber & Poggio, 1999) or learned directly from image sets by choosing S2 units to be tuned to randomly selected patches of the C1 activations that result from presentations of images from a training set (Serre, Oliva, & Poggio, 2007; Serre, Wolf, et al., 2007). This latter procedure results in S2 feature dictionaries better matched to the complexity of natural images (Serre, Wolf, et al., 2007).

In layer four, C2 units receive input from all S2 units that are tuned to some particular visual feature (again using a max function). As a result, C2 units respond at the same level as the most active S2 unit that has the same preferred feature, but regardless of the feature's size or spatial location. C2 units are similar to V4 neurons in the primate visual system (Cadieu et al., 2007). Finally, these C2 units can then be combined to serve as inputs to "view-tuned" units selective for particular views of a complex object (e.g., a human face or a hand-held tool) to mimic neurons that might be found, for example, in inferotemporal cortex.

### 11.2.2 DECISION PROCESSES USED BY THE EXPLICIT-REASONING SYSTEM

The cognitive and neural systems that implement decisional separability and other independent-decisions strategies are frequently referred to as the explicit-reasoning system since the Boolean algebra that describes these strategies is thought to underlie much of human reasoning. The most complete description is provided by COVIS (Ashby & Valentin, 2017; Ashby et al., 1998; Kovacs et al., 2021). The COVIS explicit system uses selective attention and working memory to learn explicit rules, and is mediated by a broad neural network that includes the prefrontal cortex, anterior cingulate, head of the caudate nucleus, and the hippocampus. The neural version of the model is illustrated in figure 11.2. It includes two main subnetworks: one that generates or selects new candidate rules, and one that maintains candidate rules in working memory during the testing process and mediates the switch from one rule to another.

The working memory maintenance and attentional switching network includes all structures in figure 11.2, except the anterior cingulate, premotor cortex, and motor cortex. The model assumes that the long-term representations of all possible rules are encoded in some neural network in sensory association cortex. These cortical units send excitatory signals to working memory units in lateral prefrontal cortex, which send recurrent excitatory signals back to the same cortical units (not shown in figure 11.2), thereby forming a reverberating loop. At the same time, the prefrontal cortex is part of a second excitatory reverberating loop through the medial dorsal nucleus of the thalamus (Alexander et al., 1986). These double reverberating loops maintain activation in the prefrontal working memory units during the time it takes to test the candidate rule. As it turns out however, the GABAergic

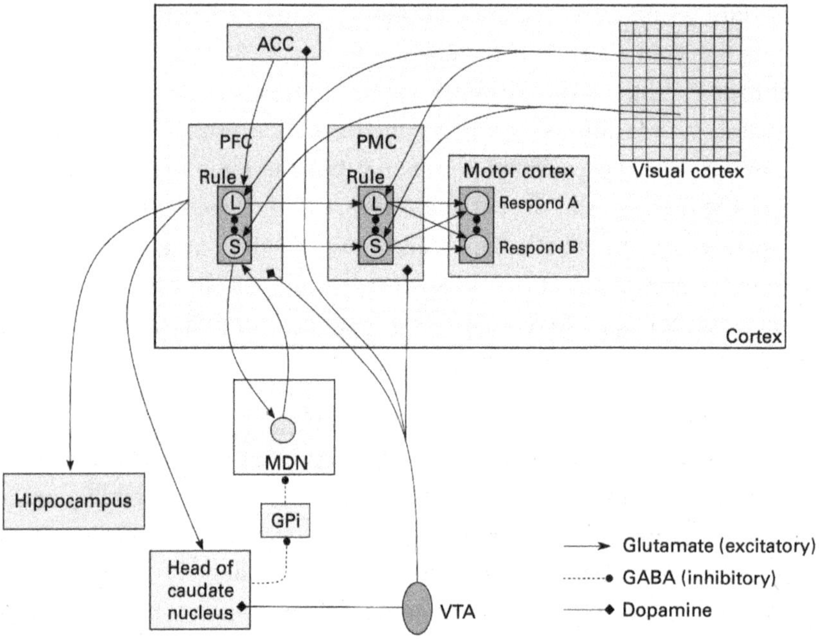


**11.2** The neural architecture of the COVIS explicit system for a two-alternative forced-choice task with responses A and B. In this example, a single one-dimensional rule is held in working memory with separate L and S units that respond when the presented stimulus has a large or small value on the attended dimension, respectively (ACC = anterior cingulate cortex; PFC = prefrontal cortex; PMC = premotor cortex; MDN = medial dorsal nucleus of the thalamus; GPi = internal segment of the globus pallidus; VTA = ventral tegmental area).

neurons in the globus pallidus have a high spontaneous firing rate that cause them to tonically inhibit the thalamus, which prevents the closing of this cortical-thalamic loop, thereby leading to the loss of information from working memory. To counteract this inhibition, the prefrontal cortex excites medium spiny neurons in the head of the caudate nucleus (Bennett & Wilson, 2000), which in turn inhibit the pallidal neurons (since medium spiny neurons are GABAergic) that are inhibiting the thalamus. Reducing the pallidal inhibition of the thalamus allows reverberation in cortical-thalamic loops, and thereby facilitates working memory maintenance. A computational (i.e., spiking neuron) version of this model successfully accounts for many behavioral and single-neuron working memory-related phenomena, such as the famous result that human working memory capacity is approximately $7 \pm 2$ items (Ashby et al., 2005).

Note that the model predicts that there are rule-sensitive units in prefrontal cortex that remain activated during the time it takes to test and either accept or reject each candidate rule. A number of studies have reported evidence for such rule-sensitive neurons in prefrontal cortex. In these studies, monkeys were trained to classify objects by applying either one rule (e.g., spatial) or another (e.g., associative), while single-unit recordings were collected from neurons in prefrontal cortex. Each trial began with a cue signaling the animal which rule to apply to the ensuing stimulus. Several studies using this paradigm reported many neurons in prefrontal cortex that showed rule-specific activity—that is, they fired during application of one of the rules but not during the other, regardless of which stimulus was shown (Asaad et al., 2000; Hoshi et al., 2000; I. M. White & Wise, 1999).

When feedback indicates that the current rule is incorrect, then a new rule must be selected and executive attention must be switched from the old rule to the new rule. Evidence suggests that these selection and switching operations are mediated by separate neural processes. The process of generating new candidate hypotheses is complex and poorly understood, and a complete model of rule selection has not yet been formulated, in part because it likely involves poorly understood phenomena such as creativity and insight.

In the computational version of the model, the anterior cingulate selects among alternative rules by enhancing the activity of the specific prefrontal working memory unit that represents a particular rule (Ashby, Valentin, & Turken, 2002). COVIS predicts that the most effortful and time-consuming processing follows error feedback, because negative feedback suggests that the current rule is incorrect. In support of this prediction, during the period of initial learning, slower response times on trials following an error predict higher accuracy in RB tasks, but not in II tasks (Tam et al., 2013). When this process leads to the selection of the correct rule, there is often a sudden transition from suboptimal to optimal performance accompanied by an abrupt transition in neural firing of prefrontal cortical neuron ensembles (Durstewitz et al., 2010; J. D. Smith & Ell, 2015).

Rule-sensitive neurons have also been reported in premotor cortex (Muhammad et al., 2006; Vallentin et al., 2012; Wallis & Miller, 2003). The figure 11.2 model assumes that these neurons and specifically the circuitry through premotor and motor cortex allow rule-based responding to eventually become automatic (Kovacs et al., 2021). The subnetwork that mediates this transition to automatic responding was briefly described in section 10.2.2.2 (e.g., compare figure 10.15 to figure 11.2). In the figure 11.2 example, each rule unit includes two simulated neurons—one that signals if the stimulus has a small value on the attended stimulus dimension (the S unit) and one that signals if the stimulus has a large value (the L unit). The idea is that neurons in visual cortex that respond to small values on the attended dimension project to the PFC-S neuron, whereas visual cortical neurons that respond to large values project to the PFC-L neuron. In this way, the S neuron responds to any stimulus with a small value and the L neuron responds to any stimulus with a large value. The model assumes that these PFC rule units develop as a result of lifelong practice with a rule. During early training sessions, the stimulus activates the appropriate PFC rule unit (i.e., S or L), which then activates the analogous PMC rule unit, which then activates the appropriate unit in motor cortex that causes the model to respond A on trials when the orientation is shallow and B when the orientation is steep. The same visual cortical units that project to the S and L PFC rule units also project to the S and L PMC rule units. Initially, however, these visual cortex-to-PMC synapses are not strong enough to activate the appropriate unit in motor cortex. Instead, PFC activation is also required.

The model also assumes that Hebbian learning will strengthen all active synapses in the figure 11.2 network. Some of these are more critical than others. First consider the synapses between visual cortex and premotor cortex. In Hebbian learning, synaptic strengthening is proportional to the product of the pre- and post-synaptic activations. During early training, much of the post-synaptic activation (i.e., the activation within the premotor units) is driven by input from prefrontal cortex. As the visual-to-premotor cortex synaptic strengths increase, they eventually become strong enough so that visual input alone is enough to cause the premotor unit to activate the appropriate target in motor cortex. The pathway through prefrontal cortex is still active, but because it is longer, it no longer controls behavior. At this point, the behavior has become automatic.

Second, consider the synapses between premotor cortex and primary motor cortex. Initially these are weak because participants have no prior association between small or large values on the attended stimulus dimension and A or B button presses (for example). But after thousands of practice trials, Hebbian learning will strengthen these associations. The model therefore predicts that early in training, when responding is controlled by prefrontal cortex, switching the location of the response buttons will have little or no effect on performance. However, after extensive practice—that is, after the connections between premotor and primary motor cortices have become strong—switching the response buttons should cause an interference. Several studies have confirmed these predictions (Ashby et al., 2003; Hélie et al., 2010).

Note that the model predicts that early rule-guided behaviors are mediated by abstract rules that are represented in prefrontal cortex, whereas automatic rule-guided behaviors are mediated by direct projections from visual cortex to the rule-sensitive units in premotor cortex that control the behavior. Because of Hebbian learning, the associations between the stimulus representations in visual cortex and the motor associations in premotor cortex eventually become strong enough to trigger the behavior without assistance from the abstract rule representations in prefrontal cortex.

Selective attention is thought to depend on the prefrontal cortex (e.g., E. K. Miller & Cohen, 2001). Thus, the ability to quickly form decision bounds that satisfy local decisional separability might be unique to species with a well-developed prefrontal cortex. Some evidence supports this prediction. First, of course, many studies have shown that healthy humans learn categories that can be separated by a one-dimensional decision bound (that satisfies decisional separability) far more quickly than categories that can be separated by a diagonal bound, even when the two category structures are exactly equated on the most widely used category-separation statistics (e.g., Ashby & Maddox, 2005; Ashby et al., 2020). Second, pigeons learn these two types of category structures equally well and at exactly the same rate (Qadri et al., 2019; J. D. Smith et al., 2011). Although pigeons do possess brain regions thought to be analogous to prefrontal cortex, such structures are not homologous, and it seems clear that pigeons lack at least some of the cognitive mechanisms recruited during rule-based learning (Berg et al., 2014; Soto & Wasserman, 2014). Third, macaque and capuchin monkeys, who both have substantial prefrontal cortices, show similar performance advantages as humans on the one-dimensional category structures (J. D. Smith et al., 2010, 2012). This result suggests that the human and monkey advantage for decisional separability is probably not language-based, even though rules that satisfy decisional separability are typically easy to describe verbally. Although much more data are needed to settle this issue, a tentative first hypothesis is that selective attention is the critical attribute that makes one-dimensional rules so easy to learn.

### 11.2.3 DECISION PROCESSES USED BY THE PROCEDURAL SYSTEM

Chapter 5 reviewed evidence that similarity-based strategies are mediated by procedural learning. The COVIS model also provides the most complete description of the procedural-learning system (Ashby & Crossley, 2011; Ashby & Waldron, 1999; Ashby et al., 1998; Cantwell et al., 2015). Figure 11.3 shows the architecture of the current version of this model. The key structure is the striatum, a major input region within the basal ganglia that includes the caudate nucleus and the putamen. In primates, all of extrastriate visual cortex projects directly to the striatum, with a cortical-striatal convergence ratio of approximately 10,000 to 1 (e.g., Wilson, 1995). Much evidence suggests that procedural learning is mediated largely within the striatum, and is facilitated by a dopamine-mediated reinforcement learning signal (Ashby & Ennis, 2006; Badgaiyan et al., 2007; Grafton et al., 1995; Knopman & Nissen, 1991; Mishkin et al., 1984; Willingham, 1998). The critical site of learning within the striatum is thought to be at synapses between pyramidal neurons projecting from cortex and striatal medium spiny neurons. The well-accepted theory is that positive feedback that follows successful behaviors increases phasic dopamine levels in the striatum, which has the effect of strengthening recently active synapses, whereas negative feedback causes dopamine levels to fall below baseline, which has the effect of weakening

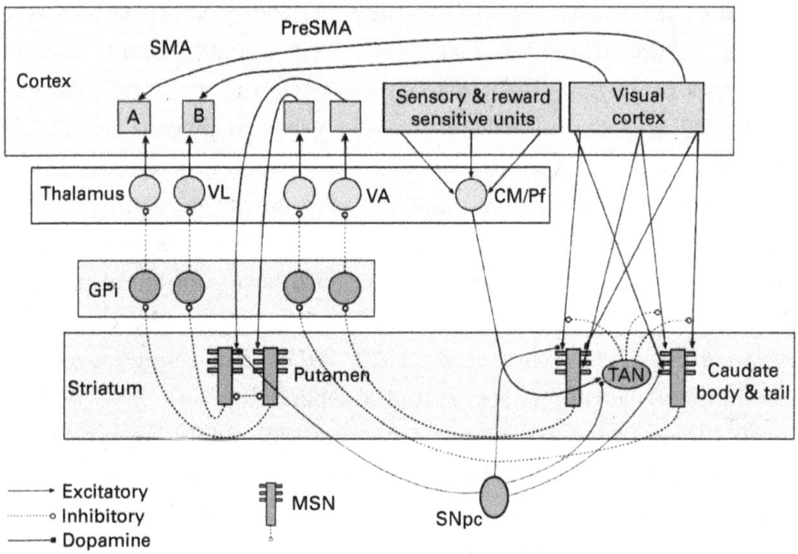


**11.3** The neural architecture of the COVIS model of procedural learning for a two-alternative forced-choice task with responses A and B (SMA = supplementary motor area; PreSMA = presupplementary motor area; VL = ventral lateral nucleus of the thalamus; VA = ventral anterior nucleus of the thalamus; CM/Pf = centromedian and parafascicular nuclei of the thalamus; GPi = internal segment of the globus pallidus; TAN = tonically active neuron; $SN_{PC}$ = substantia nigra pars compacta; MSN = medium spiny neuron of the striatum).

recently active synapses (Houk et al., 1995). In this way, units in visual cortex that are activated by the stimulus cause activation in units in premotor cortex (e.g., supplementary motor area) that encode a motor goal (e.g., press the button on the left). As a result, the procedural system learns associations between each presented stimulus (via activation of visual cortical neurons) and the relevant motor response (Ashby & Waldron, 1999; Ashby et al., 2007). The COVIS procedural-learning model (and the SPC) is a formal instantiation of these ideas.

Note that the model includes two loops through the basal ganglia (Cantwell et al., 2015). One loop projects from visual cortex through the body and tail of the caudate nucleus and terminates in presupplementary motor area, and the second loop projects from presupplementary motor area through the putamen and terminates in supplementary motor area. Because this second loop terminates in premotor cortex, COVIS predicts that the associations that are learned are between stimuli and motor goals. Both loops rely on reinforcement learning at cortical-striatal synapses. The first loop learns which stimuli are associated with the same response and the second loop learns what motor response is associated with each of these stimulus clusters. In a novel task, both types of learning are required. However, note that if we train agents to make accurate classification responses and then switch the responses associated with the two stimulus classes, then the classes remain unchanged—only the response mappings must be relearned. So COVIS predicts that reversing the locations of the response keys will interfere with procedural classification performance, but that recovery from such a reversal should be easier than novel classification learning—a prediction that has been supported in several studies (Cantwell et al., 2015; Kruschke, 1996; Maddox et al., 2010; Sanders, 1971; Wills et al., 2006).

One strong prediction of this model is that synaptic strengthening can only occur when the visual trace of the stimulus and the post-synaptic effects of dopamine overlap in time. More specifically, synaptic plasticity in the striatum is strongest when the intracellular signaling cascades driven by NMDA receptor activation and dopamine D1 receptor activation coincide (Lisman et al., 2002; Rudy, 2020). The further apart in time these two cascades peak, the less effect dopamine will have on synaptic plasticity. For example, Yagishita et al. (2014) reported that synaptic plasticity was best (i.e., greatest increase in spine volume on striatal medium spiny neurons) when dopamine neurons were stimulated 600 ms after medium spiny neurons. When the dopamine neurons were stimulated before or 5 s after the medium spiny neurons, then no evidence of any plasticity was observed. In a task mediated by procedural learning, activation of the medium spiny neurons should occur just before the motor response, and activation of the dopamine neurons should occur just after the feedback. So COVIS predicts that feedback delays during procedural learning should have effects that are similar to those observed by Yagishita et al. (2014). In fact, many studies have confirmed this prediction in II category-learning tasks (Dunn et al., 2012; Maddox et al., 2003; Maddox & Ing, 2005; Worthy et al., 2013). Valentin et al. (2014) showed that the COVIS procedural-learning model can accurately account for the effects of all these feedback delays. In contrast, the same studies showed that delays up to 10 s have no effect on RB category learning, which is thought to be mediated by the explicit-reasoning system.

Ashby and Crossley (2011) proposed that the striatal cholinergic interneurons serve as a context-sensitive gate between cortex and the striatum (see also Crossley, Horvitz, et al., 2016; Crossley et al., 2013, 2014). The idea, which is supported by a wide variety of neuroscience evidence, is that the striatal cholinergic interneurons tonically inhibit cortical input to striatal medium spiny neurons (e.g., Apicella et al., 1997; Pakhotin & Bracci, 2007). The striatal cholinergic interneurons are driven by neurons in the centremedian–parafascicular nuclei of the thalamus, which in turn are broadly tuned to features of the environment. In rewarding environments, the cholinergic interneurons learn to pause to stimuli that predict reward, which releases the cortical input to the striatum from inhibition. This allows striatal output neurons to respond to excitatory cortical input, thereby facilitating cortical-striatal plasticity. In this way, cholinergic interneuron pauses facilitate the learning and expression of striatal-dependent behaviors. When rewards are no longer available, the cholinergic interneurons cease to pause, which prevents striatal-dependent responding and protects striatal learning from decay.

Extending the COVIS procedural-learning system to include striatal cholinergic interneurons allows the model to account for many new phenomena—some of which have posed difficult challenges for previous learning theories. One of these is that the reacquisition of an instrumental behavior after it has been extinguished is considerably faster than during original acquisition (Ashby & Crossley, 2011). The model accounts for this ubiquitous phenomenon because the withholding of rewards during the extinction period causes the cholinergic interneurons to stop pausing to sensory cues in the conditioning environment (since they are no longer associated with reward). This closes the gate between cortex and the striatum, which prevents further weakening of the cortical-striatal synapses. When the rewards are reintroduced, the cholinergic interneurons relearn to pause, and the behavior immediately reappears because of the preserved synaptic strengths.

Ashby et al. (2007) extended this model to account for responding that has been practiced so much that it occurs automatically. Specifically, they proposed that skills learned procedurally are mediated entirely within cortex after they become automatized, and that the development of automaticity is associated with a gradual transfer of control from the striatum to cortical-cortical projections from the relevant sensory areas directly to the premotor areas that initiate the behavior. So in figure 11.3, the cortical-cortical projections from visual cortex to the supplementary motor area eventually mediate the expression of automatic behaviors without any assistance from the subcortical loops through the basal ganglia. Therefore, according to this account, a critical function of the basal ganglia is to train purely cortical representations of automatic behaviors. Note that this is similar to the Kovacs et al. (2021) proposal that rule-guided behaviors are automatized via a process in which the prefrontal cortex trains the cortical circuits that implement the automatic behaviors.

The Ashby et al. (2007) model was motivated by the observation that because cortical synaptic plasticity follows Hebbian, rather than reinforcement, learning rules, the purely cortical circuits are incapable of learning any behavior that requires trial-by-trial feedback. Such behaviors require the reinforcement-learning plasticity of the basal ganglia. Ashby et al. (2007) proposed that the basal ganglia use dopamine-mediated reinforcement learning (i.e., at cortical-striatal synapses) to gradually activate the correct postsynaptic targets in supplementary motor area, which thereby enables Hebbian plasticity at cortical-cortical synapses to learn the correct associations (i.e., because there will be more postsynaptic activation at the correct synapses than at synapses leading to incorrect responses). As a result, in the full version of the figure 11.3 model, plasticity at cortical-striatal synapses is modeled via reinforcement learning rules, whereas plasticity at cortical-cortical synapses is modeled via Hebbian rules.

This model accounts for many results that are problematic for other theories of automaticity. For example, it correctly predicts that people with Parkinson's disease, who have dopamine reductions and striatal dysfunction, are impaired in initial procedural learning (Soliveri et al., 1997; Thomas-Ollivier et al., 1999) but relatively normal in producing automatic skills (Asmus et al., 2008). It also correctly predicts that blocking all striatal output to cortical motor and premotor targets does not disrupt the ability of monkeys to fluidly produce an overlearned motor sequence (Desmurget & Turner, 2010). Similarly, a neuroimaging study reported that activation in the putamen was correlated with performance of a procedural skill early in training but not after automaticity developed (Waldschmidt & Ashby, 2011). Instead, automatic performance was only correlated with activity in cortical areas (i.e., presupplementary motor area and supplementary motor area).

## 11.3 EXTENDING GRT ANALYSES TO NEURAL DATA

When GRT was first developed in 1986, the only widely available method for studying human brain activity was electroencephalography (EEG). Now there are many available technologies—the most widely used of which include functional magnetic resonance imaging (fMRI), positron emission tomography (PET), functional near-infrared spectroscopy (fNIRS), and magnetoencephalography (MEG). In addition, the emergence of noninvasive brain stimulation technologies, such as transcranial magnetic stimulation (TMS) and transcranial direct current stimulation (TDCS), allow researchers to study the contributions of specific regions of the human cortex to almost any behavior of interest. Furthermore, because of advances in the surgical treatment of epilepsy and other neurological disorders (e.g., Parkinson's disease), single-neuron recordings in awake, behaving humans are now available to a handful of (fortunate) cognitive neuroscientists. All of these new methods give researchers access to human brain activity at levels that seemed like science fiction in 1986. Not surprisingly though, the development of these new technologies has greatly outpaced the development of new methods of data analysis that can exploit the power of these technologies. The most work has been done on methods of analyzing fMRI data (e.g., Ashby, 2019), but even here, new methods are needed. Statistical decision theory can play an important role in this exploding new field by providing novel data-analysis methods that are capable of addressing questions that are beyond the reach of existing tools. Only a few initial steps have been taken along this path. These are described in the next two subsections, followed by some brief remarks about possible future directions of research.

### 11.3.1 EEG

Wenger and Rhoten (2020) extended GRT to EEG recordings by showing that it is possible to use the timing of an event-related brain potential to draw inferences about perceptual independence and separability. Specifically, in a study of visual perceptual learning, they used the onset time of the lateralized readiness potential (LRP) as a substitute for RT in the timed versions of marginal response invariance and report independence that were discussed in chapter 6. The LRP is a negative-trending waveform that is widely interpreted as a signal that a movement is about to occur. It is recorded at central electrodes that are contralateral to the motor response that it precedes. The onset time of the LRP is strongly correlated with observable RT, which is necessarily slower than the LRP onset. Wenger and Rhoten (2020) reported that using the LRP as a proxy for RT in the timed versions of marginal response invariance and report independence led to conclusions that supported the inferences drawn from the response frequencies.

### 11.3.2 FUNCTIONAL MRI

Soto et al. (2018) extended the GRT concepts of perceptual separability and perceptual independence to neuroimaging data. When a visual stimulus is presented to an observer, it causes activation in many areas within the visual system. The GRT perceptual distributions are meant to model the perceptual representations that observers use to make decisions in the task under study. Therefore, these distributions most likely model activations that occur in some higher-level visual area. If these representations violate perceptual separability (or perceptual independence), then an obvious and important question to ask is: When and where was separability (or independence) first violated within the processing stream? Soto et al. (2018) proposed methods for using fMRI to answer this question.

FMRI experiments provide measures of the blood-oxygen-level-dependent (BOLD) signal while a participant completes an experimental task. The BOLD signal is an indirect measure of neural activity that depends in a complicated way on concentrations of oxygenated and deoxygenated hemoglobin, cerebral blood flow, and venous blood volume (Buxton, 2013). The important point for most fMRI users is that active brain areas consume more oxygen than inactive areas. To compensate, the vascular system rushes oxygenated hemoglobin into the active brain area. The result of these actions is that the BOLD signal rises quickly after neural activation is initiated until eventually reaching a peak after around six seconds and then it slowly decays back to baseline levels over a period of 20–25 seconds. FMRI scanners record the BOLD signal at discrete times and spatial locations throughout the duration of the experiment. The time between successive recordings is a parameter called the TR (for repetition time) that is set during experimental design. The spatial resolution is determined by the voxel size, which is another parameter set during experimental design. Current state-of-the-art scanners of the type found in many fMRI research centers are capable of voxels of approximately 8–10 $mm^3$ and TRs of less than one second. For a typical human brain, this means the scanner might record more than 150,000 BOLD responses every TR. If recordings are collected throughout a 30-minute experimental task, the end result might be as many as 400 million recorded BOLD values.

A major challenge in almost all fMRI research is that the BOLD response is only an indirect measure of neural activity. As a result, Soto et al. (2018) distinguished between what they called encoding separability and independence and decoding separability and independence. *Encoding separability* holds for stimulus dimensions 1 and 2 in some brain region of interest if the neural encoding of dimension 1 in that region does not change when the level of dimension 2 changes. Similarly, *encoding independence* holds in a region of interest if the neural representations of two stimulus dimensions are statistically independent. Encoding separability and encoding independence are properties of neural activity and, therefore, are not directly observable in fMRI experiments. As a result, Soto et al. (2018) also defined decoding separability and independence. *Decoding separability* holds for stimulus dimensions 1 and 2 in some region of interest if the BOLD response associated with dimension 1 does not change when the level of dimension 2 changes. Similarly, *decoding independence* holds in a region if the BOLD responses associated with two stimulus dimensions are statistically independent.

The obvious next question is: How are encoding and decoding separability related? Soto et al. (2018) showed that encoding separability implies decoding separability—that is, if encoding separability holds in some ROI, then decoding separability is guaranteed to also hold. The converse however, is not true. Specifically, it is logically possible that decoding separability could hold even if encoding separability is violated. As a result, an empirical demonstration that decoding separability fails is strong evidence that encoding separability is violated in this brain region. However, a finding that decoding separability holds does not guarantee that encoding separability is satisfied. Furthermore, these same logical relationships hold for encoding and decoding independence. In particular, a finding that decoding independence fails is strong evidence against encoding independence, but a finding that decoding independence holds does not guarantee that encoding independence is satisfied.

Next, Soto et al. (2018) proposed empirical tests of these constructs that are based on summary statistics derived from applying machine-learning methods to fMRI data. As a concrete example, consider the $2 \times 2$ factorial identification experiment that Soto et al. (2018) used to illustrate their approach. In this experiment, the four stimuli $A_1B_1$, $A_1B_2$, $A_2B_1$, and $A_2B_2$ were the human faces shown in figure 11.5. Note that component A refers to emotional expression, with $A_1$ and $A_2$ denoting neutral and sad expressions, respectively. In contrast, component B refers to facial identity with $B_1$ and $B_2$ denoting two different people. In the Soto et al. (2018) experiment, 21 participants completed one session of identification training with these stimuli outside of the scanner that included trial-by-trial feedback. Next, they each completed a second session without feedback inside the fMRI scanner.

The remainder of this section provides an intuitive description of the methods Soto et al. (2018) used to analyze their scanning data. For a detailed description of each step in this process, see Ashby (2019).

After preprocessing the data, the next step is to choose the search region. The standard approach is to perform a searchlight analysis in either the whole brain or in some collection of regions of interest (e.g., visual cortex). A searchlight is a spherical collection of voxels—typically some number between 20 and 50 (Kriegeskorte et al., 2006). The analysis is performed on the voxels within the current searchlight, and the results are typically attributed to the single voxel at the center of the searchlight. Next the searchlight is moved to a new center location, often just a single voxel away from the old center, and the analysis is repeated. Each analysis results in a single score—for example, a number that increases with violations of separability. In this way, a separability index can be assigned to every voxel in the search region.

The BOLD response to each stimulus is a time series because the presentation of a stimulus is expected to alter the BOLD response in stimulus-sensitive voxels for a duration of 30 seconds or so. BOLD responses are recorded every TR seconds, so the length of each time series generated by every stimulus presentation is approximately 30/TR. The next step in data analysis is to convert this time series to a single value, using one of several available methods (Ashby, 2019). If there are $N_V$ voxels in the searchlight, then after reducing each time series to a single value, we are left with a vector of $N_V$ values, called the activity vector, which summarizes the BOLD responses of all voxels in the searchlight to presentation of the stimulus. After this step, if there are $N_S$ stimulus presentations in the experiment, then there will be $N_S$ activity vectors, each of order $N_V \times 1$.

Now consider an $N_V$-dimensional space that has a dimension for every voxel in the searchlight. Each activity vector can be represented by a point in this space. Figure 11.5A shows a hypothetical example of this voxel space when the searchlight contains only three voxels. The black dots denote activity vectors on trials when stimulus $A_1B_1$ was presented, and the gray dots denote activity vectors on trials when stimulus $A_2B_1$ was presented. In the figure 11.4 experiment where each stimulus is a face, the

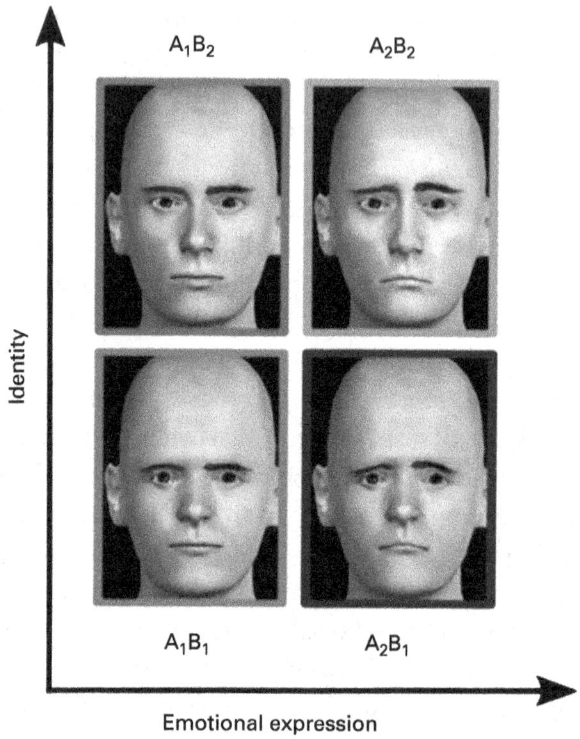


**11.4** The four faces used in the 2 × 2 factorial identification experiment reported by Soto et al. (2018).

black and gray dots describe searchlight activity for pictures of the same person (i.e., same facial identity) displaying two different emotional expressions (i.e., neutral and sad). If the voxels in this searchlight are sensitive to emotional expression, then they should respond differently to neutral and sad expressions, and, as a result, the points associated with a neutral expression (i.e., $A_1$) should cluster in a different region of voxel space than the points associated with a sad expression (i.e., $A_2$).

The statistical methods most commonly used in the neuroimaging literature to determine whether the activity vectors are clustering in different regions of voxel space are collectively referred to as multivoxel pattern analysis (MVPA). An especially common approach uses support vector machines (SVMs), which are conceptually similar to the GLC that was described in section 5.1.4. The idea is illustrated in figure 11.5A. An SVM is a machine-learning algorithm that attempts to find the plane or hyperplane that best separates points in a multidimensional space that belong to different classes (for details see Ashby, 2019).

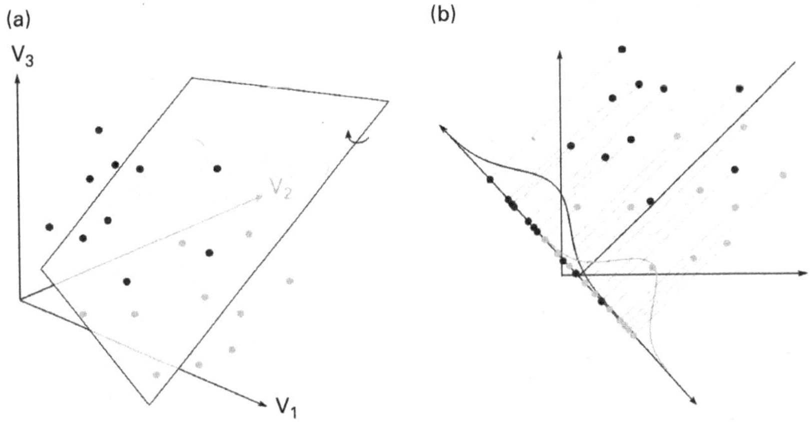


**11.5** An illustration of how a support vector machine might be used to test for decoding separability in some brain region of interest. (a) Activity vectors from 20 trials of an identification task. Black dots denote BOLD responses on trials when stimulus $A_1B_1$ was presented, and gray dots denote BOLD responses on trials when stimulus $A_2B_1$ was presented. The gray plane is the support vector machine that best separates the $A_1B_1$ and $A_2B_1$ points. (b) Same as in panel (a), except the space has been rotated so that the view is orthogonal to the classification plane.

A standard application might proceed as follows. First, all $N_S$ activity vectors are divided arbitrarily into two sets—one called the training set and the other called the validation set. For example, data from even-numbered trials could form the training set, and data from odd-numbered trials could form the validation set. The second step, called training the classifier, is to estimate the parameters of the classifier by using the data in the training set. This estimation process uses both the coordinate values of each point in the training set (i.e., within voxel space) and the identity of the stimulus that was presented when that data point was collected (i.e., whether the stimulus face displayed emotional expression $A_1$ or $A_2$). In the case of the SVM, the free parameters identify the hyperplane that best separates the $A_1$ and $A_2$ activity vectors.

The third step in a standard MVPA analysis, called validation, is to test the SVM by examining its ability to identify the stimulus label (i.e., $A_1$ versus $A_2$) of the activity vectors that comprise the validation set. An important restriction is that during this testing process, all parameters are left fixed at their final training values. If the classifier correctly classifies more activity vectors in the validation set than would be expected by chance, then we conclude that this brain region is responding differently to the two stimulus classes.

In a brain region that is sensitive to emotional expression, the classification plane represents BOLD responses that are equivocal with respect to emotion. Points on one side are suggestive of emotion $A_1$ and points on the other side are suggestive of emotion $A_2$, and the further a point is from the plane, the stronger it signals one emotional expression over the other. As a result, the single dimension in voxel space that carries the most information about emotional expression is the one orthogonal to the classification plane. Soto et al. (2018) suggested testing for encoding separability by estimating the marginal distributions associated with each stimulus on this dimension and then assessing how much the $A_1$ distribution changes when component B changes from $B_1$ to $B_2$, and similarly how much the $A_2$ distribution changes when component B changes from $B_1$ to $B_2$. These marginal distributions are illustrated in figure 11.5B for stimuli $A_1B_1$ and $A_2B_1$.

Following this approach, Soto et al. (2018) estimated all the relevant marginal distributions using a nonparametric kernel density estimator. Let $\hat{f}_{ij}(x)$ denote the pdf estimate for stimulus $A_iB_j$ at distance $x$ from the classification plane. Note that if emotional expression satisfies decoding separability, then it must be true that

$$\hat{f}_{11}(x) = \hat{f}_{12}(x), \text{ for all } x, \tag{11.1}$$

and

$$\hat{f}_{21}(x) = \hat{f}_{22}(x), \text{ for all } x. \tag{11.2}$$

Therefore, to test whether emotional expression satisfies decoding separability in any given searchlight, Soto et al. (2018) proposed the statistic

$$DDS = \sum_{i=1}^{2} \int |\hat{f}_{i1}(x) - \hat{f}_{i2}(x)| \, dx. \tag{11.3}$$

An analogous statistic can be constructed to test whether facial identity satisfies decoding separability. In either case, the larger the statistic, the more likely that decoding separability is violated. Soto et al. (2018) determined critical values for statistical significance for these statistics via a bootstrapping approach.

Using these methods, Soto et al. (2018) computed two values of the $DDS$ statistic in every voxel of the brain—one value that computed a decoding separability score for emotional expression (i.e., $DDS_E$) and one that computed a score for facial identity (i.e., $DDS_I$). Next, as is common in fMRI data analysis, they tested for the statistical significance of the two contrasts: $DDS_E - DDS_I$ and $DDS_I - DDS_E$. The results found no voxels anywhere in the brain in which the violations of decoding separability were greater for emotional expression than identity. In other words, there was no brain region that changed its response to either emotional expression when the identity changed more than it changed its response to an identity when the emotion it was displaying changed. In contrast, there were regions in which the violations of decoding separability were greater for identity than for emotional expression, suggesting that in these regions a change in emotional expression appeared to change facial identity. These regions included parts of the left superior temporal sulcus, which has been implicated in a variety of social processes, including theory-of-mind (e.g., Beauchamp, 2015).

Note that these results reinforce conclusions that were drawn from the application of GRT-wIND that was discussed in section 7.7. The experiment described there used different participants and slightly different stimuli, but that experiment also used a $2 \times 2$ identification task with faces that varied on facial identity and emotional expression (neutral versus sad). GRT-wIND accounted for 99.52% of the variance in the confusion matrices of the 24 participants, and overall it provided a better fit to the data than when a unique GRT model was fit to each of the 24 individual confusion matrices. The best-fitting version of GRT-wIND indicated that emotional expression was perceptually separable from facial identity but facial identity was not perceptually separable from emotional expression. In other words, GRT-wIND suggested that changing the emotional expression from neutral to sad changed the apparent identity of the face, but changing the facial identity had little or no effect on perceived emotional expression. This is exactly the same conclusion that was suggested by testing for decoding separability, but an advantage of the decoding separability analysis was the identification of the superior temporal sulcus as the critical brain region for these effects. This is important because the absence of violations of decoding separability in earlier visual areas suggests that the reason facial identity is not perceptually separable from emotional expression may not be primarily perceptual, but rather due to some social effect. Another advantage of the encoding separability test is that, unlike any application to behavioral data, no assumptions are required about any decision processes.

Soto et al. (2018) did not propose a test of decoding independence, so developing such a test should be a goal of future research. The challenge will be to obtain information about the joint distribution of neural activity driven by the two stimulus dimensions. The Soto et al. (2018) MVPA-based method illustrated in figure 11.5 estimates marginal distributions, but it is not clear how this approach could be extended to joint distributions.

### 11.3.3 FUTURE DIRECTIONS

Given reliable statistical methods for testing encoding separability and encoding independence, an obvious and important problem is to track these properties as the stimulus representations move from the retina, through the visual cortex, and eventually cause activation in the premotor and motor cortices. In cases where the behavioral data are characterized by violations of perceptual separability and/or perceptual independence, we would expect encoding separability and encoding independence to be satisfied in earlier brain regions on this pathway, and violated only in some downstream region(s). This is because the stimulus components in a $2 \times 2$ factorial identification experiment are constructed by separately crossing two levels of the two components, and therefore at the stimulus level, the components would typically be expected to not interact. Identifying the region where either of these encoding properties are first violated should yield some deep psychological insights into how the stimulus dimensions are processed. For example, Soto et al. (2018) identified the left superior temporal sulcus as the likely first site where the encoding of facial identity was not separable from the encoding of emotional expression. As previously mentioned, the fact that this region has been implicated in theory-of-mind and in the selective processing of social interactions (e.g., Beauchamp, 2015; Masson & Isik, 2021) suggests that the violations of perceptual separability observed in identification experiments that cross facial identity with emotional expression are not purely perceptual, but rather have a social component. Such an inference is beyond the reach of a traditional GRT analysis of identification data.

Such analyses could also address other questions that seem impossible with traditional behavioral approaches. For example, consider figure 7.4, which shows results of a $3 \times 3$ factorial identification experiment in which each stimulus was a semicircle with an interior radial line. Panel (a) shows the nine stimuli that were created by crossing 3 radial-line orientations with three circle sizes. Panels (b) and (c) show best-fitting GRT models for representative observers in two different experiments that used these stimuli. Note that in both cases, the best-fitting models suggest that perceived orientation tends to increase with circle size. A traditional GRT analysis would conclude that there must therefore be interactions between how humans perceptually process size and orientation. But note that when size increases with these stimuli, the length of the radial line also increases, and it is not unreasonable to think that orientation is easier to judge with longer lines than with shorter lines. In contrast, increasing orientation does not change circle size, so there appears to be an asymmetry in the information provided by the two stimulus components. In other words, it seems plausible that a confusion matrix from an ideal observer might also display an asymmetric perceptual integrality with these stimuli, and therefore that the asymmetric perceptual integrality exhibited in figures 7.4b and 7.4c is more a property of the stimuli than of the human perceptual system. How could one test this hypothesis? One possibility might be to run this identification experiment inside an fMRI scanner and then analyze the BOLD data using the methods of Soto et al. (2018). If the integrality is an inherent property of the stimuli then an asymmetric violation of decoding separability should be found in the earliest of visual regions, before much, if any, perceptual processing has occurred (e.g., in the lateral geniculate nucleus of the thalamus). In contrast, if the integrality occurs because of perceptual processing, then the first violations of decoding separability should occur within some downstream visual area.

The cognitive neuroscience revolution offers exciting opportunities for statistical decision theory researchers. The development of technologies that allow unparalleled access to human brain activity has far outpaced the development of appropriate methods to analyze all these new data. Statistical decision theory can make important and unprecedented contributions to this field by providing unique methods of data analysis that address problems of fundamental importance. Developing these new methods is likely to be an area of intense activity within the statistical decision theory literature during the next few decades.

# APPENDIX A: THE BIVARIATE NORMAL DISTRIBUTION

A *random vector* is a vector in which every entry is a random variable. For example, consider the vector

$$\underline{X} = \begin{bmatrix} X_1 \\ X_2 \end{bmatrix}. \tag{A.1}$$

If $\underline{X}$ is a random vector, then $X_1$ and $X_2$ are random variables. Let $g_i(x)$ denote the probability density function (pdf) of $X_i$. This function specifies the likelihood that a random sample drawn from the $X_i$ population exactly equals any specific numerical value $x$. If $X_i$ is normally distributed, then $g_i(x)$ is the familiar bell-shaped curve defined by

$$g_i(x) = \frac{1}{\sigma_i\sqrt{2\pi}} e^{-\frac{1}{2}\left(\frac{x-\mu_i}{\sigma_i}\right)^2}, \tag{A.2}$$

for any value of $x$, and where $\mu_i$ and $\sigma_i$ are the mean and standard deviation of the $X_i$ population, respectively. With respect to the random vector $\underline{X}$, the $g_i(x)$ pdfs are known as the *marginal distributions*.

The marginal distributions provide much information about the sampling behavior of $\underline{X}$, but they do not tell us everything. In particular, they provide no information about any statistical relationships that might exist between $X_1$ and $X_2$. Complete information about $\underline{X}$ is cataloged in the joint probability density function (or joint distribution) $f_{12}(x_1, x_2)$, which gives the likelihood that a random sample of $\underline{X}$ produces the exact pair of values $(x_1, x_2)$.

If there is no statistical relationship between $x_1$ and $x_2$, then $f_{12}(x_1, x_2)$ follows the usual multiplication rule for independent events, and therefore

$$f_{12}(x_1, x_2) = g_1(x_1)g_2(x_2), \tag{A.3}$$

for all values of $x_1$ and $x_2$. In this case, the random variables $X_1$ and $X_2$ are *statistically independent*, which means that the likelihood of observing the sample $(x_1, x_2)$ equals the likelihood of observing $x_1$ [i.e., $g_1(x_1)$] times the likelihood of observing $x_2$ [i.e., $g_2(x_2)$]. If equation A.3 is violated for any values of $x_1$ or $x_2$, then the random variables are *statistically dependent.*

As its name suggests, the multivariate normal distribution is a generalization of the well-known normal (or Gaussian) distribution. The multivariate normal distribution is, by far, the most widely used multivariate distribution in statistics. For example, it serves as the error model in the general linear model (GLM) and as the model that underlies principal component analysis (PCA). A multivariate normal random vector that is two-dimensional is said to have a bivariate normal distribution.

The bivariate normal distribution is defined by two assumptions: (1) the marginal distributions are both normal, so their pdfs have the form of equation A.2; and (2) $X_1$ and $X_2$ are either statistically independent or are linearly related. Note that it is logically possible that (1) holds, but (2) fails. For example, $X_1$ and $X_2$ might both be normally distributed but have an inverted U-shaped relationship to each other, rather than a linear relationship (i.e., so that a scatterplot of $(x_1, x_2)$ samples looks like an inverted U, rather than a line). In this case, the joint distribution of $(X_1, X_2)$ is not bivariate normal, even though both marginal distributions are normal.

The well-known Pearson correlation coefficient measures linear relationships between pairs of variables. This is the model of statistical dependence that underlies the bivariate normal distribution. Uncorrelated random variables have no linear relationship, but they could have a nonlinear relationship—for example, an inverted U-shaped relationship—in which case they would not be statistically independent (i.e., equation A.3 would not hold). For example, if a scatterplot of $X_1$ and $X_2$ samples has an inverted U shape, then knowledge of $X_1$ allows one to predict the value of $X_2$, so $X_1$ and $X_2$ are not independent, but the best-fitting regression line has a slope of 0, and therefore $X_1$ and $X_2$ are uncorrelated. So statistical independence implies zero correlation, but uncorrelated random variables are not necessarily statistically independent. So a correlation of 0 means there is no linear relationship, but it does not mean that there is no nonlinear relationship. In a bivariate normal distribution, however, the only possible relationships are linear, so uncorrelated is equivalent to independent.

In the bivariate normal distribution, $X_1$ and $X_2$ each have a mean and variance along with a correlation coefficient that specifies the strength of their linear relationship. If $\mu_i$ and $\sigma_i^2$ denote the mean and variance of $X_i$, respectively, then the correlation between $X_1$ and $X_2$ is defined as the standardized covariance:

$$\rho = \frac{\text{cov}_{12}}{\sigma_1 \sigma_2} = \frac{E\left[(X_1 - \mu_1)(X_2 - \mu_2)\right]}{\sigma_1 \sigma_2}, \tag{A.4}$$

where $E$ denotes expected value. So the correlation is positive, for example, if whenever $X_1$ is greater than its mean then $X_2$ also tends to be greater than its mean (because then the product $(X_1 - \mu_1)(X_2 - \mu_2)$ is positive), whereas a negative correlation occurs if values of $X_1 > \mu_1$ tend to co-occur with values of $X_2 < \mu_2$ (because then the product $(X_1 - \mu_1)(X_2 - \mu_2)$ is negative). If the means and variances are known, then note that it makes no difference whether we characterize the association of a bivariate normal distribution in terms of the correlation between $X_1$ and $X_2$ or the covariance. From either one, equation A.4 allows us to solve for the other.

So any bivariate normal distribution is characterized by five free parameters: $\mu_1, \mu_2, \sigma_1, \sigma_2$, and either $\rho$ or $\text{cov}_{12}$. A common convention is to catalog these in a mean vector

$$\underline{\mu} = \begin{bmatrix} \mu_1 \\ \mu_2 \end{bmatrix} \tag{A.5}$$

and a variance-covariance matrix

$$\Sigma = \begin{bmatrix} \sigma_1^2 & \text{cov}_{12} \\ \text{cov}_{21} & \sigma_2^2 \end{bmatrix}. \tag{A.6}$$

Note that this matrix is always symmetric because $\text{cov}_{12} = \text{cov}_{21}$. It is also positive semidefinite (i.e., both eigenvalues must be nonnegative).

Once numerical values are specified for the means, variances, and covariance, then the likelihood of sampling any pair $(x_1, x_2)$ can be computed from the bivariate normal pdf:

$$f_{12}(x_1,x_2)=\frac{1}{2\pi\sigma_1\sigma_2\sqrt{1-\rho^2}}$$
$$\times\exp\left\{-\frac{1}{2(1-\rho^2)}\left[\left(\frac{x_1-\mu_1}{\sigma_1}\right)^2-2\rho\left(\frac{x_1-\mu_1}{\sigma_1}\right)\left(\frac{x_2-\mu_2}{\sigma_2}\right)+\left(\frac{x_2-\mu_2}{\sigma_2}\right)^2\right]\right\}. \tag{A.7}$$

More generally, if an $r$-dimensional random vector $\underline{X}$ is multivariate normally distributed, then its pdf can be written using matrix notation as

$$f(\underline{x})=\frac{1}{(2\pi)^{r/2}|\Sigma|^{1/2}}\exp\left[-\frac{1}{2}\left(\underline{x}-\underline{\mu}\right)'\Sigma^{-1}\left(\underline{x}-\underline{\mu}\right)\right], \tag{A.8}$$

where $|\Sigma|$ is the determinant of $\Sigma$ and $'$ signifies matrix transpose. If $r=2$ and the indicated matrix operations are performed, then equation A.8 reduces to equation A.7.

Figure A.1 shows an example of the bivariate normal distribution. Panel (a) shows the pdf of a bivariate normal with mean vector and variance-covariance matrix

$$\underline{\mu}=\begin{bmatrix}40\\60\end{bmatrix} \text{ and } \Sigma=\begin{bmatrix}225 & 175\\175 & 225\end{bmatrix}. \tag{A.9}$$

Therefore, in this distribution, the correlation coefficient equals $\rho=0.778$ (i.e., $175/\sqrt{225^2}$). The z-axis specifies the likelihood that a random sample equals any point on the $(x_1,x_2)$ plane. As expected, note that the most likely sample is the population mean.

Three-dimensional plots like those in panel (a) are awkward to produce and interpret. For example, note that the strong positive correlation between $X_1$ and $X_2$ that characterizes this distribution is barely evident in panel (a). As a result, a common strategy is to reduce three-dimensional plots like those shown in panel (a) to two dimensions. This is done as in panel (b) by passing a plane through the distribution (i.e., that is parallel to the $(x_1,x_2)$ plane) and looking down at the result from above. The result is the ellipse shown in panel (c). Every point on this ellipse is associated with the same height above the $(x_1,x_2)$ plane, and therefore all points on this ellipse have the same likelihood of being sampled from this distribution. As a result, contours of this type are called *contours of equal likelihood.*

In bivariate normal distributions, the contours of equal likelihood are always circles or ellipses. They are always centered on the distribution mean, and their shape describes the shape of the scatterplot that results when random samples are taken from the distribution. Furthermore, all contours have the same shape, and differ only in overall size. The higher the plane that cuts through the distribution is above

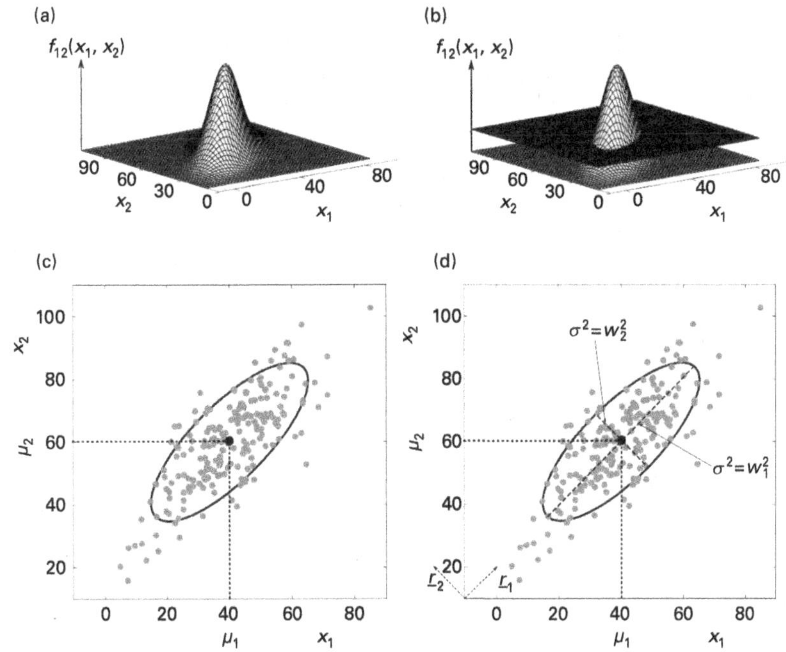


**A.1** (a) A bivariate normal pdf. (b) Same as in (a), except with the addition of a plane that is parallel to the $(x_1, x_2)$ plane. (c) A scatterplot of 200 random samples from the pdf shown in (a). The ellipse is the contour of equal likelihood created by the plane in (b), and the black circle denotes the distribution mean. (d) The same contour of equal likelihood as in panel (c) with the two eigenvectors of $\Sigma$. The eigenvalues equal the variances along these two directions.

the floor, the smaller the contour. So the size of the contour does not convey any useful information. But the center is always the distribution mean, and the relative widths of the contour in the horizontal and vertical directions are measures of the relative magnitudes of the $X_1$ and $X_2$ standard deviations, respectively. In figure A.1c, the width of the ellipse is equal in the horizontal and vertical directions, signaling that this distribution has equal $X_1$ and $X_2$ standard deviations. Finally, the orientation of the ellipse provides information about the correlation coefficient. Ellipses in which the major axis has a positive slope signal a positive correlation, ellipses in which the major axis has a negative slope signal a negative correlation, and ellipses in which the major axis is horizontal or vertical signal a zero correlation. As a result, although some information is lost when the three-dimensional plot in figure A.1a is reduced to two dimensions, much important information about the distribution is retained.

One important property of multivariate normal distributions, which is heavily used in PCA, is that the variance-covariance matrix of any multivariate normal distribution can always be rewritten in its so-called diagonal form:

$$\Sigma = \begin{bmatrix} \sigma_1^2 & \text{cov} \\ \text{cov} & \sigma_2^2 \end{bmatrix} = \begin{bmatrix} r_{11} & r_{12} \\ r_{21} & r_{22} \end{bmatrix} \begin{bmatrix} w_1^2 & 0 \\ 0 & w_2^2 \end{bmatrix} \begin{bmatrix} r_{11} & r_{12} \\ r_{21} & r_{22} \end{bmatrix}'$$

$$= \begin{bmatrix} \underline{r}_1 & \underline{r}_2 \end{bmatrix} \begin{bmatrix} w_1^2 & 0 \\ 0 & w_2^2 \end{bmatrix} \begin{bmatrix} \underline{r}_1 & \underline{r}_2 \end{bmatrix}'. \tag{A.10}$$

As it turns out, the $2 \times 1$ vectors $\underline{r}_1$ and $\underline{r}_2$ are the eigenvectors of $\Sigma$ and $w_1^2$ and $w_2^2$ are the corresponding eigenvalues. Variance-covariance matrices are always positive semidefinite, which means that $w_1^2$ and $w_2^2$ must both be greater than or equal to zero (as required by any variance).

This diagonal form of $\Sigma$ is helpful because the eigenvalues and eigenvectors of $\Sigma$ have a straightforward and highly useful geometric interpretation, which is illustrated in figure A.1d. The eigenvectors of $\Sigma$ are parallel to the major and minor axes of the ellipses that define the distribution's contours of equal likelihood. The eigenvector corresponding to the larger eigenvalue is parallel to the major axis, and the eigenvector corresponding to the smaller eigenvalue is parallel to the minor axis.

The diagonal representation is useful for constructing bivariate normal distributions that have any desired orientation, shape, and location. This is how we designed RB and II categorization tasks in section 5.6 that are simple rotations of each other in stimulus space. The idea, which is described in detail there, is to select directions for the major and minor axes, and then set the desired variance along each of these directions. These choices fix the eigenvectors and eigenvalues, and then equation A.10 is used to compute the variance-covariance matrix that has these properties.

The diagonal representation of $\Sigma$ also provides the basis for PCA. The contours of equal likelihood for all multivariate normal distributions are always ellipses (or hyperellipses when there are more than two dimensions). The major axis of any ellipse is defined as the direction through the ellipse with the greatest length, and therefore the direction along which the distribution has greatest variance. PCA exploits this fact by rotating the data so that the eigenvectors become the new coordinate axes. In PCA, the eigenvectors of $\Sigma$ are called the components, and the eigenvector with the largest eigenvalue is the principal component. If we wanted to reduce the dimensionality of the scatterplot in figure A.1d to one dimension, then it makes sense to choose to describe each $(x_1, x_2)$ ordered pair by its position along the principal component

$\underline{r}_1$ (e.g., with the mean $(\mu_1, \mu_2)$ defining the 0 point on this axis). This scheme for reducing dimensionality is optimal in the sense that it minimizes the sum of squared information losses, but only under the assumption that the data are multivariate normally distributed.

A special case of the bivariate normal distribution that is widely used when GRT is applied to categorization experiments assumes that the two variables are independent and that both variances are equal (i.e., homogeneity of variance). In this case, note that

$$\Sigma = \begin{bmatrix} \sigma^2 & 0 \\ 0 & \sigma^2 \end{bmatrix} = \sigma^2 \begin{bmatrix} 1 & 0 \\ 0 & 1 \end{bmatrix} = \sigma^2 \mathrm{I}, \tag{A.11}$$

where I is the identity matrix. The bivariate z-distribution is a special case of this in which the mean vector equals $\underline{0}$ and the variance-covariance matrix equals I. Therefore, in the bivariate z distribution, the marginal distributions are both z distributions and the two variables are statistically independent.

One common problem in probability and statistics, which arose several times in this book, is to compute the distribution of a linear transformation of some random vector (e.g., see equation 7.9). Specifically, suppose $\underline{\mathrm{X}}$ is an $r \times 1$ random vector, A is an $m \times r$ matrix of constants, and $\underline{\mathrm{b}}$ is an $m \times 1$ vector of constants. Now suppose we are interested in the distribution of the $m \times 1$ random vector

$$\underline{\mathrm{Y}} = \mathrm{A}\underline{\mathrm{X}} + \underline{\mathrm{b}}. \tag{A.12}$$

It turns out that no matter the distribution of $\underline{\mathrm{X}}$, it is always true that

$$\underline{\mu}_{\underline{\mathrm{Y}}} = \mathrm{A}\underline{\mu}_{\underline{\mathrm{X}}} + \underline{\mathrm{b}} \tag{A.13}$$

and

$$\Sigma_{\underline{\mathrm{Y}}} = \mathrm{A}\Sigma_{\underline{\mathrm{X}}}\mathrm{A}'. \tag{A.14}$$

Furthermore, if $\underline{\mathrm{X}}$ has a multivariate normal distribution, then so will $\underline{\mathrm{Y}}$ (because linear transformations of normal random variables are always normal).

# APPENDIX B: MATLAB CODE

## B.1 GENERATE STIMULI IN MATCHING RB AND II CATEGORIES

```
N = 330;                                   % Number of samples to draw for each category
N_final = 300;                             % Final number of stimuli in each category
mu_RB_A = [50; 65];                        % Mean of category A in the RB condition
mu_RB_B = [50; 35];                        % Mean of category B in the RB condition
mu_II_A = [39.4; 60.6];                    % Mean of category A in the II condition
mu_II_B = [60.6; 39.4];                    % Mean of category B in the II condition
Sig_RB = [225 0; 0 41.6];                  % Variance-covariance matrix in the RB condition
Sig_II = [133.3 91.7; 91.7 133.3];         % Variance-covariance matrix in the II condition
RB_A = mvnrnd(mu_RB_A,Sig_RB,N);           % Cat A raw stimuli in RB condition
RB_B = mvnrnd(mu_RB_B,Sig_RB,N);           % Cat B raw stimuli in RB condition

% Discard outliers
j=1; k=1;
for i=1:N
    dA=sqrt((RB_A(i,:)'-mu_RB_A)'*inv(Sig_RB)*(RB_A(i,:)'-mu_RB_A));
    if dA>3
Out_A(j)=i; j=j+1;
    end
    dB=sqrt((RB_B(i,:)'-mu_RB_B)'*inv(Sig_RB)*(RB_B(i,:)'-mu_RB_B));
    if dB>3
```

```
Out_B(k)=i; k=k+1;
    end
end
if length(Out_A)>.5                      % make sure there is at least 1 outlier
    RB_A(Out_A,:) = [];
end
if length(Out_B)>.5                      % make sure there is at least 1 outlier
    RB_B(Out_B,:) = [];
RB_A(Out_A,:) = [];
RB_B(Out_B,:) = [];

% Trim to exactly N_final stimuli in each category
RB_A = RB_A(1:N_final,:);
RB_B = RB_B(1:N_final,:);

% Transform samples to have population mean and variance-covariance matrix
S_A = cov(RB_A);          % Sample variance-covariance matrix of category A stimuli
S_B = cov(RB_B);          % Sample variance-covariance matrix of category B stimuli
P_RB = (chol(Sig_RB))';                        % Cholesky matrix for RB condition
Q_A = (chol(S_A))';                                  % Q matrix for category A
Q_B = (chol(S_B))';                                  % Q matrix for category B
for i=1:N_final
    RB_A_pop(i,:) = (P_RB*inv(Q_A)*(RB_A(i,:)'-mean(RB_A)'))' + mu_RB_A';
    RB_B_pop(i,:) = (P_RB*inv(Q_B)*(RB_B(i,:)'-mean(RB_B)'))' + mu_RB_B';
end

% Plot the RB stimuli
scatter(RB_A_pop(:,1),RB_A_pop(:,2));
axis([0 100 0 100]); axis square; hold on;
scatter(RB_B_pop(:,1),RB_B_pop(:,2));

% Rotate the RB stimuli 45 degrees to create the II stimuli
RB_A_zero=RB_A_pop-(ones(300,1)*mean(RB_A_pop));  %Zero-mean the RB stimuli
RB_B_zero=RB_B_pop-(ones(300,1)*mean(RB_B_pop));
R=[cos(pi/4) -sin(pi/4); sin(pi/4) cos(pi/4)];               %Rotation matrix
II_A_zero = (R * RB_A_zero')';                    %Rotate the zero-mean stimuli
II_B_zero = (R * RB_B_zero')';
II_A_pop=II_A_zero + (ones(300,1)*mu_II_A');                  %Set the II means
II_B_pop=II_B_zero + (ones(300,1)*mu_II_B');
```

```
% Plot the II stimuli
figure;
scatter(II_A_pop(:,1),II_A_pop(:,2));
axis([0 100 0 100]); axis square; hold on;
scatter(II_B_pop(:,1),II_B_pop(:,2));
```

## B.2 FIT THE ONE-DIMENSIONAL RULE MODEL TO CATEGORIZATION DATA

```
x0 = [40,10];                                    % Starting guess at the solution
lb = [-Inf; 0];                                  % lower bounds for all parameters
ub = [Inf; Inf];                                 % upper bounds for all parameters
[param,fval] = fmincon(@OneD,x0,[],[],[],[],lb,ub)  % call the minimization routine

% create the function that computes 1D model predictions
function BIC = OneD(param)
filename = 'Data.txt';
D = importdata(filename);                        % input data matrix

%Use the following line to fit a 1D rule on dimension 1 in which the A
%response region is on the left and the B region is on the right
Z=inv(param(2))*(param(1)*ones(length(D),1)-D(:,2));

%Use the following line to fit a 1D rule on dimension 2 in which the A
%response region is above the bound and the B region is below
%Z=-inv(param(2))*(D(:,3)-param(1)*ones(length(D),1));

PC=normcdf(Z,0,1);
for i=1:length(D)
    if D(i,4)>1.5
        PC(i)=1-PC(i);
    end
end

lnL=sum(log(PC));                                % compute log L* (Eq. 7.13)
BIC=2*log(length(D))-2*lnL;                      % compute BIC

end
```

## B.3 FIT THE GLC TO CATEGORIZATION DATA

```
x0 = [1; -1; 0; 10];                          % starting guesses for all parameters, which
                                              % assume the A response region is above
                                              % the bound and the B region is below
lb = [-Inf; -Inf; -Inf; 0];                   % lower bounds for all parameters
ub = [Inf; Inf; Inf; Inf];                    % upper bounds for all parameters
[param,fval] = fmincon(@GLC,x0,[],[],[],[],lb,ub)  % call the minimization routine

% create the function that computes GLC predictions
function BIC = GLC(param)
filename = 'Data.txt';
D = importdata(filename);                     % input data matrix

Z=inv(param(4))*(param(1)*D(:,2)+param(2)*D(:,3)+param(3)*ones(length(D),1));
PC=normcdf(Z,0,1);
for i=1:length(D)
    if D(i,4)>1.5
        PC(i)=1-PC(i);
    end
end

lnL=sum(log(PC));                             % compute log L* (Eq. 7.13)
BIC=3*log(length(D))-2*lnL;                   % compute BIC

end
```

# NOTES

## 1 INTRODUCTION

1. https://psmag.com/social-justice/newtons-needle-scientific-self-experimentation-86583/

2. For a variety of reasons, this follow-up experiment appeared in print two years before the earlier study.

## 2 SIGNAL DETECTION THEORY

1. Note that $X$ is a random variable and $x$ is a specific number. In other words, $X$ represents any possible sensory magnitude that could be observed in the experiment, so stating that $X = x$ on some trial is equivalent to stating that the sensory magnitude on that trial had the numerical value of $x$.

2. Specifically, Killeen and Taylor (2004) showed that the ROC is symmetric if the SN cumulative distribution function is generated by applying a strictly decreasing involution to the survivor function of the N distribution. An involution is a transformation that is its own inverse. So, for example, if $T$ is an involution, then $T\{T[1 - F(x)]\} = 1 - F(x)$.

## 3 SDT MODELS OF RECOGNITION MEMORY, VISUAL SEARCH, EYEWITNESS IDENTIFICATION, AND GROUP DECISION-MAKING

1. Thanks to Michael Miller for pointing this difficulty out to me.

2. Note from equation 3.3 that

$$\begin{aligned} 1 - P(\text{NEW}|\text{lure}) &= 1 - [q + (1 - q)\Phi(X_C)] \\ &= (1 - q)[1 - \Phi(X_C)] \\ &= (1 - q)\Phi(-X_C). \end{aligned}$$

3. For example, the cumulative Z distribution and its most closely approximating logistic distribution differ by less than .01 over the entire Z axis (Bowling et al., 2009).

4. This conclusion requires the assumption that the sensory distributions are normally distributed. In general, two uncorrelated random variables need not be statistically independent. However, in the bivariate normal distribution, the only possible relationship is linear. Pearson correlation measures the strength of a linear relationship, so in the bivariate normal distribution, uncorrelated random variables are statistically independent. See appendix A for details.

5. The derivation is as follows. If $\rho=0$, then

$$d'_{\text{GPD}}=\sqrt{r}[\text{Var}(d'_i)+(\bar{d}'_i)^2]^{1/2}=\sqrt{r}\left\{\left[\frac{1}{r}\sum_{i=1}^{r}(d'_i-\bar{d}_i)^2\right]+(\bar{d}'_i)^2\right\}^{1/2}$$

$$=\sqrt{r}\left\{\left[\frac{1}{r}\sum_{i=1}^{r}(d'_i)^2\right]-(\bar{d}'_i)^2+(\bar{d}'_i)^2\right\}^{1/2}=\sqrt{\sum_{i=1}^{r}(d'_i)^2}. \tag{B.1}$$

6. It should also be noted that this model does not necessarily assume that the $X_i$ in the structural model $Y_i=X_i+X_G$ has the same distributions on N and SN trials as it would on trials when group member $i$ performs the task in isolation.

## 5 DECISION PROCESSES WITH MULTIDIMENSIONAL STIMULI

1. The only real exception occurs with the normal, unequal-variance SDT model illustrated in figure 2.6 since, in this case, the optimal strategy is to create three response regions via two separate criteria.

2. The other dissociations of this type that were noted by Ashby et al. (2020) are (1) that manipulations that interfere with feedback processing impair RB learning more than II learning; (2) that pressure to perform impairs performance in RB tasks, and may even improve performance in II tasks; and (3) that increasing the number of contrasting categories slows RB learning but not II learning.

3. In the special case illustrated in figure 5.8c, a one-dimensional rule on either dimension will lead to the same accuracy. In both cases, the bound will bisect the category means on the relevant dimension.

## 6 THE SUMMARY STATISTICS APPROACH TO TESTING FOR PERCEPTUAL INTERACTIONS

1. Due to a typesetting error, the Kadlec and Townsend (1992a) article omitted the division by 2, and, as a result, the $d'$ denominator in that article was listed as the square root of the sum of the two relevant variances. Even so, it is important to note that all results in Kadlec and Townsend (i.e., in both 1992a and 1992b) are valid for either $d'$ definition.

## 7 FITTING THE GAUSSIAN GRT MODEL TO IDENTIFICATION DATA

1. Any variance-covariance matrix $\Sigma$ can be expressed via its diagonal representation as $\Sigma=\text{QDQ}'$ (see appendix A for details). Therefore,

$$\Sigma=\text{QDQ}'=\text{QD}^{1/2}\text{D}^{1/2}\text{Q}'=(\text{QD}^{1/2})(\text{QD}^{1/2})'. \tag{B.2}$$

2. The volume equals the product of the marginal values because $z_1$ and $z_2$ are statistically independent in the bivariate z distribution.

3. The fact that these experiments each only included two participants is powerful evidence that publication standards were far different 20 years ago than they are today. Currently, these experiments would probably need 20 or more participants to publish the results in these same journals.

4. Because bivariate normal distributions have positive likelihood everywhere, GRT never predicts a zero entry in a confusion matrix (so long as there is a response region assigned to that response). However, when accuracy is high, it does predict that some entries will be small. Maximum likelihood methods require precise computation of these small predicted frequencies. For example, suppose some cell of an empirical confusion matrix has an observed value of 1 (i.e., so $r_{ij} = 1$ in equation 7.3 for some $i$ and $j$), and we calculate that the model we are fitting predicts that the probability of this response is 0.01 (so $P(R_j|S_i) = 0.01$ in equation 7.3). However, if the true predicted probability of the model for this cell is really 0.0001, then the contribution of this cell to the overall fit will more than double according to maximum likelihood (i.e., according to equation 7.3). Therefore, when there are many near-zero entries in a confusion matrix and numerical integration is used to calculate model predictions, it is vital to use a numerical integration algorithm that is accurate for small predicted probabilities. The accuracy of the Cholesky factorization algorithm described in section 7.3 can be adjusted to any desired degree by selection of the preloaded z values.

5. AIC is similar to BIC. When comparing non-nested models, in both cases, the model with the smallest goodness-of-fit score is selected. Recall from equation 7.4 that $BIC = r \ln N - 2 \ln L^*$, where $N$ is the sample size and $r$ is the number of free parameters. In contrast, $AIC = 2r - 2 \ln L^*$. Note that both measures penalize a model for extra parameters. When $N \geq 8$, the BIC penalty is greater than the AIC penalty, and so, in these cases, BIC penalizes overfitting more than AIC.

6. INDSCAL parameterizes this model differently. In the INDSCAL model, equation 7.20 is replaced by

$$\delta_{ik} = \sqrt{w_1 (x_{i1} - x_{k1})^2 + w_2 (x_{i2} - x_{k2})^2}, \tag{B.3}$$

where $w_1$ and $w_2$ are free parameters and therefore not constrained to sum to 1. The two models are equivalent though when $w_1 = c\,w$ and $w_2 = c\,(1 - w)$.

## 9 GRT ACCOUNTS OF SIMILARITY, PREFERENCE, AND SAME-DIFFERENT JUDGMENTS

1. For example, suppose the contour is defined as the set of all points that are 3 Mahalanobis distance units from (0,0), and that the vertical Euclidean distance from (0,0) to the contour is 1 inch, whereas the horizontal distance is 3 inches. Then note that moving 1 inch vertically increases Mahalanobis distance by 3, whereas moving 1 inch horizontally only increases Mahalanobis distance by 1.

## 10 MODELING RESPONSE TIMES

1. It turns out that the magnitude of this bias increases with the size of the smoothing constant.

2. In the case of GRT, these would be the marginal distributions along the direction orthogonal to the decision bound.

3. This is because the drift-diffusion model assumes that all samples are independent and identically distributed.

4. These latter two conditions can always be met because multiplying all entries in $\underline{b}$ and $c$ by –1 does not change the points that lie on the bound, but it does reverse the sign of all other points.

5. The figure 10.8a bound satisfies the equation $x_2 = x_1$, which becomes $x_2 = x_1 - \Delta_c$ if the intercept is decreased by $\Delta_c$ (where $\Delta_c$ is some positive value). Note that this bound is consistent with the discriminant function $h(x_1, x_2) = x_1 - x_2 - \Delta_c$ and also with the discriminant function $h(x_1, x_2) = -x_1 + x_2 + \Delta_c$. However, only the latter function associates the A response region with positive discriminant values. As a result, decreasing the intercept of the figure 10.8a bound occurs when the numerical value of the $c$ parameter is increased.

6. The sample paths in panel (b) are less variable than the sample paths in panel (a) because, for illustrative purposes, the variance of the diffusion distribution was set to a smaller value in panel (b).

7. From an analytic perspective, elliptical barriers are much better than star-shaped barriers. An ellipse is radially symmetric if Mahalanobis distance is used rather than Euclidean distance. For example, suppose we replace the circle in figure 10.12c with an ellipse. Note that there is a linear transformation that converts the ellipse to a circle. This transformation would also change the perceptual distributions (i.e., in the figure 10.12c example, the variances on the two dimensions would become more equal). But predictions for this transformed model would follow from the model's standard equations. The problem though is that the transformed distributions would no longer describe the observer's perceptual experience with the stimuli.

8. Ashby (1989) proposed the model in discrete time. These equations are the continuous-time analogues of the discrete-time model.

9. A linear system is stable if activity in the channels decays to zero after all inputs are turned off. This occurs if and only if all the eigenvalues of the matrix A in equation 10.27 have negative real parts (e.g., C. T. Chen, 1970).

10. A bivariate normal Z distribution is a bivariate normal distribution with mean vector (0,0) and variance-covariance matrix equal to the identity. Note also that if a random variable $\epsilon$ has a Z distribution, then so does $-\epsilon$.

11. In Hebbian learning, the amount a synapse is strengthened is proportional to the product of the pre- and postsynaptic activations, and the prefrontal input will cause this product to be larger in the unit that leads to the correct response.

12. Khodadadi and Townsend (2015) showed that this failure of simple race models to mimic the push-pull dynamics of the drift-diffusion model is due to the assumption that the response barriers in the race model are constant over time. More specifically, they showed that a simple race model can mimic the drift-diffusion model perfectly if the response thresholds in the race model are allowed to vary over time (i.e., within a trial).

13. On the other hand, note that a nontarget response could be made on the basis of a self-terminating search, since a nontarget response is always correct if either stimulus component is at level 0. Therefore, an analysis of the RTs on nontarget trials would typically include the question of whether search was self-terminating or exhaustive.

14. In reliability theory, the derivative of $\log[F_{i.}(t)]$ is known as the reverse hazard function (e.g., Chechile, 2011).

15. Note that in any practical application, each $F(t)$ in Eq. 10.40 is constrained to the interval (0,1), and therefore $\log[F(t)] < 0$. As a result, if $F_{AB}(t) < F_A(t)\,F_B(t)$, as expected when capacity is limited, then $\log[F_{AB}(t)] > \log[F_A(t)\,F_B(t)]$, and as a result, the equation 10.40 capacity coefficient will be less than 1.0.

## 11 STATISTICAL DECISION THEORY AND NEUROSCIENCE

1. Section 10.1 of this chapter is based on a review published by Ashby and Soto (2016).

2. There are two possible exceptions. One is the simple reflex (e.g., the patellar reflex) that is mediated entirely within the spinal cord. Another are automatic behaviors that develop after thousands of trials of practice (i.e., see Sections 11.2.2 and 11.2.3). Statistical decision theory might therefore not provide the best account of either of these behavior types.

# REFERENCES


Aho, A.-C., Donner, K., Hydén, C., Larsen, L. O., & Reuter, T. (1988). Low retinal noise in animals with low body temperature allows high visual sensitivity. *Nature, 334*(6180), 348–350.

Alexander, G. E., DeLong, M. R., & Strick, P. L. (1986). Parallel organization of functionally segregated circuits linking basal ganglia and cortex. *Annual Review of Neuroscience, 9*(1), 357–381.

Aminoff, E. M., Clewett, D., Freeman, S., Frithsen, A., Tipper, C., Johnson, A., Grafton, S. T., & Miller, M. B. (2012). Individual differences in shifting decision criterion: A recognition memory study. *Memory & Cognition, 40,* 1016–1030.

Anderson, J. R. (1991). The adaptive nature of human categorization. *Psychological Review, 98*(3), 409–429.

Anderson, J. R., & Fincham, J. M. (1996). Categorization and sensitivity to correlation. *Journal of Experimental Psychology: Learning, Memory, and Cognition, 22*(2), 259–277.

Apicella, P., Legallet, E., & Trouche, E. (1997). Responses of tonically discharging neurons in the monkey striatum to primary rewards delivered during different behavioral states. *Experimental Brain Research, 116*(3), 456–466.

Aristotle. (1905). *Aristotle's politics.* Clarendon Press.

Asaad, W. F., Rainer, G., & Miller, E. K. (2000). Task-specific neural activity in the primate prefrontal cortex. *Journal of Neurophysiology, 84*(1), 451–459.

Ashby, F. G. (1983). A biased random walk model for two choice reaction times. *Journal of Mathematical Psychology, 27*(3), 277–297.

Ashby, F. G. (1988). Estimating the parameters of multidimensional signal detection theory from simultaneous ratings on separate stimulus components. *Perception & Psychophysics, 44*(3), 195–204.

Ashby, F. G. (1989). Stochastic general recognition theory. In D. Vickers & P. L. Smith (Eds.), *Human information processing: Measures, mechanisms, and models*, (pp. 435–457). Elsevier.

Ashby, F. G. (1992). Multidimensional models of categorization. In F. G. Ashby (Ed.), *Multidimensional models of perception and cognition*, (pp. 449–483). Lawrence Erlbaum Associates, Inc.

Ashby, F. G. (2000). A stochastic version of general recognition theory. *Journal of Mathematical Psychology, 44*(2), 310–329.

Ashby, F. G. (2019). *Statistical analysis of fMRI data*, (2nd ed.). MIT Press.

Ashby, F. G. (2023). Length of the state trace: A method for partitioning model complexity. *Journal of Mathematical Psychology, 113*, 102755.

Ashby, F. G. (2024). On using the fixed-point property of binary mixtures to discriminate among models of recognition memory. *Journal of Mathematical Psychology, 123*, 102889.

Ashby, F. G., & Alfonso-Reese, L. A. (1995). Categorization as probability density estimation. *Journal of Mathematical Psychology, 39*(2), 216–233.

Ashby, F. G., Alfonso-Reese, L. A., Turken, A. U., & Waldron, E. M. (1998). A neuropsychological theory of multiple systems in category learning. *Psychological Review, 105*(3), 442–481.

Ashby, F. G., & Bamber, D. (2022). State trace analysis: What it can and cannot do. *Journal of Mathematical Psychology, 108*, 102655.

Ashby, F. G., Boynton, G., & Lee, W. W. (1994). Categorization response time with multidimensional stimuli. *Perception & Psychophysics, 55*(1), 11–27.

Ashby, F. G., & Crossley, M. J. (2011). A computational model of how cholinergic interneurons protect striatal-dependent learning. *Journal of Cognitive Neuroscience, 23*(6), 1549–1566.

Ashby, F. G., & Ell, S. W. (2001). The neurobiology of human category learning. *Trends in Cognitive Sciences, 5*(5), 204–210.

Ashby, F. G., Ell, S. W., Valentin, V. V., & Casale, M. B. (2005). FROST: A distributed neurocomputational model of working memory maintenance. *Journal of Cognitive Neuroscience, 17*(11), 1728–1743.

Ashby, F. G., Ell, S. W., & Waldron, E. M. (2003). Procedural learning in perceptual categorization. *Memory & Cognition, 31*(7), 1114–1125.

Ashby, F. G., & Ennis, D. M. (2002). A Thurstone-Coombs model of concurrent ratings with sensory and liking dimensions. *Journal of Sensory Studies, 17*(1), 43–59.

Ashby, F. G., & Ennis, J. M. (2006). The role of the basal ganglia in category learning. *Psychology of Learning and Motivation, 46*, 1–36.

Ashby, F. G., Ennis, J. M., & Spiering, B. J. (2007). A neurobiological theory of automaticity in perceptual categorization. *Psychological Review, 114*(3), 632–656.

Ashby, F. G., & Gott, R. E. (1988). Decision rules in the perception and categorization of multidimensional stimuli. *Journal of Experimental Psychology: Learning, Memory, and Cognition, 14*(1), 33–53.

Ashby, F. G., & Lee, W. W. (1991). Predicting similarity and categorization from identification. *Journal of Experimental Psychology: General, 120*(2), 150–172.

Ashby, F. G., & Lee, W. W. (1993). Perceptual variability as a fundamental axiom of perceptual science. In S. C. Masin (Ed.), *Foundations of perceptual theory* (pp. 369–399, Vol. 99). Elsevier.

Ashby, F. G., Maddox, W. T., & Bohil, C. J. (2002). Observational versus feedback training in rule-based and information-integration category learning. *Memory & Cognition, 30*, 666–677.

Ashby, F. G., & Maddox, W. T. (1991). A response time theory of perceptual independence. In J.-P. Doignon & J.-C. Falmagne (Eds.), *Mathematical psychology: Current developments* (pp. 389–413). Springer-Verlag.

Ashby, F. G., & Maddox, W. T. (1992). Complex decision rules in categorization: Contrasting novice and experienced performance. *Journal of Experimental Psychology: Human Perception and Performance, 18*(1), 50–71.

Ashby, F. G., & Maddox, W. T. (1993). Relations between prototype, exemplar, and decision bound models of categorization. *Journal of Mathematical Psychology, 37*(3), 372–400.

Ashby, F. G., & Maddox, W. T. (1994). A response time theory of separability and integrality in speeded classification. *Journal of Mathematical Psychology, 38*(4), 423–466.

Ashby, F. G., & Maddox, W. T. (2005). Human category learning. *Annual Review of Psychology, 56*, 149–178.

Ashby, F. G., & Perrin, N. A. (1988). Toward a unified theory of similarity and recognition. *Psychological Review, 95*, 124–150.

Ashby, F. G., Queller, S., & Berretty, P. M. (1999). On the dominance of unidimensional rules in unsupervised categorization. *Perception & Psychophysics, 61*(6), 1178–1199.

Ashby, F. G., & Rosedahl, L. (2017). A neural interpretation of exemplar theory. *Psychological Review, 124*(4), 472–482.

Ashby, F. G., Smith, J. D., & Rosedahl, L. A. (2020). Dissociations between rule-based and information-integration categorization are not caused by differences in task difficulty. *Memory & Cognition, 48*(4), 541–552.

Ashby, F. G., & Soto, F. A. (2015). Multidimensional signal detection theory. In J. R. Busemeyer, Z. Wang, J. T. Townsend, & A. Eidels (Eds.), *Oxford handbook of computational and mathematical psychology*, (pp. 13–34). Oxford University Press.

Ashby, F. G., & Soto, F. A. (2016). The neural basis of general recognition theory. In J. Houpt & L. M. Blaha (Eds.), *Mathematical models of perception and cognition: Volume II, A Festschrift for James T. Townsend*, (pp. 1–31). Routledge.

Ashby, F. G., Tein, J.-Y., & Balakrishnan, J. (1993). Response time distributions in memory scanning. *Journal of Mathematical Psychology, 37*(4), 526–555.

Ashby, F. G., & Townsend, J. T. (1986). Varieties of perceptual independence. *Psychological Review, 93*(2), 154–179.

Ashby, F. G., & Valentin, V. V. (2017). Multiple systems of perceptual category learning: Theory and cognitive tests. In H. Cohen & C. Lefebvre (Eds.), *Handbook of categorization in cognitive science* (2nd ed., pp. 157–188). Elsevier.

Ashby, F. G., & Valentin, V. V. (2018). The categorization experiment: Experimental design and data analysis. In E. J. Wagenmakers & J. T. Wixted (Eds.), *Stevens' handbook of experimental psychology and cognitive neuroscience: Methodology* (4th ed., Vol. 5, pp. 307–348). Wiley.

Ashby, F. G., Valentin, V. V., & Turken, A. (2002). The effects of positive affect and arousal on working memory and executive attention: Neurobiology and computational models. In S. Moore & M. Oaksford (Eds.), *Emotional cognition: From brain to behaviour* (pp. 245–287). John Benjamins Publishing Company.

Ashby, F. G., & Waldron, E. M. (1999). On the nature of implicit categorization. *Psychonomic Bulletin & Review, 6*(3), 363–378.

Ashby, F. G., Waldron, E. M., Lee, W. W., & Berkman, A. (2001). Suboptimality in human categorization and identification. *Journal of Experimental Psychology: General, 130*(1), 77–96.

Ashby, F. G., & Wenger, M. J. (2023). Statistical decision theory. In F. G. Ashby, H. Colonius, & E. Dzhafarov (Eds.), *The new handbook of mathematical psychology* (Vol. 3, pp. 265–310). Cambridge University Press.

Asmus, F., Huber, H., Gasser, T., & Schöls, L. (2008). Kick and rush: Paradoxical kinesia in Parkinson disease. *Neurology, 71*(9), 695.

Austen-Smith, D., & Banks, J. S. (1996). Information aggregation, rationality, and the Condorcet jury theorem. *American Political Science Review, 90*(1), 34–45.

Badgaiyan, R. D., Fischman, A. J., & Alpert, N. M. (2007). Striatal dopamine release in sequential learning. *NeuroImage, 38*(3), 549–556.

Balakrishnan, J. D. (1999). Decision processes in discrimination: Fundamental misrepresentations of signal detection theory. *Journal of Experimental Psychology: Human Perception and Performance, 25*(5), 1189–1206.

Balakrishnan, J. D., & Ashby, F. G. (1992). Subitizing: Magical numbers or mere superstition? *Psychological Research, 54*(2), 80–90.

Balakrishnan, J. D., & Ratcliff, R. (1996). Testing models of decision making using confidence ratings in classification. *Journal of Experimental Psychology: Human Perception and Performance, 22*(3), 615–633.

Bamber, D. (1979). State-trace analysis: A method of testing simple theories of causation. *Journal of Mathematical Psychology, 19*(2), 137–181.

Barlow, H. B. (1957). Increment thresholds at low intensities considered as signal/noise discriminations. *Journal of Physiology, 136*(3), 469–488.

Barlow, H. B. (1977). Retinal and central factors in human vision limited by noise. In H. B. Barlow & P. Fatt (Eds.), *Vertebrate photoreception* (pp. 337–358). Academic Press.

Barlow, H. B. (1956). Retinal noise and absolute threshold. *Journal of the Optical Society of America, 46*(8), 634–639.

Barlow, R. E., Marshall, A. W., & Proschan, F. (1963). Properties of probability distributions with monotone hazard rate. *The Annals of Mathematical Statistics, 34*(2), 375–389.

Batchelder, W. H., & Romney, A. K. (1986). The statistical analysis of a general Condorcet model for dichotomous choice situations. In B. Grofman & G. Owen (Eds.), *Information pooling and group decision making: Proceedings of the second University of California, Irvine, conference on political economy* (pp. 103–112). JAI Press Greenwich, CT.

Beauchamp, M. S. (2015). The social mysteries of the superior temporal sulcus. *Trends in Cognitive Sciences, 19*(9), 489–490.

Bennett, B. D., & Wilson, C. J. (2000). Synaptology and physiology of neostriatal neurones. In R. Miller & J. R. Wickens (Eds.), *Brain dynamics and the striatal complex* (pp. 123–152). CRC Press.

Berg, M. E., Ward, M. D., Dai, Z., Arantes, J., & Grace, R. C. (2014). Comparing performance of humans and pigeons in rule-based visual categorization tasks. *Learning and Motivation, 45*, 44–58.

Bernbach, H. A. (1967). Decision processes in memory. *Psychological Review, 74*(6), 462–480.

Bogacz, R., Brown, E., Moehlis, J., Holmes, P., & Cohen, J. D. (2006). The physics of optimal decision making: A formal analysis of models of performance in two-alternative forced-choice tasks. *Psychological Review, 113*(4), 700–765.

Bowling, S. R., Khasawneh, M. T., Kaewkuekool, S., & Cho, B. R. (2009). A logistic approximation to the cumulative normal distribution. *Journal of Industrial Engineering and Management, 2*(1), 114–127.

Boynton, G. M., Engel, S. A., Glover, G. H., & Heeger, D. J. (1996). Linear systems analysis of functional magnetic resonance imaging in human V1. *Journal of Neuroscience, 16*(13), 4207–4221.

Britten, K. H., Newsome, W. T., Shadlen, M. N., Celebrini, S., & Movshon, J. A. (1996). A relationship between behavioral choice and the visual responses of neurons in macaque MT. *Visual Neuroscience, 13*(1), 87–100.

Brody, C. D., Hernández, A., Zainos, A., & Romo, R. (2003). Timing and neural encoding of somatosensory parametric working memory in macaque prefrontal cortex. *Cerebral Cortex, 13*(11), 1196–1207.

Broschard, M. B., Kim, J., Love, B. C., Wasserman, E. A., & Freeman, J. H. (2019). Selective attention in rat visual category learning. *Learning & Memory, 26*(3), 84–92.

Burbeck, S. L., & Luce, R. D. (1982). Evidence from auditory simple reaction times for both change and level detectors. *Perception & Psychophysics, 32*(2), 117–133.

Buxton, R. B. (2013). The physics of functional magnetic resonance imaging (fMRI). *Reports on Progress in Physics, 76*(9), 096601.

Cadieu, C., Kouh, M., Pasupathy, A., Connor, C. E., Riesenhuber, M., & Poggio, T. (2007). A model of V4 shape selectivity and invariance. *Journal of Neurophysiology, 98*(3), 1733–1750.

Canini, K. R., Griffiths, T. L., Vanpaemel, W., & Kalish, M. L. (2014). Revealing human inductive biases for category learning by simulating cultural transmission. *Psychonomic Bulletin & Review, 21*(3), 785–793.

Cantwell, G., Crossley, M. J., & Ashby, F. G. (2015). Multiple stages of learning in perceptual categorization: Evidence and neurocomputational theory. *Psychonomic Bulletin & Review, 22*(6), 1598–1613.

Carroll, J. D., & Chang, J.-J. (1970). Analysis of individual differences in multidimensional scaling via an n-way generalization of "Eckart-Young" decomposition. *Psychometrika, 35*(3), 283–319.

Casale, M. B., Roeder, J. L., & Ashby, F. G. (2012). Analogical transfer in perceptual categorization. *Memory & Cognition, 40*(3), 434–449.

Chechile, R. A. (2011). Properties of reverse hazard functions. *Journal of Mathematical Psychology, 55*(3), 203–222.

Chen, C. T. (1970). *Introduction to linear system theory*. Holt, Rinehart; Winston.

Chen, M.-Y., Jimura, K., White, C. N., Maddox, W. T., & Poldrack, R. A. (2015). Multiple brain networks contribute to the acquisition of bias in perceptual decision-making. *Frontiers in Neuroscience, 9*, 63.

Chernev, A. (2003). When more is less and less is more: The role of ideal point availability and assortment in consumer choice. *Journal of Consumer Research, 30*(2), 170–183.

Clark, S. E. (2003). A memory and decision model for eyewitness identification. *Applied Cognitive Psychology, 17*(6), 629–654.

Clark, S. E., Erickson, M. A., & Breneman, J. (2011). Probative value of absolute and relative judgments in eyewitness identification. *Law and Human Behavior, 35*, 364–380.

Coombs, C. H. (1964). *A theory of data*. Wiley.

Cox, D. R., & Miller, H. D. (1977). *The theory of stochastic processes* (Vol. 134). CRC press.

Cox, P. H., & Riesenhuber, M. (2015). There is a "U" in clutter: Evidence for robust sparse codes underlying clutter tolerance in human vision. *Journal of Neuroscience, 35*(42), 14148–14159.

Creelman, C. D. (2015). Signal detection theory, history of. In J. D. Wright (Ed.), *International Encyclopedia of the Social & Behavioral Sciences* (2nd ed., pp. 952–956). Elsevier.

Crossley, M. J., & Ashby, F. G. (2015). Procedural learning during declarative control. *Journal of Experimental Psychology: Learning, Memory, and Cognition, 41*(5), 1388–1403.

Crossley, M. J., Ashby, F. G., & Maddox, W. T. (2013). Erasing the engram: The unlearning of procedural skills. *Journal of Experimental Psychology: General, 142*(3), 710–741.

Crossley, M. J., Ashby, F. G., & Maddox, W. T. (2014). Context-dependent savings in procedural category learning. *Brain & Cognition, 92*, 1–10.

Crossley, M. J., Horvitz, J. C., Balsam, P. D., & Ashby, F. G. (2016). Expanding the role of striatal cholinergic interneurons and the midbrain dopamine system in appetitive instrumental conditioning. *Journal of Neurophysiology, 115*, 240–254.

Crossley, M. J., Paul, E. J., Roeder, J. L., & Ashby, F. G. (2016). Declarative strategies persist under increased cognitive load. *Psychonomic Bulletin & Review, 23*(1), 213–222.

Davison, M. L. (1979). Testing a unidimensional, qualitative unfolding model for attitudinal or developmental data. *Psychometrika, 44*(2), 179–194.

De Baene, W., Ons, B., Wagemans, J., & Vogels, R. (2008). Effects of category learning on the stimulus selectivity of macaque inferior temporal neurons. *Learning & Memory, 15*(9), 717–727.

De Condorcet, N. (2014). *Essai sur l'application de l'analyse à la probabilité des décisions rendues à la pluralité des voix*. Cambridge University Press.

De Soete, G., Carroll, J. D., & DeSarbo, W. S. (1986). The wandering ideal point model: A probabilistic multidimensional unfolding model for paired comparisons data. *Journal of Mathematical Psychology, 30*(1), 28–41.

DeAngelis, G. C., Cumming, B. G., & Newsome, W. T. (1998). Cortical area MT and the perception of stereoscopic depth. *Nature, 394*(6694), 677–680.

de Beeck, H. O., Wagemans, J., & Vogels, R. (2001). Inferotemporal neurons represent low-dimensional configurations of parameterized shapes. *Nature Neuroscience, 4*(12), 1244–1252.

DeCarlo, L. T. (2002). Signal detection theory with finite mixture distributions: Theoretical developments with applications to recognition memory. *Psychological Review, 109*(4), 710–721.

DeCarlo, L. T. (2010). On the statistical and theoretical basis of signal detection theory and extensions: Unequal variance, random coefficient, and mixture models. *Journal of Mathematical Psychology, 54*(3), 304–313.

de Micheaux, P. L., & de Micheaux, M. P. L. (2017). Package "CompQuadForm." *CRAN Repository*.

DeSarbo, W. S., & Rao, V. R. (1984). GENFOLD2: A set of models and algorithms for the GENeral unFOLDing analysis of preference/dominance data. *Journal of Classification, 1*(1), 147–186.

DeSarbo, W. S., Young, M. R., & Rangaswamy, A. (1997). A parametric multidimensional unfolding procedure for incomplete nonmetric preference/choice set data in marketing research. *Journal of Marketing Research, 34*(4), 499–516.

Desmurget, M., & Turner, R. S. (2010). Motor sequences and the basal ganglia: Kinematics, not habits. *Journal of Neuroscience, 30*(22), 7685–7690.

DeValois, R. L., & DeValois, K. K. (1988). *Spatial vision*. Oxford University Press.

Dubois, B. (1975). Ideal point versus attribute models of brand preference: A comparison of predictive validity. *ACR North American Advances*.

Dunn, J. C., Kaesler, M., & Semmler, C. (2022). A model of position effects in the sequential lineup. *Journal of Memory and Language, 122*, 104297.

Dunn, J. C., Newell, B. R., & Kalish, M. L. (2012). The effect of feedback delay and feedback type on perceptual category learning: The limits of multiple systems. *Journal of Experimental Psychology: Learning, Memory, and Cognition, 38*(4), 840–859.

Dunnington, G. W., Gray, J., & Dohse, F. E. (2004). *Carl Friedrich Gauss: Titan of science*. MAA.

Durstewitz, D., Vittoz, N. M., Floresco, S. B., & Seamans, J. K. (2010). Abrupt transitions between prefrontal neural ensemble states accompany behavioral transitions during rule learning. *Neuron, 66*(3), 438–448.

Edmunds, C. E., Milton, F., & Wills, A. J. (2015). Feedback can be superior to observational training for both rule-based and information-integration category structures. *Quarterly Journal of Experimental Psychology, 68*(6), 1203–1222.

Edmunds, C. E., Milton, F., & Wills, A. J. (2018). Due process in dual process: Model-recovery simulations of decision-bound strategy analysis in category learning. *Cognitive Science, 42*, 833–860.

Egan, J. P. (1958). Recognition memory and the operating characteristic. *USAF Operational Applications Laboratory Technical Note*.

Eichenbaum, H., & Cohen, N. J. (2001). *From conditioning to conscious recollection: Memory systems of the brain*. Oxford University Press.

Ell, S. W., Ashby, F. G., & Hutchinson, S. (2012). Unsupervised category learning with integral-dimension stimuli. *Quarterly Journal of Experimental Psychology, 65*(8), 1537–1562.

Emmerich, D. S., Gray, J. L., Watson, C. S., & Tanis, D. C. (1972). Response latency, confidence, and ROCs in auditory signal detection. *Perception & Psychophysics, 11*(1), 65–72.

Ennis, D. M. (1988). Confusable and discriminable stimuli: Comment on Nosofsky (1986) and Shepard (1986). *Journal of Experimental Psychology: General, 117*(4), 408–411.

Ennis, D. M. (1993). A single multidimensional model for discrimination, identification and preferential choice. *Acta Psychologica, 84*(1), 17–27.

Ennis, D. M., & Ashby, F. G. (2003). Fitting decision bound models to identification or categorization data. Unpublished manuscript. Available at https://labs.psych.ucsb.edu/ashby/gregory/sites/labs.psych.ucsb.edu.ashby.gregory/files/pubs/ennisashby2003_0.pdf.

Ennis, D. M., & Johnson, N. L. (1993). Thurstone-Shepard similarity models as special cases of moment generating functions. *Journal of Mathematical Psychology, 37*(1), 104–110.

Ennis, D. M., & Johnson, N. L. (1994). A general model for preferential and triadic choice in terms of central F distribution functions. *Psychometrika, 59*(1), 91–96.

Ennis, D. M., & Mullen, K. (1986). A multivariate model for discrimination methods. *Journal of Mathematical Psychology, 30*(2), 206–219.

Ennis, D. M., Mullen, K., & Frijters, J. E. (1988). Variants of the method of triads: Unidimensional Thurstonian models. *British Journal of Mathematical and Statistical Psychology, 41*(1), 25–36.

Ennis, D. M., Palen, J. J., & Mullen, K. (1988). A multidimensional stochastic theory of similarity. *Journal of Mathematical Psychology, 32*(4), 449–465.

Erev, I., Gopher, D., Itkin, R., & Greenshpan, Y. (1995). Toward a generalization of signal detection theory to N-person games: The example of two-person safety problem. *Journal of Mathematical Psychology, 39*(4), 360–375.

Estes, W. (2002). Traps in the route to models of memory and decision. *Psychonomic Bulletin & Review, 9*(1), 3–25.

Faisal, A. A. (2012). Noise in neurons and other constraints. In N. L. Novère (Ed.), *Computational systems neurobiology* (pp. 227–257). Springer.

Faisal, A. A., & Laughlin, S. B. (2007). Stochastic simulations on the reliability of action potential propagation in thin axons. *PLoS Computational Biology, 3*(5), e79.

Faisal, A. A., Selen, L. P., & Wolpert, D. M. (2008). Noise in the nervous system. *Nature Reviews Neuroscience, 9*(4), 292–303.

Falmagne, J.-C. (1968). Note on a simple fixed-point property of binary mixtures. *British Journal of Mathematical and Statistical Psychology, 21*(1), 131–132.

Farah, M. J., Monheit, M. A., & Wallace, M. A. (1991). Unconscious perception of "extinguished" visual stimuli: Reassessing the evidence. *Neuropsychologia, 29*(10), 949–958.

Farah, M. J., Wilson, K. D., Drain, M., & Tanaka, J. N. (1998). What is "special" about face perception? *Psychological Review, 105*(3), 482–498.

Farell, B. (1985). "Same"–"different" judgments: A review of current controversies in perceptual comparisons. *Psychological Bulletin, 98*(3), 419–456.

Fechner, G. T. (1860). *Elements of psychophysics* (H. E. Adler, Trans.). Breitkopf; Hartel (Holt, Rinehart, Winston).

Feldman, J. (2000). Minimization of Boolean complexity in human concept learning. *Nature, 407*(6804), 630–633.

Fisher, R. A. (1922). On the mathematical foundations of theoretical statistics. *Philosophical Transactions of the Royal Society of London. Series A, 222*(594-604), 309–368.

Fletcher, H. (1940). Auditory patterns. *Reviews of Modern Physics, 12*(1), 47–66.

Foroni, F., & Rothbart, M. (2011). Category boundaries and category labels: When does a category name influence the perceived similarity of category members? *Social Cognition, 29*(5), 547–576.

Forstmann, B. U., Anwander, A., Schäfer, A., Neumann, J., Brown, S., Wagenmakers, E.-J., Bogacz, R., & Turner, R. (2010). Cortico-striatal connections predict control over speed and accuracy in perceptual decision making. *Proceedings of the National Academy of Sciences, 107*(36), 15916–15920.

Forstmann, B. U., Dutilh, G., Brown, S., Neumann, J., Von Cramon, D. Y., Ridderinkhof, K. R., & Wagenmakers, E.-J. (2008). Striatum and pre-SMA facilitate decision-making under time pressure. *Proceedings of the National Academy of Sciences, 105*(45), 17538–17542.

Freedman, D. J., Riesenhuber, M., Poggio, T., & Miller, E. K. (2003). A comparison of primate prefrontal and inferior temporal cortices during visual categorization. *Journal of Neuroscience, 23*(12), 5235–5246.

Frijters, J. (1979). Variations of the triangular method and the relationship of its unidimensional probabilistic models to three-alternative forced-choice signal detection theory models. *British Journal of Mathematical and Statistical Psychology, 32*(2), 229–241.

Fujita, I., Tanaka, K., Ito, M., & Cheng, K. (1992). Columns for visual features of objects in monkey inferotemporal cortex. *Nature, 360*(6402), 343–346.

Fukunaga, K. (2013). *Introduction to statistical pattern recognition* (2nd ed.). Academic Press.

Fullerton, G., & Cattell, J. (1892). On the perception of small differences. University of Pennsylvania Philosophy Series, No. 2. University of Pennsylvania.

Garner, W. R. (1974). *The processing of information and structure.* Erlbaum.

Garner, W. R., & Morton, J. (1969). Perceptual independence: Definitions, models, and experimental paradigms. *Psychological Bulletin, 72,* 233–259.

Gehrlein, W. V., & Lepelley, D. (2010). *Voting paradoxes and group coherence: The Condorcet efficiency of voting rules.* Springer Science & Business Media.

Geisler, W. S. (1989). Sequential ideal-observer analysis of visual discriminations. *Psychological Review, 96*(2), 267–314.

Geisler, W. S. (2011). Contributions of ideal observer theory to vision research. *Vision Research, 51*(7), 771–781.

Gescheider, G. A., Wright, J. H., Weber, B. J., Kirchner, B. M., & Milligan, E. A. (1969). Reaction time as a function of the intensity and probability of occurrence of vibrotactile signals. *Perception & Psychophysics, 5,* 18–20.

Gilbert, C. D., & Sigman, M. (2007). Brain states: Top-down influences in sensory processing. *Neuron, 54*(5), 677–696.

Glanzer, M., Kim, K., Hilford, A., & Adams, J. K. (1999). Slope of the receiver-operating characteristic in recognition memory. *Journal of Experimental Psychology: Learning, Memory, and Cognition, 25*(2), 500–513.

Gold, J. I., & Ding, L. (2013). How mechanisms of perceptual decision-making affect the psychometric function. *Progress in Neurobiology, 103,* 98–114.

Gold, J. I., & Shadlen, M. N. (2007). The neural basis of decision making. *Annual Review of Neuroscience, 30,* 535–574.

Grafton, S. T., Hazeltine, E., & Ivry, R. (1995). Functional mapping of sequence learning in normal humans. *Journal of Cognitive Neuroscience, 7*(4), 497–510.

Graham, N. V. S. (1989). *Visual pattern analysers*. Oxford University Press.

Green, D. M., & Swets, J. A. (1966). *Signal detection theory and psychophysics*. Wiley.

Green, D. M., & Swets, J. A. (1964). General prediction relating yes-no and forced-choice results. *Journal of the Acoustical Society of America, 36*(5), 1042–1042.

Hanks, T. D., Ditterich, J., & Shadlen, M. N. (2006). Microstimulation of macaque area LIP affects decision-making in a motion discrimination task. *Nature Neuroscience, 9*(5), 682–689.

Hecht, S., Shlaer, S., & Pirenne, M. H. (1942). Energy, quanta, and vision. *Journal of General Physiology, 25*(6), 819–840.

Heekeren, H. R., Marrett, S., Bandettini, P. A., & Ungerleider, L. G. (2004). A general mechanism for perceptual decision-making in the human brain. *Nature, 431*(7010), 859–862.

Heitz, R. P. (2014). The speed-accuracy tradeoff: History, physiology, methodology, and behavior. *Frontiers in Neuroscience, 8*, 150.

Hélie, S., & Ashby, F. G. (2012). Learning and transfer of category knowledge in an indirect categorization task. *Psychological Research, 76*(3), 292–303.

Hélie, S., Turner, B. O., Crossley, M. J., Ell, S. W., & Ashby, F. G. (2017). Trial-by-trial identification of categorization strategy using iterative decision-bound modeling. *Behavior Research Methods, 49*(3), 1146–1162.

Hélie, S., Waldschmidt, J. G., & Ashby, F. G. (2010). Automaticity in rule-based and information-integration categorization. *Attention, Perception, & Psychophysics, 72*(4), 1013–1031.

Helmholtz, H. v. (1867). *Handbuch der physiologischen Optik* (Vol. 9). Voss.

Henry, R. A. (1993). Group judgment accuracy: Reliability and validity of postdiscussion confidence judgments. *Organizational Behavior and Human Decision Processes, 56*(1), 11–27.

Hernández, A., Zainos, A., & Romo, R. (2000). Neuronal correlates of sensory discrimination in the somatosensory cortex. *Proceedings of the National Academy of Sciences, 97*(11), 6191–6196.

Hess, R. F., & Plant, G. T. (1985). Temporal frequency discrimination in human vision: Evidence for an additional mechanism in the low spatial and high temporal frequency region. *Vision Research, 25*(10), 1493–1500.

Hirsch, J., Hylton, R., & Graham, N. (1982). Simultaneous recognition of two spatial-frequency components. *Vision Research, 22*(3), 365–375.

Hohle, R. H. (1965). Inferred components of reaction times as functions of foreperiod duration. *Journal of Experimental Psychology, 69*(4), 382–386.

Holden, J. G., Van Orden, G. C., & Turvey, M. T. (2009). Dispersion of response times reveals cognitive dynamics. *Psychological Review, 116*(2), 318–342.

Homa, D., Sterling, S., & Trepel, L. (1981). Limitations of exemplar-based generalization and the abstraction of categorical information. *Journal of Experimental Psychology: Human Learning and Memory, 7*(6), 418–439.

Hoshi, E., Shima, K., & Tanji, J. (2000). Neuronal activity in the primate prefrontal cortex in the process of motor selection based on two behavioral rules. *Journal of Neurophysiology, 83*(4), 2355–2373.

Houk, J., Adams, J., & Barto, A. (1995). A model of how the basal ganglia generate and use neural signals that predict reinforcement. In J. C. Houk, J. L. Davis, & D. G. Beiser (Eds.), *Models of information processing in the basal ganglia* (pp. 249–270). MIT Press.

Hubel, D. H., & Wiesel, T. N. (1962). Receptive fields, binocular interaction and functional architecture in the cat's visual cortex. *Journal of Physiology, 160*(1), 106–154.

Hung, C. P., Kreiman, G., Poggio, T., & DiCarlo, J. J. (2005). Fast readout of object identity from macaque inferior temporal cortex. *Science, 310*(5749), 863–866.

Inglis, J. B., Bird, J., & Ashby, F. G. (2022). A general recognition theory model for identifying an ideal stimulus. *Attention, Perception, & Psychophysics, 84*(7), 2408–2421.

Izhikevich, E. M. (2003). Simple model of spiking neurons. *IEEE Transactions on Neural Networks, 14*(6), 1569–1572.

James, W. (1890). *The principles of psychology*. Holt.

Jiang, X., Rosen, E., Zeffiro, T., VanMeter, J., Blanz, V., & Riesenhuber, M. (2006). Evaluation of a shape-based model of human face discrimination using fMRI and behavioral techniques. *Neuron, 50*(1), 159–172.

Jones, M., & Dzhafarov, E. N. (2014a). Analyzability, ad hoc restrictions, and excessive flexibility of evidence-accumulation models: Reply to two critical commentaries. *Psychological Review, 121*(4), 689–695.

Jones, M., & Dzhafarov, E. N. (2014b). Unfalsifiability and mutual translatability of major modeling schemes for choice reaction time. *Psychological Review, 121*(1), 1–32.

Kadlec, H. (1995). Multidimensional signal detection analyses (MSDA) for testing separability and independence: A PASCAL program. *Behavior Research Methods, Instruments, & Computers, 4*, 442–458.

Kadlec, H., & Townsend, J. T. (1992a). Implications of marginal and conditional detection parameters for the separabilities and independence of perceptual dimensions. *Journal of Mathematical Psychology, 36*(3), 325–374.

Kadlec, H., & Townsend, J. T. (1992b). Signal detection analysis of dimensional interactions. In F. G. Ashby (Ed.), *Multidimensional models of perception and cognition* (pp. 181–228). Erlbaum.

Kaesler, M., Dunn, J. C., Ransom, K., & Semmler, C. (2020). Do sequential lineups impair underlying discriminability? *Cognitive Research: Principles and Implications, 5*, 1–21.

Kaku, M., (2014, August 20). The golden age of neuroscience has arrived. *Wall Street Journal*. https://www.wsj.com/articles/michio-kaku-the-golden-age-of-neuroscience-has-arrived-1408577023?msockid=010e7be3472b66ef3dd16fa346b967d8

Kanwisher, N., & Yovel, G. (2006). The fusiform face area: A cortical region specialized for the perception of faces. *Philosophical Transactions of the Royal Society B: Biological Sciences, 361*(1476), 2109–2128.

Kastner, S., & Ungerleider, L. G. (2000). Mechanisms of visual attention in the human cortex. *Annual Review of Neuroscience, 23*(1), 315–341.

Kelly, S. P., & O'Connell, R. G. (2013). Internal and external influences on the rate of sensory evidence accumulation in the human brain. *Journal of Neuroscience, 33*(50), 19434–19441.

Kerr, N. L., & Tindale, R. S. (2004). Group performance and decision making. *Annual Review of Psychology, 55*, 623–655.

Khodadadi, A., & Townsend, J. T. (2015). On mimicry among sequential sampling models. *Journal of Mathematical Psychology, 68–69*, 37–48.

Kiani, R., Hanks, T. D., & Shadlen, M. N. (2008). Bounded integration in parietal cortex underlies decisions even when viewing duration is dictated by the environment. *Journal of Neuroscience, 28*(12), 3017–3029.

Killeen, P. R., & Taylor, T. J. (2004). Symmetric receiver operating characteristics. *Journal of Mathematical Psychology, 48*(6), 432–434.

Kleppe, I. C., & Robinson, H. P. (2006). Correlation entropy of synaptic input-output dynamics. *Physical Review E, 74*(4), 041909.

Knopman, D., & Nissen, M. J. (1991). Procedural learning is impaired in Huntington's disease: Evidence from the serial reaction time task. *Neuropsychologia, 29*(3), 245–254.

Koen, J. D., Barrett, F. S., Harlow, I. M., & Yonelinas, A. P. (2017). The ROC Toolbox: A toolbox for analyzing receiver-operating characteristics derived from confidence ratings. *Behavior Research Methods, 49*(4), 1399–1406.

Kovacs, P., Hélie, S., Tran, A. N., & Ashby, F. G. (2021). A neurocomputational theory of how rule-guided behaviors become automatic. *Psychological Review, 128*(3), 488–508.

Kriegeskorte, N., Goebel, R., & Bandettini, P. (2006). Information-based functional brain mapping. *Proceedings of the National Academy of Sciences, 103*(10), 3863–3868.

Krueger, L. E. (1978). A theory of perceptual matching. *Psychological Review, 85*(4), 278–304.

Krumhansl, C. L. (1978). Concerning the applicability of geometric models to similarity data: The interrelationship between similarity and spatial density. *Psychological Review, 85*(5), 445–463.

Kruschke, J. K. (1992). ALCOVE: An exemplar-based connectionist model of category learning. *Psychological Review, 99*(1), 22–44.

Kruschke, J. K. (1996). Dimensional relevance shifts in category learning. *Connection Science, 8*(2), 225–247.

Kruskal, J. B. (1964). Multidimensional scaling by optimizing goodness of fit to a nonmetric hypothesis. *Psychometrika, 29*(1), 1–27.

Kruskal, J. B., & Wish, M. (1978). *Multidimensional scaling* (Vol. 11). Sage.

Lara, F., Hahn, A., Yu, N.-Y., & Yamauchi, T. (2012). Arbitrary category labels can change similarity judgments of human faces. In *Proceedings of the Annual Meeting of the Cognitive Science Society*, 34(34) ISSN: 1069-7977.

Lattman, P. (2007, September 27). The origins of Justice Stewart's "I know it when I see it." *Wall Street Journal*. LawBlog at *The Wall Street Journal Online*. URL is: https://www.wsj.com/articles/BL-LB-4558?msockid=010e7be3472b66ef3dd16fa346b967d8

Lauwereyns, J., Watanabe, K., Coe, B., & Hikosaka, O. (2002). A neural correlate of response bias in monkey caudate nucleus. *Nature, 418*(6896), 413–417.

Layher, E., Dixit, A., & Miller, M. B. (2020). Who gives a criterion shift? A uniquely individualistic cognitive trait. *Journal of Experimental Psychology: Learning, Memory, and Cognition, 46*(11), 2075–2105.

Le Pelley, M. E., Newell, B. R., & Nosofsky, R. M. (2019). Deferred feedback does not dissociate implicit and explicit category-learning systems: Commentary on Smith et al. (2014). *Psychological Science, 30*(9), 1403–1409.

Lee, J., & Penrod, S. D. (2019). New signal detection theory-based framework for eyewitness performance in lineups. *Law and Human Behavior, 43*(5), 436–454.

Lee, M. D., Steyvers, M., De Young, M., & Miller, B. (2012). Inferring expertise in knowledge and prediction ranking tasks. *Topics in Cognitive Science, 4*(1), 151–163.

Link, S. W. (1992). *The wave theory of difference and similarity*. Lawrence Erlbaum Associates.

Link, S. W. (1994). Rediscovering the past: Gustav Fechner and signal detection theory. *Psychological Science, 5*(6), 335–340.

Link, S. W., & Heath, R. A. (1975). A sequential theory of psychological discrimination. *Psychometrika, 40*(1), 77–105.

Lisman, J., Schulman, H., & Cline, H. (2002). The molecular basis of CaMKII function in synaptic and behavioural memory. *Nature Reviews Neuroscience, 3*(3), 175–190.

Liu, Y., Townsend, J. T., & Wenger, M. J. (2023). Don't be a square: The processing mechanisms characterising the elemental dimensions of width and height. *Quarterly Journal of Experimental Psychology, 76*(4), 792–826.

Lockhead, G. R. (1966). Effects of dimensional redundancy on visual discrimination. *Journal of Experimental Psychology, 72*(1), 95–104.

Luce, R. D. (1959). *Individual choice behavior: A theoretical analysis*. Wiley.

Luce, R. D. (1963). Detection and recognition. In R. D. Luce, R. R. Bush, & Galanter, E. (Eds.), *Handbook of mathematical psychology* (Vol. 1, pp. 103–190). Wiley.

Luce, R. D. (1986). *Response times: Their role in inferring elementary mental organization*. Oxford University Press, USA.

MacKay, D. B. (2001). Probabilistic multidimensional scaling using a city-block metric. *Journal of Mathematical Psychology, 45*(2), 249–264.

Macmillan, N. A., & Creelman, C. D. (2004). *Detection theory: A user's guide*. Psychology Press.

Maddox, W. T., Ashby, F. G., & Bohil, C. J. (2003). Delayed feedback effects on rule-based and information-integration category learning. *Journal of Experimental Psychology: Learning, Memory, and Cognition, 29*(4), 650–662.

Maddox, W. T. (1992). Perceptual and decisional separability. In F. G. Ashby (Ed.), *Multidimensional models of perception and cognition* (pp. 147–180). Lawrence Erlbaum Associates.

Maddox, W. T., & Ashby, F. G. (1996). Perceptual separability, decisional separability, and the identification-speeded classification relationship. *Journal of Experimental Psychology: Human Perception and Performance, 22*(4), 795–817.

Maddox, W. T., Ashby, F. G., & Gottlob, L. R. (1998). Response time distributions in multidimensional perceptual categorization. *Perception & Psychophysics, 60*(4), 620–637.

Maddox, W. T., Ashby, F. G., Ing, A. D., & Pickering, A. D. (2004). Disrupting feedback processing interferes with rule-based but not information-integration category learning. *Memory & Cognition, 32*(4), 582–591.

Maddox, W. T., Ashby, F. G., & Waldron, E. M. (2002). Multiple attention systems in perceptual categorization. *Memory & Cognition, 30*(3), 325–339.

Maddox, W. T., Bohil, C. J., & Ing, A. D. (2004). Evidence for a procedural-learning-based system in perceptual category learning. *Psychonomic Bulletin & Review, 11*(5), 945–952.

Maddox, W. T., Glass, B. D., O'Brien, J. B., Filoteo, J. V., & Ashby, F. G. (2010). Category label and response location shifts in category learning. *Psychological Research, 74*(2), 219–236.

Maddox, W. T., & Ing, D. A. (2005). Delayed feedback disrupts the procedural-learning system but not the hypothesis-testing system in perceptual category learning. *Journal of Experimental Psychology: Learning, Memory, and Cognition, 31*(1), 100–107.

Mahon, B. Z., & Caramazza, A. (2009). Concepts and categories: A cognitive neuropsychological perspective. *Annual Review of Psychology, 60*(1), 27–51.

Marcum, J. I. (1947). *A statistical theory of target detection by pulsed radar* (tech. rep.). Rand Corporation. https://www.rand.org/content/dam/rand/pubs/research_memoranda/2006/RM754.pdf.

Marr, D. (1982). *Vision: A computational investigation into the human representation and processing of visual information*. Freeman.

Masson, H. L., & Isik, L. (2021). Functional selectivity for social interaction perception in the human superior temporal sulcus during natural viewing. *NeuroImage, 245*, 118741.

Medin, D. L., Altom, M. W., Edelson, S. M., & Freko, D. (1982). Correlated symptoms and simulated medical classification. *Journal of Experimental Psychology: Learning, Memory, and Cognition, 8*(1), 37–50.

Medin, D. L., & Schwanenflugel, P. J. (1981). Linear separability in classification learning. *Journal of Experimental Psychology: Human Learning and Memory*, *7*(5), 355–368.

Menne, N. M., Winter, K., Bell, R., & Buchner, A. (2022). A validation of the two-high threshold eyewitness identification model by reanalyzing published data. *Scientific Reports*, *12*(1), 13379.

Menneer, T., Wenger, M., & Blaha, L. (2010). Inferential challenges for general recognition theory: Mean-shift integrality and perceptual configurality. *Journal of Vision*, *10*(7), 1211–1211.

Miller, D. R., & Singpurwalla, N. D. (1980). Failure rate estimation using random smoothing. *Sankhyā: The Indian Journal of Statistics, Series B*, *42*(3–4), 217–228.

Miller, E. K., & Cohen, J. D. (2001). An integrative theory of prefrontal cortex function. *Annual Review of Neuroscience*, *24*(1), 167–202.

Miller, M. B., Handy, T. C., Cutler, J., Inati, S., & Wolford, G. L. (2001). Brain activations associated with shifts in response criterion on a recognition test. *Canadian Journal of Experimental Psychology*, *55*(2), 162–173.

Miller, M. B., & Kantner, J. (2020). Not all people are cut out for strategic criterion shifting. *Current Directions in Psychological Science*, *29*(1), 9–15.

Milner, A. D. (2017). How do the two visual streams interact with each other? *Experimental Brain Research*, *235*(5), 1297–1308.

Mishkin, M., Malamut, B., & Bachevalier, J. (1984). Memories and habits: Two neural systems. In G. Lynch, J. L. McGaugh, & N. M. Weinberger (Eds.), *Neurobiology of human learning and memory* (pp. 65–77). Guilford Press.

Muhammad, R., Wallis, J. D., & Miller, E. K. (2006). A comparison of abstract rules in the prefrontal cortex, premotor cortex, inferior temporal cortex, and striatum. *Journal of Cognitive Neuroscience*, *18*(6), 974–989.

Mullen, K., Ennis, D., Doncker, E., & Kapenga, J. (1988). Multivariate models for the triangular and duo-trio methods. *Biometrics*, *44*(116), 1169–1175.

Mullen, K., & Ennis, D. M. (1991). A simple multivariate probabilistic model for preferential and triadic choices. *Psychometrika*, *56*(1), 69–75.

Murdock, B. B. (1985). An analysis of the strength-latency relationship. *Memory & Cognition*, *13*(6), 511–521.

Murdock, B. B., & Anderson, R. E. (1975). Encoding, storage, and retrieval of item information. In R. L. Solso (Ed.), *Information processing and cognition: The Loyola symposium* (pp. 145–194). Erlbaum.

Murphy, J., Gray, K. L. H., & Cook, R. (2017). The composite face illusion. *Psychonomic Bulletin & Review*, *24*(2), 245–261.

Myung, J. I., Cavagnaro, D. R., & Pitt, M. A. (2016). Model evaluation and selection. In W. H. Batchelder, H. Colonius, E. Dzhafarov, & J. I. Myung (Eds.), *New handbook of mathematical psychology, Volume 1: Measurement and methodology* (pp. 552–598). Cambridge University Press.

Neyman, J., & Pearson, E. S. (1933). IX. On the problem of the most efficient tests of statistical hypotheses. *Philosophical Transactions of the Royal Society of London. Series A, Containing Papers of a Mathematical or Physical Character, 231*(694–706), 289–337.

Noles, N. S., & Gelman, S. A. (2012). Effects of categorical labels on similarity judgments: A critical analysis of similarity-based approaches. *Developmental Psychology, 48*(3), 890–896.

Nosofsky, R. M. (1986). Attention, similarity, and the identification-categorization relationship. *Journal of Experimental Psychology: General, 115*(1), 39–57.

Nosofsky, R. M. (1991). Stimulus bias, asymmetric similarity, and classification. *Cognitive Psychology, 23*(1), 94–140.

Nosofsky, R. M., Stanton, R. D., & Zaki, S. R. (2005). Procedural interference in perceptual classification: Implicit learning or cognitive complexity? *Memory & Cognition, 33*(7), 1256–1271.

O'Connell, R. G., Dockree, P. M., & Kelly, S. P. (2012). A supramodal accumulation-to-bound signal that determines perceptual decisions in humans. *Nature Neuroscience, 15*(12), 1729–1735.

Olzak, L. A. (1986). Widely separated spatial frequencies: Mechanism interactions. *Vision Research, 26*(7), 1143–1153.

O'Toole, A. J., Wenger, M. J., & Townsend, J. T. (2001). Quantitative models of perceiving and remembering faces: Precedents and possibilities. In M. J. Wenger, & J. T. Townsend (Eds.), *Computational, geometric, and process perspectives on facial cognition: Contexts and challenges* (pp. 1–38). Erlbaum.

Owen, D. B. (1956). Tables for computing bivariate normal probabilities. *The Annals of Mathematical Statistics, 27*(4), 1075–1090.

Pachella, R. G. (2021). The interpretation of reaction time in information-processing research. In B. H. Kantowitz (Ed.), *Human information processing: Tutorials in performance and cognition* (pp. 41–82). Erlbaum.

Pajani, A., Kok, P., Kouider, S., & Lange, F. P. (2015). Spontaneous activity patterns in primary visual cortex predispose to visual hallucinations. *Journal of Neuroscience, 35*(37), 12947–12953.

Pakhotin, P., & Bracci, E. (2007). Cholinergic interneurons control the excitatory input to the striatum. *Journal of Neuroscience, 27*(2), 391–400.

Palmer, J., Verghese, P., & Pavel, M. (2000). The psychophysics of visual search. *Vision Research, 40*(10–12), 1227–1268.

Parzen, E. (1962). On estimation of a probability density function and mode. *Annals of Mathematical Statistics, 33*(3), 1065–1076.

Perrin, N. A. (1992). Uniting identification, similarity and preference: General recognition theory. In F. G. Ashby (Ed.), *Multidimensional models of perception and cognition* (pp. 123–146). Lawrence Erlbaum Associates.

Perrin, N. A., & Ashby, F. G. (1991). A test for perceptual independence with dissimilarity data. *Applied Psychological Measurement, 15*(1), 79–93.

Perry, L., Blaha, L., & Townsend, J. (2008). Reassessing the architecture of same-different face judgments. *Journal of Vision, 8*(6), 889.

Peterson, W. W., Birdsall, T., & Fox, W. (1954). The theory of signal detectability. *Transactions of the IRE Professional Group on Information Theory, 4*(4), 171–212.

Peterson, W. W., & Birdsall, T. G. (1953). The theory of signal detectability: Part I. The general theory. Electronic Defense Group (tech. rep.). Technical Report 13, June 1953. University of Michigan.

Piepers, D., & Robbins, R. (2012). A review and clarification of the terms "holistic," "configural," and "relational" in the face perception literature. *Frontiers in Psychology, 3*, 559.

Posner, M. I., & Keele, S. W. (1968). On the genesis of abstract ideas. *Journal of Experimental Psychology, 77*(3p1), 353–363.

Purushothaman, G., & Bradley, D. C. (2005). Neural population code for fine perceptual decisions in area MT. *Nature Neuroscience, 8*(1), 99–106.

Qadri, M. A., Ashby, F. G., Smith, J. D., & Cook, R. G. (2019). Testing analogical rule transfer in pigeons (Columba livia). *Cognition, 183*, 256–268.

Raftery, A. E. (1995). Bayesian model selection in social research. *Sociological Methodology, 25*, 111–163.

Ratcliff, R. (1978). A theory of memory retrieval. *Psychological Review, 85*(2), 59–108.

Ratcliff, R. (1985). Theoretical interpretations of the speed and accuracy of positive and negative responses. *Psychological Review, 92*(2), 212–225.

Ratcliff, R., & Rouder, J. N. (1998). Modeling response times for two-choice decisions. *Psychological Science, 9*(5), 347–356.

Ratcliff, R., Sheu, C.-F., & Gronlund, S. D. (1992). Testing global memory models using ROC curves. *Psychological Review, 99*(3), 518–535.

Ratcliff, R., Smith, P. L., Brown, S. D., & McKoon, G. (2016). Diffusion decision model: Current issues and history. *Trends in Cognitive Sciences, 20*(4), 260–281.

Ratcliff, R., & Tuerlinckx, F. (2002). Estimating parameters of the diffusion model: Approaches to dealing with contaminant reaction times and parameter variability. *Psychonomic Bulletin & Review, 9*(3), 438–481.

Ratcliff, R., Van Zandt, T., & McKoon, G. (1999). Connectionist and diffusion models of reaction time. *Psychological Review, 106*(2), 261–300.

Ress, D., & Heeger, D. J. (2003). Neuronal correlates of perception in early visual cortex. *Nature Neuroscience, 6*(4), 414–420.

Reynolds, J. H., Pasternak, T., & Desimone, R. (2000). Attention increases sensitivity of V4 neurons. *Neuron, 26*(3), 703–714.

Richler, J. J., & Gauthier, I. (2014). A meta-analysis and review of holistic face processing. *Psychological Bulletin, 140*(5), 1281–1302.

Richler, J. J., & Gauthier, I. (2013). When intuition fails to align with data: A reply to Rossion (2013). *Visual Cognition, 21*(2), 254–276.

Richler, J. J., Gauthier, I., Wenger, M. J., & Palmeri, T. J. (2008). Holistic processing of faces: Perceptual and decisional components. *Journal of Experimental Psychology: Learning, Memory, and Cognition, 34*(2), 328–342.

Richler, J. J., Palmeri, T. J., & Gauthier, I. (2012). Meanings, mechanisms, and measures of holistic processing. *Frontiers in Psychology, 3*, 553.

Riesenhuber, M., & Poggio, T. (1999). Hierarchical models of object recognition in cortex. *Nature Neuroscience, 2*(11), 1019–1025.

Roberts, J. S., Donoghue, J. R., & Laughlin, J. E. (2000). A general item response theory model for unfolding unidimensional polytomous responses. *Applied Psychological Measurement, 24*(1), 3–32.

Robson, J. G. (1975). Receptive fields: Neural representation of the spatial and intensive attributes of the visual image. In E. C. Carterette & M. P. Friedman (Eds.), *Handbook of perception, volume 5: Vision* (pp. 81–116). Academic Press.

Roitman, J. D., & Shadlen, M. N. (2002). Response of neurons in the lateral intraparietal area during a combined visual discrimination reaction time task. *Journal of Neuroscience, 22*(21), 9475–9489.

Rolls, E. T. (2009). The neurophysiology and computational mechanisms of object representation. In S. Dickinson, M. J. Tarr, A. Leonardis, & B. Schiele (Eds.), *Object categorization: Computer and human vision perspectives* (pp. 257–287). Citeseer.

Rolls, E. T., Judge, S. J., & Sanghera, M. K. (1977). Activity of neurones in the inferotemporal cortex of the alert monkey. *Brain Research, 130*(2), 229–238.

Romo, R., Hernández, A., & Zainos, A. (2004). Neuronal correlates of a perceptual decision in ventral premotor cortex. *Neuron, 41*(1), 165–173.

Rose, A. (1942). The relative sensitivities of television pickup tubes, photographic film, and the human eye. *Proceedings of the IRE, 30*(6), 293–300.

Rose, A. (1948). The sensitivity performance of the human eye on an absolute scale. *Journal of the Optical Society of America, 38*(2), 196–208.

Rosedahl, L. A., & Ashby, F. G. (2019). A difficulty predictor for perceptual category learning. *Journal of Vision, 19*(6), 20.

Rosedahl, L. A., Serota, R., & Ashby, F. G. (2021). When instructions don't help: Knowing the optimal strategy facilitates rule-based but not information-integration category learning. *Journal of Experimental Psychology: Human Perception and Performance, 47*(9), 1226–1236.

Rossion, B. (2013). The composite face illusion: A whole window into our understanding of holistic face perception. *Visual Cognition, 21*(2), 139–253.

Rotello, C. M., Macmillan, N. A., & Reeder, J. A. (2004). Sum-difference theory of remembering and knowing: A two-dimensional signal-detection model. *Psychological Review, 111*(3), 588–616.

Rousseau, B., Ennis, D. M., & Rossi, F. (2012). Internal preference mapping and the issue of satiety. *Food Quality and Preference, 24*(1), 67–74.

Rudy, J. W. (2020). *The neurobiology of learning and memory* (3rd ed.). Oxford University Press.

Salin, P.-A., & Bullier, J. (1995). Corticocortical connections in the visual system: Structure and function. *Physiological Reviews, 75*(1), 107–154.

Sanders, B. (1971). Factors affecting reversal and nonreversal shifts in rats and children. *Journal of Comparative and Physiological Psychology, 74*(2), 192–202.

Scheffé, H. (1999). *The analysis of variance.* John Wiley & Sons.

Schönemann, P. H., & Wang, M. M. (1972). An individual difference model for the multidimensional analysis of preference data. *Psychometrika, 37*(3), 275–309.

Schwarz, G. (1978). Estimating the dimension of a model. *Annals of Statistics, 6*(2), 461–464.

Serre, T., Oliva, A., & Poggio, T. (2007). A feedforward architecture accounts for rapid categorization. *Proceedings of the National Academy of Sciences, 104*(15), 6424–6429.

Serre, T., Wolf, L., Bileschi, S., Riesenhuber, M., & Poggio, T. (2007). Robust object recognition with cortex-like mechanisms. *IEEE Transactions on Pattern Analysis and Machine Intelligence, 29*(3), 411–426.

Shadlen, M. N., & Newsome, W. T. (1998). The variable discharge of cortical neurons: Implications for connectivity, computation, and information coding. *Journal of Neuroscience, 18*(10), 3870–3896.

Shadlen, M. N., & Newsome, W. T. (2001). Neural basis of a perceptual decision in the parietal cortex (area LIP) of the rhesus monkey. *Journal of Neurophysiology, 86*(4), 1916–1936.

Shaw, M. L. (1980). Identifying attentional and decision-making components in information processing. In R. S. Nickerson (Ed.), *Attention and performance VIII* (pp. 106–121). Erlbaum.

Shaw, M. L. (1982). Attending to multiple sources of information: I. The integration of information in decision making. *Cognitive Psychology, 14*(3), 353–409.

Shepard, R. N. (1957). Stimulus and response generalization: A stochastic model relating generalization to distance in psychological space. *Psychometrika, 22*(4), 325–345.

Shepard, R. N. (1962). The analysis of proximities: Multidimensional scaling with an unknown distance function. I. *Psychometrika, 27*(2), 125–140.

Shepard, R. N. (1964). Attention and the metric structure of the stimulus space. *Journal of Mathematical Psychology, 1*(1), 54–87.

Shepard, R. N. (1987). Toward a universal law of generalization for psychological science. *Science, 237*(4820), 1317–1323.

Shepard, R. N., & Chang, J.-J. (1963). Stimulus generalization in the learning of classifications. *Journal of Experimental Psychology, 65*(1), 94–102.

Shepard, R. N., Hovland, C. I., & Jenkins, H. M. (1961). Learning and memorization of classifications. *Psychological Monographs: General and Applied, 75*(13), 1–42.

Sigala, N. (2004). Visual categorization and the inferior temporal cortex. *Behavioural Brain Research, 149*(1), 1–7.

Sigala, N., & Logothetis, N. K. (2002). Visual categorization shapes feature selectivity in the primate temporal cortex. *Nature, 415*(6869), 318–320.

Silbert, N. H., & Hawkins, R. X. D. (2016). A tutorial on general recognition theory. *Journal of Mathematical Psychology, 73*, 94–109.

Silbert, N. H., & Thomas, R. D. (2013). Decisional separability, model identification, and statistical inference in the general recognition theory framework. *Psychonomic Bulletin & Review, 20*(1), 1–20.

Sloutsky, V. M., & Fisher, A. V. (2004). Induction and categorization in young children: A similarity-based model. *Journal of Experimental Psychology: General, 133*(2), 166–188.

Smith, J. D., Ashby, F. G., Berg, M. E., Murphy, M. S., Spiering, B., Cook, R. G., & Grace, R. C. (2011). Pigeons' categorization may be exclusively nonanalytic. *Psychonomic Bulletin & Review, 18*(2), 414–421.

Smith, J. D., Beran, M. J., Crossley, M. J., Boomer, J. T., & Ashby, F. G. (2010). Implicit and explicit category learning by macaques (macaca mulatta) and humans (homo sapiens). *Journal of Experimental Psychology: Animal Behavior Processes, 36*, 54–65.

Smith, J. D., Boomer, J., Zakrzewski, A. C., Roeder, J. L., Church, B. A., & Ashby, F. G. (2014). Deferred feedback sharply dissociates implicit and explicit category learning. *Psychological Science, 25*(2), 447–457.

Smith, J. D., Crossley, M. J., Boomer, J., Church, B. A., Beran, M. J., & Ashby, F. G. (2012). Implicit and explicit category learning by capuchin monkeys (Cebus apella). *Journal of Comparative Psychology, 126*(3), 294–304.

Smith, J. D., & Ell, S. W. (2015). One giant leap for categorizers: One small step for categorization theory. *PloS One, 10*(9), e0137334.

Smith, J. D., Jamani, S., Boomer, J., & Church, B. A. (2018). One-back reinforcement dissociates implicit-procedural and explicit-declarative category learning. *Memory & Cognition, 46*, 261–273.

Smith, J. D., & Minda, J. P. (1998). Prototypes in the mist: The early epochs of category learning. *Journal of Experimental Psychology: Learning, Memory, and Cognition, 24*(6), 1411–1436.

Smith, J. K. (1992). Alternative biased choice models. *Mathematical Social Sciences, 23*(2), 199–219.

Smith, P. L. (2016). Diffusion theory of decision making in continuous report. *Psychological Review, 123*(4), 425–451.

Smith, P. L. (2019). Linking the diffusion model and general recognition theory: Circular diffusion with bivariate-normally distributed drift rates. *Journal of Mathematical Psychology, 91*, 145–158.

Smith, P. L., & Corbett, E. A. (2019). Speeded multielement decision-making as diffusion in a hypersphere: Theory and application to double-target detection. *Psychonomic Bulletin & Review, 26*(1), 127–162.

Smith, P. L., Garrett, P. M., & Zhou, J. (2023). Obtaining stable predicted distributions of response times and decision outcomes for the circular diffusion model. *Computational Brain & Behavior, 6*, 543–555.

Soliveri, P., Brown, R. G., Jahanshahi, M., Caraceni, T., & Marsden, C. D. (1997). Learning manual pursuit tracking skills in patients with Parkinson's disease. *Brain, 120*(8), 1325–1337.

Sorkin, R. D., & Dai, H. (1994). Signal detection analysis of the ideal group. *Organizational Behavior and Human Decision Processes, 60*(1), 1–13.

Sorkin, R. D., Hays, C. J., & West, R. (2001). Signal-detection analysis of group decision making. *Psychological Review, 108*(1), 183–203.

Sorkin, R. D., West, R., & Robinson, D. E. (1998). Group performance depends on the majority rule. *Psychological Science, 9*(6), 456–463.

Soto, F. A., Vucovich, L., Musgrave, R., & Ashby, F. G. (2015). General recognition theory with individual differences: A new method for examining perceptual and decisional interactions with an application to face perception. *Psychonomic Bulletin & Review, 22*(1), 88–111.

Soto, F. A., Vucovich, L. E., & Ashby, F. G. (2018). Linking signal detection theory and encoding models to reveal independent neural representations from neuroimaging data. *PLoS Computational Biology, 14*(10), e1006470.

Soto, F. A., & Wasserman, E. A. (2014). Mechanisms of object recognition: What we have learned from pigeons. *Frontiers in Neural Circuits, 8*, 122.

Soto, F. A., Zheng, E., Fonseca, J., & Ashby, F. G. (2017). Testing separability and independence of perceptual dimensions with general recognition theory: A tutorial and new R package (grtools). *Frontiers in Psychology, 8*, 696.

Squire, L. R. (2004). Memory systems of the brain: A brief history and current perspective. *Neurobiology of Learning and Memory, 82*(3), 171–177.

Starns, J. J., Cohen, A. L., & Rotello, C. M. (2023). A complete method for assessing the effectiveness of eyewitness identification procedures: Expected information gain. *Psychological Review, 130*(3), 677–719.

Sternberg, S. (1966). High-speed scanning in human memory. *Science, 153*(3736), 652–654.

Stevens, S. S. (1957). On the psychophysical law. *Psychological Review, 64*(3), 153–181.

Stone, M. (1960). Models for choice-reaction time. *Psychometrika, 25*(3), 251–260.

Swensson, R. G., & Judy, P. F. (1981). Detection of noisy visual targets: Models for the effects of spatial uncertainty and signal-to-noise ratio. *Perception & Psychophysics*, *29*(6), 521–534.

Swets, J. A. (1961). Is there a sensory threshold? *Science*, *134*(3473), 168–177.

Tam, H., Maddox, W. T., & Huang-Pollock, C. L. (2013). Posterror slowing predicts rule-based but not information-integration category learning. *Psychonomic Bulletin & Review*, *20*, 1343–1349.

Tanaka, K. (1996). Inferotemporal cortex and object vision. *Annual Review of Neuroscience*, *19*(1), 109–139.

Tanner, W. (1956). Theory of recognition. *Journal of the Acoustical Society of America*, *28*(5), 882–888.

Tanner Jr, W. P., & Swets, J. A. (1954). A decision-making theory of visual detection. *Psychological Review*, *61*(6), 401–409.

Thomas, E., Van Hulle, M. M., & Vogel, R. (2001). Encoding of categories by noncategory-specific neurons in the inferior temporal cortex. *Journal of Cognitive Neuroscience*, *13*(2), 190–200.

Thomas, E.A.C. (1971). Sufficient conditions for monotone hazard rate: An application to latency-probability curves. *Journal of Mathematical Psychology*, *8*(3), 303–332.

Thomas, R. D. (1996). Separability and independence of dimensions within the same–different judgment task. *Journal of Mathematical Psychology*, *40*(4), 318–341.

Thomas, R. D. (1999). Assessing sensitivity in a multidimensional space: Some problems and a definition of a general *d*′. *Psychonomic Bulletin & Review*, *6*, 224–238.

Thomas, R. D. (2001). Characterizing perceptual interactions in face identification using multidimensional signal detection theory. In M. J. Wenger, & J. T. Townsend (Eds.), *Computational, geometric, and process perspectives on facial cognition: Contexts and challenges* (pp. 193–228). Erlbaum.

Thomas-Ollivier, V., Reymann, J., Le Moal, S., Schück, S., Lieury, A., & Allain, H. (1999). Procedural memory in recent-onset Parkinson's disease. *Dementia and Geriatric Cognitive Disorders*, *10*(2), 172–180.

Thorpe, S. J., & Fabre-Thorpe, M. (2001). Seeking categories in the brain. *Science*, *291*(5502), 260–263.

Thurstone, L. L. (1927a). A law of comparative judgment. *Psychological Review*, *34*(4), 273–286.

Thurstone, L. L. (1927b). The method of paired comparisons for social values. *Journal of Abnormal and Social Psychology*, *21*(4), 384–400.

Thurstone, L. L. (1927c). Psychophysical analysis. *American Journal of Psychology*, *38*(3), 368–389.

Tindale, R. S., & Winget, J. R. (2019). Group decision-making. *Oxford Research Encyclopedia of Psychology*. https://doi.org/10.1093/acrefore/9780190236557.013.262

Tolhurst, D. J., Movshon, J. A., & Dean, A. F. (1983). The statistical reliability of signals in single neurons in cat and monkey visual cortex. *Vision Research*, *23*(8), 775–785.

Torgerson, W. S. (1952). Multidimensional scaling: I. Theory and method. *Psychometrika, 17*(4), 401–419.

Townsend, J. T. (1971). Theoretical analysis of an alphabetic confusion matrix. *Perception & Psychophysics, 9*(1A), 40–50.

Townsend, J. T., & Altieri, N. (2012). An accuracy–response time capacity assessment function that measures performance against standard parallel predictions. *Psychological Review, 119*(3), 500–516.

Townsend, J. T., & Ashby, F. G. (1978). Methods of modeling capacity in simple processing systems. In N. J. Castellan & F. Restle (Eds.), *Cognitive theory* (Vol. 3, pp. 199–239). Psychology Press.

Townsend, J. T., & Ashby, F. G. (1982). Experimental test of contemporary mathematical models of visual letter recognition. *Journal of Experimental Psychology: Human Perception and Performance, 8*(6), 834–864.

Townsend, J. T., & Ashby, F. G. (1983). *Stochastic modeling of elementary psychological processes.* Cambridge University Press.

Townsend, J. T., Houpt, J. W., & Silbert, N. H. (2012). General recognition theory extended to include response times: Predictions for a class of parallel systems. *Journal of Mathematical Psychology, 56*(6), 476–494.

Townsend, J. T., Hu, G. G., & Ashby, F. G. (1980). A test of visual feature sampling independence with orthogonal straight lines. *Bulletin of the Psychonomic Society, 15*(3), 163–166.

Townsend, J. T., & Landon, D. E. (1983). Mathematical models of recognition and confusion in psychology. *Mathematical Social Sciences, 4*(1), 25–71.

Townsend, J. T., & Nozawa, G. (1995). Spatio-temporal properties of elementary perception: An investigation of parallel, serial, and coactive theories. *Journal of Mathematical Psychology, 39*(4), 321–359.

Townsend, J. T., & Wenger, M. J. (2004). A theory of interactive parallel processing: New capacity measures and predictions for a response time inequality series. *Psychological Review, 111*(4), 1003–1035.

Townsend, J. T., & Wenger, M. J. (2015). On the dynamic perceptual characteristics of Gestalten: Theory-based methods. In J. Wagemans (Ed.), *The Oxford handbook of perceptual organization* (pp. 948–968). Oxford.

Treutwein, B., Rentschler, I., & Caelli, T. (1989). Perceptual spatial frequency-orientation surface: Psychophysics and line element theory. *Biological Cybernetics, 60*(4), 285–295.

Tsunoda, K., Yamane, Y., Nishizaki, M., & Tanifuji, M. (2001). Complex objects are represented in macaque inferotemporal cortex by the combination of feature columns. *Nature Neuroscience, 4*(8), 832–838.

Tucker, L. R. (1972). Relations between multidimensional scaling and three-mode factor analysis. *Psychometrika, 37*(1), 3–27.

Tversky, A. (1977). Features of similarity. *Psychological Review, 84*(4), 327–352.

Tversky, A., & Gati, I. (1982). Similarity, separability, and the triangle inequality. *Psychological Review, 89*(2), 123–154.

Tversky, A., & Kahneman, D. (1974). Judgment under uncertainty: Heuristics and biases. *Science, 185*(4157), 1124–1131.

Uka, T., & DeAngelis, G. C. (2004). Contribution of area MT to stereoscopic depth perception: Choice-related response modulations reflect task strategy. *Neuron, 42*(2), 297–310.

Valentin, V. V., Maddox, W. T., & Ashby, F. G. (2014). A computational model of the temporal dynamics of plasticity in procedural learning: Sensitivity to feedback timing. *Frontiers in Psychology, 5*(643).

Vallentin, D., Bongard, S., & Nieder, A. (2012). Numerical rule coding in the prefrontal, premotor, and posterior parietal cortices of macaques. *Journal of Neuroscience, 32*(19), 6621–6630.

Van Meter, D., & Middleton, D. (1954). Modern statistical approaches to reception in communication theory. *Transactions of the IRE Professional Group on Information Theory, 4*(4), 119–145.

Van Zandt, T. (2002). Analysis of response time distributions. In J. Wixted (Ed.), *Stevens' handbook of experimental psychology, methodology in experimental psychology* (3rd ed., Vol. 4, pp. 461–516). Wiley.

Vandekerckhove, J., & Tuerlinckx, F. (2007). Fitting the Ratcliff diffusion model to experimental data. *Psychonomic Bulletin & Review, 14*(6), 1011–1026.

van Maanen, L., Couto, J., & Lebreton, M. (2016). Three boundary conditions for computing the fixed-point property in binary mixture data. *PLoS ONE, 11*(11), e0167377.

van Maanen, L., Jong, R., & Rijn, H. (2014). How to assess the existence of competing strategies in cognitive tasks: A primer on the fixed-point property. *PLoS ONE, 9*(8), e106113.

van Vugt, M. K., Simen, P., Nystrom, L. E., Holmes, P., & Cohen, J. D. (2012). EEG oscillations reveal neural correlates of evidence accumulation. *Frontiers in Neuroscience, 6*, 106.

Vazquez, A. L., & Noll, D. C. (1998). Nonlinear aspects of the BOLD response in functional MRI. *NeuroImage, 7*(2), 108–118.

Vogels, R. (1999). Categorization of complex visual images by rhesus monkeys. Part 2: single-cell study. *European Journal of Neuroscience, 11*(4), 1239–1255.

Waldron, E. M., & Ashby, F. G. (2001). The effects of concurrent task interference on category learning: Evidence for multiple category learning systems. *Psychonomic Bulletin & Review, 8*(1), 168–176.

Waldschmidt, J. G., & Ashby, F. G. (2011). Cortical and striatal contributions to automaticity in information-integration categorization. *NeuroImage, 56*(3), 1791–1802.

Wallis, J. D., & Miller, E. K. (2003). From rule to response: Neuronal processes in the premotor and prefrontal cortex. *Journal of Neurophysiology, 90*(3), 1790–1806.

Wandell, B., & Brainard, D. (2023). Principles and consequences of the initial visual encoding. In F. G. Ashby, H. Colonius, & E. N. Dzhafarov (Eds.), *New handbook of mathematical psychology, Volume III: Perceptual and cognitive processes* (pp. 1–41). Cambridge.

Wang, G., Tanifuji, M., & Tanaka, K. (1998). Functional architecture in monkey inferotemporal cortex revealed by in vivo optical imaging. *Neuroscience Research, 32*(1), 33–46.

Watkins, S., Shams, L., Tanaka, S., Haynes, J.-D., & Rees, G. (2006). Sound alters activity in human V1 in association with illusory visual perception. *NeuroImage, 31*(3), 1247–1256.

Watson, A. B., & Robson, J. G. (1981). Discrimination at threshold: Labelled detectors in human vision. *Vision Research, 21*(7), 1115–1122.

Wells, G. L. (1984). The psychology of lineup identifications. *Journal of Applied Social Psychology, 14*(2), 89–103.

Wenger, M. J., & Ingvalson, E. M. (2002). A decisional component of holistic encoding. *Journal of Experimental Psychology: Learning, Memory, and Cognition, 28*(5), 872–892.

Wenger, M. J., & Ingvalson, E. M. (2003). Preserving informational separability and violating decisional separability in facial perception and recognition. *Journal of Experimental Psychology: Learning, Memory, and Cognition, 29*(6), 1106–1118.

Wenger, M. J., & Rhoten, S. E. (2020). Perceptual learning produces perceptual objects. *Journal of Experimental Psychology: Learning, Memory, and Cognition, 46*(3), 455–475.

Wenger, M. J., & Townsend, J. T. (2006). On the costs and benefits of faces and words: Process characteristics of feature search in highly meaningful stimuli. *Journal of Experimental Psychology: Human Perception and Performance, 32*(3), 755–779.

White, I. M., & Wise, S. P. (1999). Rule-dependent neuronal activity in the prefrontal cortex. *Experimental Brain Research, 126*(3), 315–335.

White, J. A., Rubinstein, J. T., & Kay, A. R. (2000). Channel noise in neurons. *Trends in Neurosciences, 23*(3), 131–137.

Wichmann, F. A., Jäkel, F., & Wixted, J. (2018). Methods in psychophysics. In E. J. Wagenmakers & J. T. Wixted (Eds.), *Stevens handbook of experimental psychology and cognitive neuroscience: Methodology* (4th ed., Vol. 5, pp. 265–306). Wiley.

Wickens, T. D. (1992). Maximum-likelihood estimation of a multivariate Gaussian rating model with excluded data. *Journal of Mathematical Psychology, 36*(2), 213–234.

Wickens, T. D. (2002). *Elementary signal detection theory*. Oxford University Press.

Wiecki, T. V., Sofer, I., & Frank, M. J. (2013). HDDM: Hierarchical Bayesian estimation of the drift-diffusion model in Python. *Frontiers in Neuroinformatics, 7*, 14.

Willingham, D. B. (1998). A neuropsychological theory of motor skill learning. *Psychological Review, 105*(3), 558–584.

Wills, A., Noury, M., Moberly, N. J., & Newport, M. (2006). Formation of category representations. *Memory & Cognition, 34*(1), 17–27.

Wilson, C. J. (1995). The contribution of cortical neurons to the firing pattern of striatal spiny neurons. In J. C. Houk, J. L. Davis, & D. G. Beiser (Eds.), *Models of information processing in the basal ganglia* (pp. 29–50). MIT Press.

Wixted, J. T. (2007). Dual-process theory and signal-detection theory of recognition memory. *Psychological Review, 114*(1), 152–176.

Wixted, J. T. (2020). The forgotten history of signal detection theory. *Journal of Experimental Psychology: Learning, Memory, and Cognition, 46*(2), 201–233.

Wixted, J. T., & Mickes, L. (2014). A signal-detection-based diagnostic-feature-detection model of eyewitness identification. *Psychological Review, 121*(2), 262–276.

Wixted, J. T., & Stretch, V. (2004). In defense of the signal detection interpretation of remember/know judgments. *Psychonomic Bulletin & Review, 11*(4), 616–641.

Wixted, J. T., Vul, E., Mickes, L., & Wilson, B. M. (2018). Models of lineup memory. *Cognitive Psychology, 105*, 81–114.

Wogalter, M. S., Malpass, R. S., & McQuiston, D. E. (2004). A national survey of U.S. police on preparation and conduct of lineups. *Psychology, Crime & Law, 10*(1), 69–82.

Wolfe, J. M. (2020). Visual search: How do we find what we are looking for? *Annual Review of Vision Science, 6*(1), 539–562.

Worthy, D. A., Markman, A. B., & Maddox, W. T. (2013). Feedback and stimulus-offset timing effects in perceptual category learning. *Brain and Cognition, 81*(2), 283–293.

Wyszecki, G., & Stiles, W. S. (1982). *Color science: Concepts and methods, quantitative data and formulas*. Wiley.

Yagishita, S., Hayashi-Takagi, A., Ellis-Davies, G. C., Urakubo, H., Ishii, S., Kasai, H. (2014). A critical time window for dopamine actions on the structural plasticity of dendritic spines. *Science, 345*(6204), 1616–1620.

Yamane, Y., Tsunoda, K., Matsumoto, M., Phillips, A. N., & Tanifuji, M. (2006). Representation of the spatial relationship among object parts by neurons in macaque inferotemporal cortex. *Journal of Neurophysiology, 96*(6), 3147–3156.

Yellott , J. I., Jr. (1971). Correction for fast guessing and the speed-accuracy tradeoff in choice reaction time. *Journal of Mathematical Psychology, 8*(2), 159–199.

Yonelinas, A. P. (1994). Receiver-operating characteristics in recognition memory: Evidence for a dual-process model. *Journal of Experimental Psychology: Learning, Memory, and Cognition, 20*(6), 1341–1354.

Yonelinas, A. P., Kroll, N. E., Dobbins, I., Lazzara, M., & Knight, R. T. (1998). Recollection and familiarity deficits in amnesia: Convergence of remember-know, process dissociation, and receiver operating characteristic data. *Neuropsychology, 12*(3), 323–339.

Young, A. W., Hellawell, D., & Hay, D. C. (1987). Configurational information in face perception. *Perception, 16*(6), 747–759.

Young, F. W., & Hamer, R. M. (2013). *Multidimensional scaling: History, theory, and applications.* Psychology Press.

Zaki, S. R., & Kleinschmidt, D. F. (2014). Procedural memory effects in categorization: Evidence for multiple systems or task complexity? *Memory & Cognition, 42*, 508–524.

Zeithamova, D., & Maddox, W. T. (2006). Dual-task interference in perceptual category learning. *Memory & Cognition, 34*(2), 387–398.

Zinnes, J. L., & Griggs, R. A. (1974). Probabilistic, multidimensional unfolding analysis. *Psychometrika, 39*(3), 327–350.

# INDEX

Publisher contact:
The MIT Press
Massachusetts Institute of Technology
77 Massachusetts Avenue, Cambridge, MA 02139
mitpress.mit.edu

EU Authorised Representative:
Easy Access System Europe, Mustamäe tee 50,
10621 Tallinn, Estonia
gpsr.requests@easproject.com

Printed by Integrated Books International,
United States of America